Trends in packaging of food, beverages and other fast-moving consumer goods (FMCG)

Related titles:

Packaging technology: fundamentals, materials and processes
(ISBN 978-1-84569-665-8)

Emerging food packaging technologies: principles and practice
(ISBN 978-1-84569-809-6)

Biaxial stretching of film: principles and applications
(ISBN 978-1-84569-675-7)

Details of these books and a complete list of titles from Woodhead Publishing can be obtained by:

- visiting our web site at www.woodheadpublishing.com
- contacting Customer Services (e-mail: sales@woodheadpublishing.com; fax: +44 (0) 1223 832819; tel.: +44 (0) 1223 499140 ext. 130; address: Woodhead Publishing Limited, 80, High Street, Sawston, Cambridge CB22 3HJ, UK)
- in North America, contacting our US office (e-mail: usmarketing@woodheadpublishing.com; tel.: (215) 928 9112; address: Woodhead Publishing, 1518 Walnut Street, Suite 1100, Philadelphia, PA 19102-3406, USA

If you would like e-versions of our content, please visit our online platform: www.woodheadpublishingonline.com. Please recommend it to your librarian so that everyone in your institution can benefit from the wealth of content on the site.

We are always happy to receive suggestions for new books from potential editors. To enquire about contributing to our food series, please send your name, contact address and details of the topic/s you are interested in to nell.holden@woodheadpublishing.com. We look forward to hearing from you.

The Woodhead team responsible for publishing this book:

Commissioning Editor: Nell Holden
Publications Co-ordinator: Adam Davies
Project Editor: Rachel Cox
Editorial and Production Manager: Mary Campbell
Production Editor: Richard Fairclough
Freelance Project Manager: Annette Wiseman
Copy-editor: Jo Egré
Proofreader: Don Cleeter
Cover Designer: Terry Callanan

Woodhead Publishing Series in Food Science, Technology and Nutrition: Number 244

Trends in packaging of food, beverages and other fast-moving consumer goods (FMCG)

Markets, materials and technologies

Edited by
Neil Farmer

WOODHEAD PUBLISHING

Oxford Cambridge Philadelphia New Delhi

Published by Woodhead Publishing Limited,
80 High Street, Sawston, Cambridge CB22 3HJ, UK
www.woodheadpublishing.com
www.woodheadpublishingonline.com

Woodhead Publishing, 1518 Walnut Street, Suite 1100, Philadelphia,
PA 19102-3406, USA

Woodhead Publishing India Private Limited, G-2, Vardaan House, 7/28 Ansari Road,
Daryaganj, New Delhi – 110002, India
www.woodheadpublishingindia.com

First published 2013, Woodhead Publishing Limited

British Library Cataloguing in Publication Data
A catalogue record for this book is available from the British Library.

Library of Congress Control Number: 2012955694

ISBN 978-0-85709-503-9 (print)
ISBN 978-0-85709-897-9 (online)
ISSN 2042-8049 Woodhead Publishing Series in Food Science, Technology and Nutrition (print)
ISSN 2042-8057 Woodhead Publishing Series in Food Science, Technology and Nutrition (online)

The publisher's policy is to use permanent paper from mills that operate a sustainable forestry policy, and which has been manufactured from pulp which is processed using acid-free and elemental chlorine-free practices. Furthermore, the publisher ensures that the text paper and cover board used have met acceptable environmental accreditation standards.

Typeset by RefineCatch Limited, Bungay, Suffolk, UK

Contents

Contributor contact details

(* = main contact)

Editor, Chapters 1 and 11

Neil Farmer
Neil Farmer Associates
Westbridge House
19 Church Lane
Alveston
Stratford-upon-Avon CV37 7QJ
UK

Email: neilfarmer.48@btinternet.com

Chapter 2

Anne Emblem
School of Management and Science
London College of Fashion
London
UK

Email: packaging@btinternet.com

Chapter 3

Jeremy Plimmer
c/o Product & Image Security Foundation
81 Houting
Dosthill
Tamworth B77 1PB
UK

Email: jeremyplimmer@aol.com

Chapters 4 and 5

Dr Terrence A. Cooper
ARGO Group International
15 Stage Road
Newark
DE 19711-4001
USA

Email: tacooper@post.com

Chapter 6

Daniel A. Abramowicz*
Crown Holdings
One Crown Way
Philadelphia
PA 19154
USA

Email: Daniel.Abramowicz@crowncork.com

Leonard Jenkins and Brian Fields
Crown Packaging Technology
Alsip Technical Center
11535 S. Central Avenue
Alsip
IL 60803
USA

Kevin Ambrose, Ian Bucklow, Terry Benge, Grahame Hughes, Adrian Noke, John Bilko, Andrew Ioannides and Chris Ramsey
Crown Packaging Technology
Wantage Technical Center
Downsview Road
Wantage
Oxfordshire OX12 9BP
UK

Chapter 7

Richard Coles
Director
Emagine
London
www.letsemagine.com

Email: richard@letsemagine.com

Chapter 8

G. Richard Inns
Director PEC Partnership Limited
PO Box 160 Ashtead
Surrey KT21 9BN
UK

Email: richard.inns@pecpartnership.com

Chapter 9

Dr Barry Park
GBP Consulting Ltd
14 Orchard Gardens
Purton
Swindon SN5 4EJ
UK

Email: barry.park@gbpconsulting.co.uk

Chapter 10

Dr Paul Butler
Packaging Materials & Technologies
1278 S Pennsylvania Ave
Winter Park
Florida 32789
USA

Email: packaginginnovation@yahoo.co.uk

Woodhead Publishing Series in Food Science, Technology and Nutrition

1 **Chilled foods: a comprehensive guide** *Edited by C. Dennis and M. Stringer*
2 **Yoghurt: science and technology** *A. Y. Tamime and R. K. Robinson*
3 **Food processing technology: principles and practice** *P. J. Fellows*
4 **Bender's dictionary of nutrition and food technology Sixth edition** *D. A. Bender*
5 **Determination of veterinary residues in food** *Edited by N. T. Crosby*
6 **Food contaminants: sources and surveillance** *Edited by C. Creaser and R. Purchase*
7 **Nitrates and nitrites in food and water** *Edited by M. J. Hill*
8 **Pesticide chemistry and bioscience: the food-environment challenge** *Edited by G. T. Brooks and T. Roberts*
9 **Pesticides: developments, impacts and controls** *Edited by G. A. Best and A. D. Ruthven*
10 **Dietary fibre: chemical and biological aspects** *Edited by D. A. T. Southgate, K. W. Waldron, I. T. Johnson and G. R. Fenwick*
11 **Vitamins and minerals in health and nutrition** *M. Tolonen*
12 **Technology of biscuits, crackers and cookies Second edition** *D. Manley*
13 **Instrumentation and sensors for the food industry** *Edited by E. Kress-Rogers*
14 **Food and cancer prevention: chemical and biological aspects** *Edited by K. W. Waldron, I. T. Johnson and G. R. Fenwick*
15 **Food colloids: proteins, lipids and polysaccharides** *Edited by E. Dickinson and B. Bergenstahl*
16 **Food emulsions and foams** *Edited by E. Dickinson*
17 **Maillard reactions in chemistry, food and health** *Edited by T. P. Labuza, V. Monnier, J. Baynes and J. O'Brien*
18 **The Maillard reaction in foods and medicine** *Edited by J. O'Brien, H. E. Nursten, M. J. Crabbe and J. M. Ames*
19 **Encapsulation and controlled release** *Edited by D. R. Karsa and R. A. Stephenson*
20 **Flavours and fragrances** *Edited by A. D. Swift*
21 **Feta and related cheeses** *Edited by A. Y. Tamime and R. K. Robinson*

22 **Biochemistry of milk products** *Edited by A. T. Andrews and J. R. Varley*
23 **Physical properties of foods and food processing systems** *M. J. Lewis*
24 **Food irradiation: a reference guide** *V. M. Wilkinson and G. Gould*
25 **Kent's technology of cereals: an introduction for students of food science and agriculture Fourth edition** *N. L. Kent and A. D. Evers*
26 **Biosensors for food analysis** *Edited by A. O. Scott*
27 **Separation processes in the food and biotechnology industries: principles and applications** *Edited by A. S. Grandison and M. J. Lewis*
28 **Handbook of indices of food quality and authenticity** *R. S. Singhal, P. K. Kulkarni and D. V. Rege*
29 **Principles and practices for the safe processing of foods** *D. A. Shapton and N. F. Shapton*
30 **Biscuit, cookie and cracker manufacturing manuals Volume 1: ingredients** *D. Manley*
31 **Biscuit, cookie and cracker manufacturing manuals Volume 2: biscuit doughs** *D. Manley*
32 **Biscuit, cookie and cracker manufacturing manuals Volume 3: biscuit dough piece forming** *D. Manley*
33 **Biscuit, cookie and cracker manufacturing manuals Volume 4: baking and cooling of biscuits** *D. Manley*
34 **Biscuit, cookie and cracker manufacturing manuals Volume 5: secondary processing in biscuit manufacturing** *D. Manley*
35 **Biscuit, cookie and cracker manufacturing manuals Volume 6: biscuit packaging and storage** *D. Manley*
36 **Practical dehydration Second edition** *M. Greensmith*
37 **Lawrie's meat science Sixth edition** *R. A. Lawrie*
38 **Yoghurt: science and technology Second edition** *A. Y. Tamime and R. K. Robinson*
39 **New ingredients in food processing: biochemistry and agriculture** *G. Linden and D. Lorient*
40 **Benders' dictionary of nutrition and food technology Seventh edition** *D. A. Bender and A. E. Bender*
41 **Technology of biscuits, crackers and cookies Third edition** *D. Manley*
42 **Food processing technology: principles and practice Second edition** *P. J. Fellows*
43 **Managing frozen foods** *Edited by C. J. Kennedy*
44 **Handbook of hydrocolloids** *Edited by G. O. Phillips and P. A. Williams*
45 **Food labelling** *Edited by J. R. Blanchfield*
46 **Cereal biotechnology** *Edited by P. C. Morris and J. H. Bryce*
47 **Food intolerance and the food industry** *Edited by T. Dean*
48 **The stability and shelf-life of food** *Edited by D. Kilcast and P. Subramaniam*
49 **Functional foods: concept to product** *Edited by G. R. Gibson and C. M. Williams*
50 **Chilled foods: a comprehensive guide Second edition** *Edited by M. Stringer and C. Dennis*
51 **HACCP in the meat industry** *Edited by M. Brown*
52 **Biscuit, cracker and cookie recipes for the food industry** *D. Manley*
53 **Cereals processing technology** *Edited by G. Owens*
54 **Baking problems solved** *S. P. Cauvain and L. S. Young*
55 **Thermal technologies in food processing** *Edited by P. Richardson*
56 **Frying: improving quality** *Edited by J. B. Rossell*
57 **Food chemical safety Volume 1: contaminants** *Edited by D. Watson*

58 **Making the most of HACCP: learning from others' experience** *Edited by T. Mayes and S. Mortimore*
59 **Food process modelling** *Edited by L. M. M. Tijskens, M. L. A. T. M. Hertog and B. M. Nicolaï*
60 **EU food law: a practical guide** *Edited by K. Goodburn*
61 **Extrusion cooking: technologies and applications** *Edited by R. Guy*
62 **Auditing in the food industry: from safety and quality to environmental and other audits** *Edited by M. Dillon and C. Griffith*
63 **Handbook of herbs and spices Volume 1** *Edited by K. V. Peter*
64 **Food product development: maximising success** *M. Earle, R. Earle and A. Anderson*
65 **Instrumentation and sensors for the food industry Second edition** *Edited by E. Kress-Rogers and C. J. B. Brimelow*
66 **Food chemical safety Volume 2: additives** *Edited by D. Watson*
67 **Fruit and vegetable biotechnology** *Edited by V. Valpuesta*
68 **Foodborne pathogens: hazards, risk analysis and control** *Edited by C. de W. Blackburn and P. J. McClure*
69 **Meat refrigeration** *S. J. James and C. James*
70 **Lockhart and Wiseman's crop husbandry Eighth edition** *H. J. S. Finch, A. M. Samuel and G. P. F. Lane*
71 **Safety and quality issues in fish processing** *Edited by H. A. Bremner*
72 **Minimal processing technologies in the food industries** *Edited by T. Ohlsson and N. Bengtsson*
73 **Fruit and vegetable processing: improving quality** *Edited by W. Jongen*
74 **The nutrition handbook for food processors** *Edited by C. J. K. Henry and C. Chapman*
75 **Colour in food: improving quality** *Edited by D MacDougall*
76 **Meat processing: improving quality** *Edited by J. P. Kerry, J. F. Kerry and D. A. Ledward*
77 **Microbiological risk assessment in food processing** *Edited by M. Brown and M. Stringer*
78 **Performance functional foods** *Edited by D. Watson*
79 **Functional dairy products Volume 1** *Edited by T. Mattila-Sandholm and M. Saarela*
80 **Taints and off-flavours in foods** *Edited by B. Baigrie*
81 **Yeasts in food** *Edited by T. Boekhout and V. Robert*
82 **Phytochemical functional foods** *Edited by I. T. Johnson and G. Williamson*
83 **Novel food packaging techniques** *Edited by R. Ahvenainen*
84 **Detecting pathogens in food** *Edited by T. A. McMeekin*
85 **Natural antimicrobials for the minimal processing of foods** *Edited by S. Roller*
86 **Texture in food Volume 1: semi-solid foods** *Edited by B. M. McKenna*
87 **Dairy processing: improving quality** *Edited by G. Smit*
88 **Hygiene in food processing: principles and practice** *Edited by H. L. M. Lelieveld, M. A. Mostert, B. White and J. Holah*
89 **Rapid and on-line instrumentation for food quality assurance** *Edited by I. Tothill*
90 **Sausage manufacture: principles and practice** *E. Essien*
91 **Environmentally-friendly food processing** *Edited by B. Mattsson and U. Sonesson*
92 **Bread making: improving quality** *Edited by S. P. Cauvain*
93 **Food preservation techniques** *Edited by P. Zeuthen and L. Bøgh-Sørensen*
94 **Food authenticity and traceability** *Edited by M. Lees*

95 **Analytical methods for food additives** *R. Wood, L. Foster, A. Damant and P. Key*
96 **Handbook of herbs and spices Volume 2** *Edited by K. V. Peter*
97 **Texture in food Volume 2: solid foods** *Edited by D. Kilcast*
98 **Proteins in food processing** *Edited by R. Yada*
99 **Detecting foreign bodies in food** *Edited by M. Edwards*
100 **Understanding and measuring the shelf-life of food** *Edited by R. Steele*
101 **Poultry meat processing and quality** *Edited by G. Mead*
102 **Functional foods, ageing and degenerative disease** *Edited by C. Remacle and B. Reusens*
103 **Mycotoxins in food: detection and control** *Edited by N. Magan and M. Olsen*
104 **Improving the thermal processing of foods** *Edited by P. Richardson*
105 **Pesticide, veterinary and other residues in food** *Edited by D. Watson*
106 **Starch in food: structure, functions and applications** *Edited by A.-C. Eliasson*
107 **Functional foods, cardiovascular disease and diabetes** *Edited by A. Arnoldi*
108 **Brewing: science and practice** *D. E. Briggs, P. A. Brookes, R. Stevens and C. A. Boulton*
109 **Using cereal science and technology for the benefit of consumers: proceedings of the 12th International ICC Cereal and Bread Congress, 24–26th May, 2004, Harrogate, UK** *Edited by S. P. Cauvain, L. S. Young and S. Salmon*
110 **Improving the safety of fresh meat** *Edited by J. Sofos*
111 **Understanding pathogen behaviour: virulence, stress response and resistance** *Edited by M. Griffiths*
112 **The microwave processing of foods** *Edited by H. Schubert and M. Regier*
113 **Food safety control in the poultry industry** *Edited by G. Mead*
114 **Improving the safety of fresh fruit and vegetables** *Edited by W. Jongen*
115 **Food, diet and obesity** *Edited by D. Mela*
116 **Handbook of hygiene control in the food industry** *Edited by H. L. M. Lelieveld, M. A. Mostert and J. Holah*
117 **Detecting allergens in food** *Edited by S. Koppelman and S. Hefle*
118 **Improving the fat content of foods** *Edited by C. Williams and J. Buttriss*
119 **Improving traceability in food processing and distribution** *Edited by I. Smith and A. Furness*
120 **Flavour in food** *Edited by A. Voilley and P. Etievant*
121 **The Chorleywood bread process** *S. P. Cauvain and L. S. Young*
122 **Food spoilage microorganisms** *Edited by C. de W. Blackburn*
123 **Emerging foodborne pathogens** *Edited by Y. Motarjemi and M. Adams*
124 **Benders' dictionary of nutrition and food technology Eighth edition** *D. A. Bender*
125 **Optimising sweet taste in foods** *Edited by W. J. Spillane*
126 **Brewing: new technologies** *Edited by C. Bamforth*
127 **Handbook of herbs and spices Volume 3** *Edited by K. V. Peter*
128 **Lawrie's meat science Seventh edition** *R. A. Lawrie in collaboration with D. A. Ledward*
129 **Modifying lipids for use in food** *Edited by F. Gunstone*
130 **Meat products handbook: practical science and technology** *G. Feiner*
131 **Food consumption and disease risk: consumer-pathogen interactions** *Edited by M. Potter*
132 **Acrylamide and other hazardous compounds in heat-treated foods** *Edited by K. Skog and J. Alexander*

167 **Shellfish safety and quality** *Edited by S. E. Shumway and G. E. Rodrick*
168 **Functional and speciality beverage technology** *Edited by P. Paquin*
169 **Functional foods: principles and technology** *M. Guo*
170 **Endocrine-disrupting chemicals in food** *Edited by I. Shaw*
171 **Meals in science and practice: interdisciplinary research and business applications** *Edited by H. L. Meiselman*
172 **Food constituents and oral health: current status and future prospects** *Edited by M. Wilson*
173 **Handbook of hydrocolloids Second edition** *Edited by G. O. Phillips and P. A. Williams*
174 **Food processing technology: principles and practice Third edition** *P. J. Fellows*
175 **Science and technology of enrobed and filled chocolate, confectionery and bakery products** *Edited by G. Talbot*
176 **Foodborne pathogens: hazards, risk analysis and control Second edition** *Edited by C. de W. Blackburn and P. J. McClure*
177 **Designing functional foods: measuring and controlling food structure breakdown and absorption** *Edited by D. J. McClements and E. A. Decker*
178 **New technologies in aquaculture: improving production efficiency, quality and environmental management** *Edited by G. Burnell and G. Allan*
179 **More baking problems solved** *S. P. Cauvain and L. S. Young*
180 **Soft drink and fruit juice problems solved** *P. Ashurst and R. Hargitt*
181 **Biofilms in the food and beverage industries** *Edited by P. M. Fratamico, B. A. Annous and N. W. Gunther*
182 **Dairy-derived ingredients: food and neutraceutical uses** *Edited by M. Corredig*
183 **Handbook of waste management and co-product recovery in food processing Volume 2** *Edited by K. W. Waldron*
184 **Innovations in food labelling** *Edited by J. Albert*
185 **Delivering performance in food supply chains** *Edited by C. Mena and G. Stevens*
186 **Chemical deterioration and physical instability of food and beverages** *Edited by L. H. Skibsted, J. Risbo and M. L. Andersen*
187 **Managing wine quality Volume 1: viticulture and wine quality** *Edited by A. G. Reynolds*
188 **Improving the safety and quality of milk Volume 1: milk production and processing** *Edited by M. Griffiths*
189 **Improving the safety and quality of milk Volume 2: improving quality in milk products** *Edited by M. Griffiths*
190 **Cereal grains: assessing and managing quality** *Edited by C. Wrigley and I. Batey*
191 **Sensory analysis for food and beverage quality control: a practical guide** *Edited by D. Kilcast*
192 **Managing wine quality Volume 2: oenology and wine quality** *Edited by A. G. Reynolds*
193 **Winemaking problems solved** *Edited by C. E. Butzke*
194 **Environmental assessment and management in the food industry** *Edited by U. Sonesson, J. Berlin and F. Ziegler*
195 **Consumer-driven innovation in food and personal care products** *Edited by S. R. Jaeger and H. MacFie*
196 **Tracing pathogens in the food chain** *Edited by S. Brul, P. M. Fratamico and T. A. McMeekin*

197 **Case studies in novel food processing technologies: innovations in processing, packaging, and predictive modelling** *Edited by C. J. Doona, K. Kustin and F. E. Feeherry*
198 **Freeze-drying of pharmaceutical and food products** *T.-C. Hua, B.-L. Liu and H. Zhang*
199 **Oxidation in foods and beverages and antioxidant applications Volume 1: understanding mechanisms of oxidation and antioxidant activity** *Edited by E. A. Decker, R. J. Elias and D. J. McClements*
200 **Oxidation in foods and beverages and antioxidant applications Volume 2: management in different industry sectors** *Edited by E. A. Decker, R. J. Elias and D. J. McClements*
201 **Protective cultures, antimicrobial metabolites and bacteriophages for food and beverage biopreservation** *Edited by C. Lacroix*
202 **Separation, extraction and concentration processes in the food, beverage and nutraceutical industries** *Edited by S. S. H. Rizvi*
203 **Determining mycotoxins and mycotoxigenic fungi in food and feed** *Edited by S. De Saeger*
204 **Developing children's food products** *Edited by D. Kilcast and F. Angus*
205 **Functional foods: concept to product Second edition** *Edited by M. Saarela*
206 **Postharvest biology and technology of tropical and subtropical fruits Volume 1: fundamental issues** *Edited by E. M. Yahia*
207 **Postharvest biology and technology of tropical and subtropical fruits Volume 2: açai to citrus** *Edited by E. M. Yahia*
208 **Postharvest biology and technology of tropical and subtropical fruits Volume 3: cocona to mango** *Edited by E. M. Yahia*
209 **Postharvest biology and technology of tropical and subtropical fruits Volume 4: mangosteen to white sapote** *Edited by E. M. Yahia*
210 **Food and beverage stability and shelf life** *Edited by D. Kilcast and P. Subramaniam*
211 **Processed meats: improving safety, nutrition and quality** *Edited by J. P. Kerry and J. F. Kerry*
212 **Food chain integrity: a holistic approach to food traceability, safety, quality and authenticity** *Edited by J. Hoorfar, K. Jordan, F. Butler and R. Prugger*
213 **Improving the safety and quality of eggs and egg products Volume 1** *Edited by Y. Nys, M. Bain and F. Van Immerseel*
214 **Improving the safety and quality of eggs and egg products Volume 2** *Edited by F. Van Immerseel, Y. Nys and M. Bain*
215 **Animal feed contamination: effects on livestock and food safety** *Edited by J. Fink-Gremmels*
216 **Hygienic design of food factories** *Edited by J. Holah and H. L. M. Lelieveld*
217 **Manley's technology of biscuits, crackers and cookies Fourth edition** *Edited by D. Manley*
218 **Nanotechnology in the food, beverage and nutraceutical industries** *Edited by Q. Huang*
219 **Rice quality: a guide to rice properties and analysis** *K. R. Bhattacharya*
220 **Advances in meat, poultry and seafood packaging** *Edited by J. P. Kerry*
221 **Reducing saturated fats in foods** *Edited by G. Talbot*
222 **Handbook of food proteins** *Edited by G. O. Phillips and P. A. Williams*

223 **Lifetime nutritional influences on cognition, behaviour and psychiatric illness** *Edited by D. Benton*
224 **Food machinery for the production of cereal foods, snack foods and confectionery** *L.-M. Cheng*
225 **Alcoholic beverages: sensory evaluation and consumer research** *Edited by J. Piggott*
226 **Extrusion problems solved: food, pet food and feed** *M. N. Riaz and G. J. Rokey*
227 **Handbook of herbs and spices Second edition Volume 1** *Edited by K. V. Peter*
228 **Handbook of herbs and spices Second edition Volume 2** *Edited by K. V. Peter*
229 **Breadmaking: improving quality Second edition** *Edited by S. P. Cauvain*
230 **Emerging food packaging technologies: principles and practice** *Edited by K. L. Yam and D. S. Lee*
231 **Infectious disease in aquaculture: prevention and control** *Edited by B. Austin*
232 **Diet, immunity and inflammation** *Edited by P. C. Calder and P. Yaqoob*
233 **Natural food additives, ingredients and flavourings** *Edited by D. Baines and R. Seal*
234 **Microbial decontamination in the food industry: novel methods and applications** *Edited by A. Demirci and M. O. Ngadi*
235 **Chemical contaminants and residues in foods** *Edited by D. Schrenk*
236 **Robotics and automation in the food industry: current and future technologies** *Edited by D. G. Caldwell*
237 **Fibre-rich and wholegrain foods: improving quality** *Edited by J. A. Delcour and K. Poutanen*
238 **Computer vision technology in the food and beverage industries** *Edited by D.-W. Sun*
239 **Encapsulation technologies and delivery systems for food ingredients and nutraceuticals** *Edited by N. Garti and D. J. McClements*
240 **Case studies in food safety and authenticity** *Edited by J. Hoorfar*
241 **Heat treatment for insect control: developments and applications** *D. Hammond*
242 **Advances in aquaculture hatchery technology** *Edited by G. Allan and G. Burnell*
243 **Open innovation in the food and beverage industry: concepts and case studies** *Edited by M. Garcia Martinez*
244 **Trends in packaging of food, beverages and other fast-moving consumer goods (FMCG)** *Edited by Neil Farmer*
245 **New analytical approaches for verifying the origin of food** *Edited by P. Brereton*
246 **Microbial production of food ingredients, enzymes and nutraceuticals** *Edited by B. McNeil, D. Archer, I. Giavasis and L. Harvey*
247 **Persistent organic pollutants and toxic metals in foods** *Edited by M. Rose and A. Fernandes*
248 **Cereal grains for the food and beverage industries** *E. Arendt and E. Zannini*
249 **Viruses in food and water: risks, surveillance and control** *Edited by N. Cook*
250 **Improving the safety and quality of nuts** *Edited by L. J. Harris*
251 **Metabolomics in food and nutrition** *Edited by B. Weimer and C. Slupsky*
252 **Food enrichment with omega-3 fatty acids** *Edited by C. Jacobsen, N. Skall Nielsen, A. Frisenfeldt Horn and A.-D. Moltke Sørensen*
253 **Instrumental assessment of food sensory quality: a practical guide** *Edited by D. Kilcast*
254 **Food microstructures: microscopy, measurement and modelling** *Edited by V. J. Morris and K. Groves*

255 **Handbook of food powders: processes and properties** *Edited by B. R. Bhandari, N. Bansal, M. Zhang and P. Schuck*
256 **Functional ingredients from algae for foods and nutraceuticals** *Edited by H. Domínguez*
257 **Satiation, satiety and the control of food intake: theory and practice** *Edited by J. E. Blundell and F. Bellisle*
258 **Hygiene in food processing: principles and practice Second edition** *Edited by H. L. M. Lelieveld, J. Holah and D. Napper*

About the editor

Neil Farmer is the owner/proprietor of Neil Farmer Associates, an international marketing, market research and business development consultancy. Established in 2001, the company provides marketing expertise, knowledge, analysis and information in all packaging markets on a global basis. Particular areas of expertise include flexible packaging, rigid plastics, bioplastics, food contact packaging, closures, active and intelligent packaging, metal packaging and folding cartons.

With 35 years' experience in the international packaging industry, Neil has held the position of director or senior manager in several leading publicly quoted packaging companies including Wace Group, Jarvis Porter, Ferry Pickering and Metal Closures Group, prior to setting up the consultancy. Neil Farmer Associates' clients include leading international management consultancies, research organisations, and multinational packaging groups.

Neil holds a BSc in Economics, a Diploma in Marketing and a postgraduate Business Studies qualification. He is a Fellow of the former Institute of Packaging (now The Packaging Society and part of The Institute of Materials, Minerals and Mining), a Fellow of the Chartered Institute of Marketing, and a Fellow of the Institute of Paper, Printing and Publishing. Neil is a Chartered Scientist, Chartered Environmentalist, a Freeman of the City of London and a Freeman of the Worshipful Company of Stationers and Newspaper Makers. He is a previous National Chairman and Trustee of The Institute of Packaging and a member of The Packaging Society Board.

Neil is a regular writer of articles and reviews on packaging and a speaker at many international packaging conferences and seminars. He is also the author of the chapter on Packaging and Marketing in the recent Woodhead Publishing title *Packaging technology: fundamentals, materials and processes.*

As an enthusiastic marathon runner, Neil has competed in 275 marathon and half marathon races around the world, including London and Los Angeles, raising money for many charities in the process.

1

Present status and trends in innovations in packaging for food, beverages and other fast-moving consumer goods

N. Farmer, Consultant, UK

DOI: 10.1533/9780857098979.1

Abstract: This chapter first considers the international fast-moving consumer goods (FMCG) packaging market in terms of major trends and innovations from 2006 to 2011. It then describe the innovations that are occurring in individual materials, technologies and market sectors and contains information about some of the leading companies in global markets, statistics and market trends. Some of the most important innovations have occurred in areas such as the environment and sustainability, material reductions, waste disposal and recycling. These areas are considered from a market trends and market analysis viewpoint. Similarly, the importance of the development of new markets and technologies in bioplastics, biodegradable materials and compostability of materials are also evaluated. The final section deals with the future of many of these materials, technologies, international companies and global markets, evaluating the trends and innovations over the coming decade and assessing their likely impact in the international FMCG packaging sector.

Key words: bioplastics, biodegradable materials, compostable materials, flexible packaging, active and intelligent packaging, globalisation of packaging markets, light-weighting of packaging, packaging material reductions, recycling and waste initiatives.

1.1 Introduction

The packaging industry is now truly a global business. In 2010, industry estimates valued the global consumer packaging market at \$395 billion (£247 billion). The UK market for food and drink packaging alone was estimated to be worth over £5 billion. These are significant numbers, reflecting the order of magnitude of an industry with massive international significance. However, there is a tremendous amount of diversity and individual influences within each of the sectors within the packaging market.

Innovations within the market are certainly one of the driving factors behind the dynamic nature of today's global industry. Between 2006 and 2011, these innovative driving forces have manifested themselves in a variety of ways, which are dealt with in this chapter.

1.2 Light-weighting, material reductions, recycling and waste initiatives

Some might argue that to seek to reduce the weight of a pack and its material content is hardly an innovation. However, in recent years the weight reduction programmes of leading consumer goods companies has galvanised the industry into re-evaluating its whole approach to innovative and cost-effective use of materials. We are probably all aware of the massive material savings in sectors such as the Easter egg market by leading producers in the UK, such as Cadbury, Nestlé, Thorntons and Mars. We are talking about material savings of between 20 and 50% in some of these types of packs.

These numbers, although considerable, are not unique to that sector. For example, Kraft Foods announced that it would eliminate £150 million of packaging material by 2011 from a base year of 2005. This would be achieved by creating more efficient packaging designs and converting to lighter weight materials. The company said it would also source materials in a more sustainable way and establish programmes to deal with the packaging at the end of the product's life. Relating to the latter, Kraft has encapsulated what has been the key driver for leading consumer goods companies in this sector, namely the holistic approach to the environment, materials, sustainability and avoidance of further depletion of the Earth's resources.

1.2.1 The holistic approach

At the current time, global players in the consumer goods market are embracing similar mantras to that of Kraft. In 2011 Unilever set far reaching Sustainable Living Plan objectives, including a quote from Jay Gouliard, then their Vice-President of Global Packaging and R&D: 'I view sustainability as the opening of a new space for innovation and the opportunity to demonstrate our collective innovation power in Unilever.' It was reported in 2011 that, as part of its plan, Unilever had become a zero landfill manufacturer in the UK, with its 11 sites not sending any non-hazardous waste to landfill. Procter and Gamble has also created a Sustainability Vision, with far reaching targets and objectives in sustainable packaging usage now and over the coming 10 years. It has pledged to reduce packaging by 20% by 2020 and to reduce its use of petroleum-derived materials by 25%. Ultimately, it plans to make 100% of its products and packaging with renewable or recycled materials.

The Coca-Cola company too has established a new global Office of Sustainability. This will integrate the company's existing system sustainability initiatives in the areas of water, climate protection, packaging and recycling, and community. In December 2011, Nestlé announced what is claimed to be the UK's lightest-weight water bottle, with 25% less material than previous. In its Sustainability and Responsibility Report 2011, Diageo, one of the world's leading drinks companies, said it had Sustainable Packaging Targets to reduce the average unit weight of packaging by 10% by 2015 and to increase to 42% the average recycled content across all packaging, also by 2015. The company intends to make all packaging 100% recyclable or reusable within the same timeframe.

Thus, what started out as material reduction programmes across major consumer goods producers has now become a global drive to save and replenish the Earth's natural resources. Efficient and cost-effective use of packaging materials is very much part of this holistic world-view.

1.2.2 Waste and recycling

In the context of material usage, reduction and light-weighting, it is important to mention the important area of waste and recycling. The *Review of Waste Policy*,[1] published by the UK's coalition government in June 2011, sought to clarify the way forward and is a good example of all stakeholders seeking to work together for the collective good. The document talked about voluntary agreements and local initiatives and broadly spoke of the benefits of recycling, energy from waste and material resource efficiency. It acknowledged the benefits of packaging and said that its carbon footprint is small compared to that of its contents. Indeed, the Courtauld Commitment Phase 2[2] from WRAP in the UK set a target of reducing the carbon impact of grocery packaging by 10% by the end of 2012, against a 2009 baseline. Courtauld 3 was being developed in the autumn of 2011 and is likely to tackle issues concerning products, product design and lifecycle impacts to reduce waste.

Waste reduction is now clearly high on the UK government's and other national governments' agendas. In terms of recycled packaging, the report identified the need to see more recycled content being utilised throughout the packaging supply chain and called for packaging that is easier to recycle, such as non-mixed materials. The issue of sorting and reprocessing capacity for non-bottle plastics is important. This area is a significant impediment to achieving increased rates of plastic packaging recycling and the government's commitment to develop and provide loans to companies to fund this activity is important. We will consider the implications of the UK government's 2013 to 2017 packaging recycling targets in Chapter 11, when looking at the future.

1.2.3 UK's first mixed plastics recycling facility

The forecast growth in recycling was aided by projects such as the official launch in the spring of 2011 of the UK's first mixed plastics recycling facility. The plant is in Redcar, North Yorkshire and owned by the Biffa Group. It is able to process a range of mixed household plastic packaging, including food trays, margarine containers and yoghurt tubs that previously could not have been recycled. The Biffa Polymers plant has been built by Greenstar WES,[22] part of the Biffa Group, and is able to process up to 15 000 tonnes of mixed plastic packaging every year. It planned a full run-rate of 20 000 tonnes a year from April 2012. The project was funded by a £1.187 million grant from Wrap,[3] which was provided to Greenstar in the spring of 2010. Some of the output will be processed through Biffa's food grade high density polyethylene (HDPE) recycling facility at the same site and it is planned that this will go back into the production of new milk bottles.

1.2.4 British Retail Consortium/retailer initiative

Other recycling initiatives in the UK also demonstrate the pace of growth of the recycling industry. In April 2011, an agreement was made by retailers and the British Retail Consortium (BRC)[4] for bread bags, cereal pack liners and the thin plastics used for carrying canned multipacks and other items, to be collected by the 4500 stores that already take plastic bags for recycling. The initiative was backed by Tesco, Sainsbury's, Asda, the Co-op, Waitrose and Morrisons supermarkets. The BRC claim that in the UK thin plastics make up 43% of plastic household waste, amounting to 645 000 tonnes per year. This initiative is another excellent example of the progressive moves to recycle plastic films by those in the packaging recycling chain. Much of thin plastic film is fully recyclable, but there have always been issues about the ways and means of recycling it.

1.2.5 Global Packaging Project

Projects such as the Global Packaging Project,[5] which seeks to establish a common language of definitions to benchmark sustainability, are also important for the future. The project involves some of the world's leading consumer goods producers and its framework is based on work produced by Europen,[6] the European Organisation for Packaging and the Environment. In September 2011, the Global Packaging Project released the Global Protocol on Packaging Sustainability (GPPS) to enable the consumer goods industry to better assess the relative sustainability of packaging. The protocol aims to help companies reduce the environmental impact of packaging by providing a common language, which consists of a framework and a measurement system to address a range of business questions about packaging sustainability.

All these positive developments are poised to play a vital part in achieving greater sustainability in Western economies now and throughout the next ten years. Inevitably it is a question of meeting the needs of all in the packaging supply chain, including the final consumer, without depleting future resources.

1.2.6 Global edible waste disposal

Care needs to be taken when considering waste disposal. According to WRAP in the 2008 Food Waste report, 23.4% of edible food by value, or nearly

18.4% by weight, is discarded by UK consumers. Taking the global picture, at the Save Food Congress at Interpack in May 2011 in Dusseldorf, the Swedish Institute for Food and Biotechnology (SIK)[7] said that research had shown that one-third of food produced for human consumption was wasted globally. It said that wastage in developing countries was concentrated early in the production chain, whereas in the developed world it was largely found at the point of consumption. Thus there is clearly much work to be done in the whole sustainability and waste disposal arena. Supporting this view, in September 2011, the Department for Food, Rural and Agricultural Affairs (DEFRA) in the UK said that it will recommend that sell-by and display-until dates be dropped from packaging. DEFRA claim these confuse consumers and are simply a stock method for retailers. However, they proposed that use-by and best-before dates would remain. It is estimated that £12 billion of edible food is thrown away in the UK each year. If this type of figure is replicated around the world, it is damaging to all future sustainability targets and needs to be substantially reduced.

1.3 The flexible packaging market

1.3.1 Overview

According to a survey by Smithers Pira International[8] in August 2011, global consumer flexible packaging sales were worth $58.3 billion in 2011, a higher figure than other surveys, as it includes other end use sectors (Chapter 11). Smithers Pira[9] forecast a compound annual growth rate of 4.1% per annum up to 2016; this is compared with a growth rate of almost 6% per annum over the previous three years. There is a marked trend in the reduction of the number of flexible packaging layers being used. The reduction in packaging usage in general terms will also impact on volumes in the future, coupled with the previously noted packaging reduction initiatives by many leading consumer goods producers. Improved barrier performance and developments in barrier resins, discussed later in this chapter, are also important factors in the market.

Converted flexible packaging demand in the USA was forecast to exceed $16 billion in 2011. Growth was attributed to cost, performance and technical advances in flexible materials, compared with alternative materials such as rigid packaging. PCI Consulting,[10] in their European Flexible Market Report 2011, estimated the European market to be worth 11 billion Euros. One of the most interesting conclusions and a strong indication of how market trends have developed and will continue to develop in the future was the statistic that, after its acquisition of Alcan Packaging in February 2010, Amcor now accounted for nearly 25% of all European flexible packaging sales. The power of the larger groups is clearly demonstrated by this fact and is a theme that will be evaluated and analysed in more detail in Chapter 11 when considering globalisation, mergers and acquisitions and the world market.

1.3.2 Steam venting systems

Some of the latest generation of innovations in flexible packaging in the period under review have been steam venting systems. Convenience and ease of use have been important factors in these developments, with demand rapidly growing in a variety of food market applications.

When looking at individual applications for steam venting systems, Heinz in the USA adopted a high-performance sealant on the interior of a laminated film, providing a steamed-in feature for their frozen mashed potatoes. This particular package was made by Alcan, now part of Bemis. In-register laser technology creates the self-venting feature, which is key to the whole concept. Amcor's Protect Valve System allows the product to be cooked in a microwave oven without the need to burst the film prior to cooking. There are now numerous applications of steam in pack microwavable vegetables, which have become highly popular because of cost effectiveness, convenience and ease of cooking, without having to tear open the pack prior to cooking. The benefits of technically outstanding sealants on the interior of the laminated film will ensure that the packaging material will perform across the range of temperatures associated with frozen food distribution and microwave cooking.

Sealed Air Cryovac's Simple Steps Microwavable Vacuum Skin packaging is another example of a concept that is self-venting, thereby obviating the requirement to pierce the pack before placing it in the microwave. In one particular application for Meralliance, a French food producer, the system allows microwave steam cooking of raw marinated fish fillets in 90 seconds. The concept allows for the formation of a pressure bubble within the pack, which ensures that the flavour and moisture of the dish are retained; it also provides a shelf life of three weeks. Similarly, it is used by food producer Mirital and retailer Metro C&C in Russia, for frozen ready meals.

All these types of packs have been, and will continue to be, successful and an important part of the growth of the high performance flexible packaging films sector.

1.3.3 Pouch packs

One of the most significant innovations in the last decade in the flexible packaging market has been the inexorable rise of the pouch pack. The global market for stand-up pouches in 2007 was estimated at 30 billion units. Subsequent reports indicate that this figure might have been ambitious. However, since then, demand for pouches has increased by 6 to 8% per annum. In Japan, 1.5 to 3 billion pouches are produced per year. Arguably stand-up pouches are one of the most important innovations to affect flexible packaging in recent years. High performance laminates have improved the pouch packs image and, with longer shelf lives, have opened up new markets. The rise of on-the-go snacking and trends towards convenience eating means that the world market for pouches will continue to rise significantly. Although pouch packs have been widely adopted in a variety of food contact applications, in other sectors, including pet foods and household products, the demand for pouches has also grown significantly.

The benefits of pouch packs are illustrated in the following example. In Europe, the Kraft Kenco brand was launched in 2009. The current Kenco brand was a screw-cap glass package. Kenco decided to introduce a new refill stand-up pouch with freeze-dried instant coffee, with 97% less packaging than the jar. It provided extra value because the consumer bought the glass package and reused it multiple times with the refill bag. The theme is once again the environment, convenience and packaging reduction, a mantra of the supply chain, which will be repeated again and again in the future.

The original standing pouch pack from Wild of Germany was for Capri Sun orange drink. This innovation, like others in the packaging material world, took many years to achieve wide international acceptance. In recent years, Capri Sun has broadened its range to gain a foothold in the teenage market, using a shaped, spouted pouch with a foil laminate material. Amcor Flexibles has the Heat Flex pouch, which has various alternative constructions, including one with good barrier properties. Amcor also has a microwavable pouch for ready-to-cook vegetables, which Tesco in the UK and others are using. Huhtamaki of Finland has an extensive range, including the Cyclero, which is a modular system with a variety of shapes and sizes. Leading producers in North America include Peel Plastic Products in Ontario, Canada and Ampac Flexibles of Cincinnati, USA.

All of these types of companies, with their innovations and expertise, will drive the pouch market forward. Further healthy gains in volume are expected due to improved barrier structures and features, such as re-sealable zipper closures, spouts and tear strips. Whilst food will continue to be the market sector offering the best continuing growth opportunities, new applications are being found on a regular basis. Certainly, pet food demand shows no sign of slowing and in personal care and toiletries, new markets are being opened up as a result of superior pack performance. A global study by Allied Development Corp,[11] entitled *Stand-up Pouches 2012–2016*, stated that stand-up pouch sales are estimated to reach $4 billion by 2016.

1.4 Active and intelligent packaging

Active packaging usually means packaging that has active functions beyond passive containment and protection of the product. Technologies within the active packaging sector, such as packaging that can actively reduce the level of oxygen or even microbial count in a pack, are of great importance. Similarly, intelligent packaging, which has the ability to sense an attribute of the product or atmosphere around it, is a key part of recent innovations in the global packaging world.

1.4.1 Overview of the market

In a 2009 report entitled *The Future of Active and Modified Atmosphere Packaging: Market Forecasts to 2014*, Smithers Pira said that the European market for active food packaging grew at an annual rate of 23.8% between 2004 and 2009. The report states that US demand for active packaging has a compound annual growth rate (CAGR) of 8.8% for the period 2009 to 2014.

Along similar lines, Freedonia Group[12] stated in 2009 that demand for smart or intelligent packaging would grow by 13% per annum to 2011. Market expansion was being driven by the development of new products with improved performance and lower costs.

However, despite these figures in the active and intelligent packaging field over the years of the economic downturn between 2008 and 2011, it is probably true that some innovative developments were put on hold, as a result of the cost versus innovation equation falling down on the side of cost savings and not further investment in technology and R&D.

However, there are signs that developments in active and intelligent packaging will grow substantially over the next decade, as world economies improve. Markets will be driven by innovations in time temperature indicators (TTIs), thermochromic inks, barrier films, coatings and additives, oxygen scavenging technology, biodegradable and compostable materials, and the long-term sustainability visions of leading companies, such as the ones previously mentioned in this chapter. Developments such as these will be dealt with from a technical viewpoint in later chapters, but are commented on in the following sections in terms of market trends.

1.4.2 Modified atmosphere packaging

The science of regulating moisture to ensure shelf stability of packaging foods has been around for many years (Chapter 2) and modified atmosphere packaging (MAP) has long been used to extend shelf life. In recent years there has been an enormous growth in the use of MAP for fresh produce in the retail sector. It is now an established technique for packaging and storing fresh meat, with life expectancy of up to 25 to 30 days. In the UK in 2009, over 80% of the raw meat sold was in case-ready packs. Case-ready developments are slower in other countries in Europe, where meat production is more fragmented. In Germany, leading supermarkets are encouraging growth; in France, MAP penetration was over 40% in 2009. However, in Italy, the introduction of MAP has been slower. New applications are regularly being found for MAP in sectors such as fruit and vegetables.

One of the issues still facing the industry is that once a pack is opened these protective benefits are lost. Thus developments that have been forthcoming in resealable packs and portion controlled packs, which allow contents to be opened and consumed incrementally, are of great significance.

1.4.3 Time temperature indicators (TTIs)

Apart from acting as a quality assurance tool, TTIs have an important marketing benefit in that they tell the consumer and the retailer whether the food and drink on sale is fresh and ready to be eaten or drunk. These, for example, can be applied as stand-alone labels on food packaging. This is particularly important when statistics by the World Trade Organisation[13] are considered, showing that there are an estimated 76 million cases of food-borne disease a year, resulting in 325 000 hospitalisations and 5000 deaths. It is also estimated that in Western countries, over 30% of all food produced is disposed of unused, as has been noted earlier in this chapter. Apart from the issues of human suffering, there are the enormous cost issues of these figures to also be considered. The simple colour changes in a pack that TTIs can achieve, to show whether the cool chain has been maintained and the food or drink is fresh, are therefore vital technologies for the packaging supply chain. Companies such as BASF with the On Vu system, Cryolog's (eO), Vitsab's Check Point, Temptime's Fresh-Check and Avery Dennison's TT sensor have all been used in recent years in this sector of the intelligent packaging market. Whilst some reluctance to adopt these types of systems has been apparent in the supply chain in recent years, due to factors such as cost, the development of TTIs over the next decade will be more dynamic, as the pace of post-recessionary innovative change quickens.

Cryolog's (eO) freshness label is a good example of TTIs at work. Based on patented technology, it uses micro-organisms to simulate the actual deterioration of food stuffs. The use of microbiology allows for the precise tracking of what is happening inside the food. Cryolog's (eO) indicates the level of freshness of a product through a change in colour. It is one of the innovative solutions to overcome our concerns over food freshness.

1.4.4 Thermochromic ink technology

Thermochromic ink technology, utilising special microcapsules, is an innovation of great significance in the intelligent packaging sector and one that will also expand over the coming decade. Originally launched in 2005, it essentially lets the consumer know when the contents of a pack have been cooked for long enough or when the contents have been chilled for long enough in the refrigerator. An example of this innovation is the Thermochromic Bevcan by Ball Packaging Europe for Miller Coors Brewing's 'Coors Light' brand. The Rocky Mountain logo on the can turns from white to shimmering blue. The consumer then knows the beer is ready to take from the fridge and drink. This particular pack is the first time that thermochromic ink has been successfully used for a design covering a large surface area. Future developments by Ball will make it possible to change colour not just from 'white' to 'blue', but from any colour to any other colour. Subsequent innovations are also planned by the company, including making the starting colour of the can change to different hues on the same can.

In the summer of 2011, Miller Coors announced its latest cold-activated technology: a new two-stage super cold activation, which makes it easy to see when the beer goes from cold to super cold. The Rocky Mountains still turn blue when the beer is cold and in addition a new cold indicator bar will turn blue at the same time. When the beer's temperature is at what is considered to be the peak of refreshment, a second super cold indicator bar turns blue. These are just two examples of where thermochromic ink technology can offer both improved functional performance and marketing and consumer benefits.

Irreversible thermochromic technologies have also impacted in the intelligent packaging sector in recent years. As microwaves and ovens differ in the amount of heat they really generate, food is sometimes overcooked or still cold on the inside when it is thought to be ready. Companies such as BASF have been developing an irreversible thermochromic technology that allows the food producer to add to the outside of the food pack a design that appears only when the product is ready to eat. Irreversible thermochromics are based on normal printing inks and can therefore be integrated into any existing food packaging. This is another innovation that is poised for further growth.

1.5 Bioplastics packaging market

1.5.1 Overview

The bioplastics for packaging market in Europe in 2009 was small, representing around 100 000 tonnes per year. However, these numbers are poised to increase rapidly over the next decade (Chapter 11). In fresh food packaging, bioplastics can enhance the product's lifespan compared to using petrochemical plastics. Whilst they are not going to be able to fulfil all purposes, they can offer longer shelf life. Leading groups are backing the R&D currently going into bioplastic materials, with companies such as Nestlé sponsoring, and indeed co-funding, chairs in bioplastics at leading international academic institutions.

1.5.2 Compostabilty and biodegradability

Compostability and biodegradability of bioplastic materials are vital for the future. Companies such as Innovia Films already have attractive product offerings in this area. They claim that their Nature Flex NK material is not only biodegradable and compostable but has the highest moisture barrier of any biopolymer film currently available. Like many materials in the bioplastics field, it is starting from a relatively low base in terms of volume, but has achieved significant growth, with considerable further potential over the next ten years. The barrier features of bioplastics materials make them an exciting innovation for the global fast-moving consumer goods (FMCG) industry. However, there are still issues, both in terms of capital and set-up costs and comparative pricing of finished material, to be overcome.

1.5.3 Plantic

The Australian company Plantic, who offer their Plantic trays, are also worthy of mention in terms of innovations in this field. The trays are made from cornstarch and are biodegradable, home compostable and compostable to both EU and US standards. In 2010, Marks and Spencer in the UK adopted Plantic's compostable trays for its range of Swiss chocolates. The trays are derived from 85% renewable sources and offer anti-static and odour barrier solutions, essential for chocolate production. The trays take about three weeks to completely break down on a garden compost heap or dissolve in minutes when put in water. Plantic claim their energy requirements are low compared to petrochemical-based polymers, but they are more expensive to produce. The Plantic material was also chosen by Plush Chocolates in 2009 for its Luxury Fairtrade English and Belgian chocolate collections.

1.5.4 Polylactic acid (PLA)

The benefits of other materials, such as polylactic acid (PLA), will continue to challenge the minds of those in the packaging supply chain. PLA is a biodegradable material derived from renewable sources such as cornstarch or sugar cane. Producers claim it has better structural properties than some petroplastics, allowing a 10 to 15% reduction in weight. Use of PLA has been limited in some applications by low melting temperatures. However, it has been claimed that recent technological advances, allowing the control of the molecular orientation and crystallisation of PLA, can produce a second generation of material with a melting temperature of 175°C.

US bioplastic producer NatureWorks is a leading company in PLA production. It is positive as to the material's performance characteristics and suitability for a wide variety of applications. In October 2011 the company, which was previously owned by US agri-business Cargill, announced that it would invest in a new PLA plant in Thailand, following the purchase of a 50% stake in the business by PTT Chemicals of Thailand. Cargill will own the remaining 50% of NatureWorks. The proposed new Ingeo plant, the brand name for the company's biopolymers made from renewable materials, would be NatureWorks' second production operation; it will come on-line in 2015. The company already has a 140 000 tonnes/year PLA facility in Blair, Nebraska, USA.

1.6 Recycled polyethelene terephthalate (PET) market developments

1.6.1 Closed Loop bottle-to-bottle recycling

The demand for recycled PET is increasing in many consumer markets. Capacity is coming on-stream to process and recycle PET bottles, but there is a shortage of raw materials coming onto the market. Closed Loop bottle-to-bottle recycling will drive the market forward and the developments in the UK by the company London Closed Loop Recycling are worthy of note. The company currently sorts, granulates, washes and cleans 35 000 tonnes of recovered plastics bottles a year. It plans to double capacity in 2012. A major driver in its growth is the demand from major FMCG producers and retailers for recycled plastics materials.

1.6.2 Coca-Cola and PepsiCo plant bottles

In the global market, the developments by Coca-Cola in the recycling and use of bioplastics are part of an important innovative trend. The key driver here is the

Plant Bottle, something which Coca-Cola has been pioneering since 2009. Up to 22.5% of the Plant Bottle is produced from plant-based PET. The plant-based part is currently made from sustainable sugar cane from Brazil. Up to 25% of the bottle is recycled PET plastic. It was initially introduced in Denmark and Western Canada in late 2009 and is now available in 20 different countries in a variety of Coca-Cola's drinks, including Coca-Cola itself. It is believed that usage of the Plant Bottle will double in 2011.

Not to be out done by its rival in the international food and beverage market, PepsiCo announced in March 2011 that it had developed the world's first 100% plant-based PET plastic bottle, which is also 100% recyclable. PepsiCo will undertake pilot production of the new bottle in 2012 and will move to full-scale commercialisation upon successful completion of the pilot launch. It was said that the development would enable the company to produce plastic containers from non-fossil base resources with a significantly reduced carbon footprint. It was further announced in March 2011 that Coca-Cola and Heinz had entered into a collaboration agreement to produce PET plastic ketchup bottles made partially from plant-based materials. The key driver for this is the Plant Bottle. For more on this development see Chapters 5 and 11.

1.6.3 Procter and Gamble-Pantene developments

In May 2011, Procter and Gamble announced that it had started producing its Pantene shampoo and conditioner bottles made primarily from plant-based materials. The new sugar cane-based bottles will be replacing traditional HDPE plastic containers for Pantene's NatureFusion collection. The packs are being initially launched in Western Europe, with the rest of the world to follow over the next two years.

These types of developments are vital parts of the sustainability programmes of major supply chain companies and are key innovations in the global packaging industry.

1.6.4 Tetra Pak-Braskem developments and Nestlé

In 2010 Tetra Pak signed an agreement with Brazilian thermoplastic resin producer Braskem, one of the world's largest producers of biopolymers, to buy small quantities of HDPE from renewable feedstock, as a raw material in the manufacture of plastic caps and closures. Braskem had announced plans to open the world's first commercial-scale environmental polyethylene plant by the end of 2010 and said it planned to deliver the first batches to Tetra Pak in 2011. It was therefore of great significance when, in August 2011, Nestlé said it had joined with Tetra Pak to launch a cap made from plant-based plastics for two of its Brazilian milk brands. This is thought to be the first polyethylene (PE) closure made out of sugar cane derivatives and a significant event, coming as it does from two leading international groups, as part of their moves towards further global reductions of greenhouse gases. Tetra Pak has said that its goal is to achieve

100% renewable packaging; Nestlé's global social responsibility platform is called 'creating shared value', a theme that is very much part of global initiatives of this kind.

Developments and partnerships of this type will become a regular part of the bioplastics market from now to the end of the decade. Bioplastics may currently be more expensive to produce compared to all petrochemical-based materials, but with economies of scale benefiting major players over the longer term and the massive rises in oil prices in 2011, they will become more cost-effective and an attractive sustainable proposition longer term. However, infrastructure and initial set-up costs will be a major hurdle to overcome in the current recessionary climate. It is probably true that the financial muscle and economic power of companies of the size mentioned in this section will be necessary to drive these innovations through to commercial success.

1.7 High performance barrier additives, materials and coatings

1.7.1 Market overview

In 2009 Smithers Pira valued the functional additives and barrier coating for the plastics market at $366 million globally. Development of high performance barrier materials, coatings and additives are a key factor in the plastic container market. Additives and barrier coatings can enhance, improve or modify barrier properties. ColorMatrix is one of the leading international companies in this sector, with operations in North America, Europe, South America and Asia. Its PET barrier technology can ensure product protection and longer shelf life in its oxygen-sensitive beverages, particularly beers, wines and juices. The company's AMOSORB SOI02 technology provides an active barrier against oxygen ingress through container walls. In July 2011, the private equity group Audax of the US announced the proposed sale of ColorMatrix, a deal that could be worth more than $400 million for Audax, according to Reuters. It was subsequently sold to PolyOne Corporation of Ohio, USA in January 2012. In October 2011, APPE (the packaging division of La Seda of Spain) announced a major collaboration with ColorMatrix and specialist mould-maker KTW to develop a new barrier technology for PET and other plastic containers. This is initially intended for the juice market, with food and other drinks applications to follow. APPE is the only company licensed to sell the technology which, it is claimed, will offer advantages over existing barrier methods, and also provide excellent material clarity and recycling benefits.

1.7.2 Development of the PET container market

The PET container market has changed considerably since its introduction over 30 years ago. Increased application for PET containers revolves around many things including, of course, economics and material prices. However, an important factor will always be barrier improvement. As has already been noted, barrier bottles are proving increasingly successful and this technology is rapidly developing.

Barrier technologies now exist across a wide range of beverage products. Leading companies are developing oxygen scavenging technology for hot-fill products. The conversion to PET containers for products such as juices and nectars, table sauces, jams and coffee and iced ready-to-drink teas is likely to increase. The usage of PET for beer still remains an area for debate and there is still reticence in traditional beer-drinking countries to adopt the PET pack. However, this is changing, as we shall see in Chapter 11.

1.7.3 Development of the market for plastic container barrier technology

As we shall see later in this chapter and in Chapter 11, there is a battle in the packaging industry between traditional forms of packaging such as glass and metal and newer technologies such as plastic containers, which incorporate barrier technology. Over the last few years, more and more companies have been evaluating plastic containers as a viable option to traditional glass packs, because of the benefits they can provide. These are many, not least the light-weighting advantages already discussed. However, the benefits of the extended shelf life as a result of the use of new technologies, is driving the market forward in a big way. RPC Group, a leading international plastics packaging producer, with operations throughout Europe and one in the USA (Chapter 11) as part of the important trend of globalisation of companies in the packaging industry, acquired Superfos Industries of Denmark for £203 million (240 million Euros) in December 2010. Superfos claim to have technologies that can double the shelf life of certain food products compared with conventional packaging. This fits perfectly with the requirement of food and drink supply chain companies who are demanding products with a longer shelf life, greater protection of flavour, good nutritional content, visual appeal and of course, weight reduction. For example, Premier Foods announced in 2010 that it was changing its Sun-Pat peanut butter packs from glass to recycled PET (rPET). The jars are said to be 90% lighter than glass and the use of 50% recycled material is also an environmental advantage. To reinforce the point about innovation by particular companies, the jars are made by RPC, Blackburn, UK.

In April 2010, Marks and Spencer became the first retailer in the UK to convert its entire range of mini (25 cl) still wine bottles to PET. In addition to saving 525 tonnes of packaging a year and achieving a weight reduction of 88% over glass bottles, the containers utilised technology by having two layers of PET with a barrier material inbetween, preventing oxygen entering the bottles.

Thus it seems the relentless rise of plastics barrier technology will further erode the glass packaging market.

1.8 Current status of the market for glass containers, plastics containers and metal packaging

The long-term implications for the market in terms of plastics packaging versus glass packaging will be dealt with in more detail in Chapter 11. As regards current status, glass packaging has many advantages and has achieved considerable recycling success, reductions in material usage and the creation of innovative new shapes, decorations and designs. Metal packaging too has many benefits in food and drink markets and has achieved much in light-weighting, particularly in cans, advanced decorative and shaping techniques, developments in printing technology (including digital), innovations in easy-opening devices and enhanced re-sealability. Glass will retain a share in up-market applications, such as added value wines and spirits and in niche and prestige food market sectors. Notwithstanding all these issues, improved shelf life, light-weighting and continuing advances in recycling of plastics will expedite the move away from glass, with further market share gains likely for plastics.

Other factors have impacted in the metal container market, including the threat from BPA in the epoxy linings in cans and drinks. The monomer is widely used in the linings of food and beverage cans and by the end of the period under review had been banned for children's food packaging in several countries throughout the world. This factor has had a negative impact on volumes. Alternative materials to BPA with greater consumer acceptance are being developed.

1.9 Innovations in paper and paperboard packaging

1.9.1 Overview

Paper and paperboard packaging has greatly benefited from its ability to provide high recycling rates, ensuring greater sustainability advantages in the use of materials. Carton-board and paper are the most highly recycled products in Europe. Indeed the European carton-board and paper industry voluntarily undertook to achieve a recycling rate of 66% by 2010. In 2011, CERES Logistics[14] estimated the percentage of recycled material in corrugated cartons in the UK was, on average, 76%. The Confederation of Paper Industries,[15] UK puts the recycling rate for corrugated board at over 80%.

1.9.2 Aseptic packaging

Innovations in aseptic packaging technology have provided consumers with greater food safety, quality and hygiene. It is estimated that 30% of all aseptic carton packs are now small-sized packs, up to 33 cl in size, compared with a figure of 20% at the start of the period under review. This development has been influenced by the trend towards on-the-go beverages. Convenience has been the driving force over the period in the carton market, with shapes and styles needing to offer enhanced ease of opening, pouring and drinking.

1.9.3 Retail-ready packs

There has been a rise in the use of retail-ready packs, which would previously have been seen purely as transit packs, into consumer-facing packs, ready for display on the supermarket shelf. The need for product differentiation at the point of purchase and the requirement for cost savings in the complete supply chain have prompted packaging supply companies, working with FMCG partners, to produce packaging that encompasses both transit and display features. As a result, corrugated transit packs have been transformed in recent years and, with attractive print and graphics, can offer consumer-facing display benefits. The demands of a 24/7 culture and, as a consequence, greater speed to market has also prompted this trend. The move towards packaging reduction and the need to provide environmental benefits in all aspects of the supply chain has encouraged the features of minimal handling, attractive display and designs that ensure easy conversion from transit to retail ready, to be incorporated into the new generation of packs.

Improved pack presentation and supply chain efficiencies have greatly aided in the rise of a new generation of corrugated packs, which look good on the shelf and also offer sustainable advantages. The advances in corrugated board grades have aided this whole process. As a result of improved pack quality and aesthetic appearance on the shelf, and the ability of corrugated packs to offer superior print performance, the conversion of transit to consumer-facing packs has gained momentum. The whole sector has also benefited from the continuing development of radio frequency identification (RFID) technology, which will be dealt with in detail in Chapter 3. Essentially, enhanced delivery systems and greater efficiencies in supply chain logistics have been major consequences of these developments.

1.10 Holographic images

1.10.1 Market overview

The whole area of enhancing and securing consumer brands is dealt with in Chapter 3, from an innovations and technical viewpoint. In terms of market trends, for many years the holographic image has been something which the packaging industry has evaluated, enthused about but, because of cost considerations, has sometimes been reluctant to embrace on a large scale. It is well known that in the drinks industry, when attempting to overcome the threat of counterfeiting and parallel trade, particularly in Europe and in the Far East, holograms on packs have proved to be extremely beneficial.

One of the important things to realise from a market trends and current market status viewpoint is the continuing rise in illegal activities such as counterfeiting, which is a major threat to world trade. In 2005 a report by the Organisation of Economic Co-operation and Development (OECD)[16] said that up to \$200 billion of international trade could be in counterfeited or illegally copied goods. In November 2009, updated estimates by the OECD said that the share of counterfeited and pirated goods in world trade had increased to almost 2% in 2007. These figures are staggering enough in their own right, but are particularly concerning when it is realised that other organisations, such as the lobby group of the International Anti-Counterfeiting Coalition,[17] estimate a figure of $600 billion is more accurate, including counterfeits bought on-line. Thus the need to find ways to overcome the counterfeiting issue are now essential, particularly as more and more counterfeited goods are appearing in the supply chain.

1.10.2 Developments in holographic foil

A recent innovative development in the holographic foil market, which could prove to be of great benefit, is the Crown Speciality Packaging Europe HoloCrown technology, which allows holographic images to be stamped directly onto decorative metal tins. Claimed to be the first for metal packaging, the technology uses the diffraction of light from the design on the tin to create a cost-effective three-dimensional image that continuously changes position and colour. This concept has the potential to reduce the risk of product counterfeiting.

1.10.3 Holographic images for mock-ups

In other developments, there are suggestions that holographic images could be of greater benefit in mock-ups and customisation of packs. For example, API Foils, a leading international holographic foil producer, in April 2011 launched a bespoke holographic product called Holonique. With this product, API is seeking to create a holographic mock-up in a quick and inexpensive way. The company claim that the mock-ups look bespoke but cost only around £250, compared to traditional bespoke holographic products, which can cost around £3000. This is certainly an important step in cost-effective innovation in recessionary times.

1.11 Nanotechnology

1.11.1 Market overview

The nanotechnology global market was estimated to be worth US$ 11.6 billion in 2007. According to BCC Research,[18] the nanofood market increased from US$ 2.6 billion in 2003 to US$ 5.3 billion in 2005. The Helmut Kaiser Consultancy[19] believes that the nanofood market could grow to US$ 20.4 billion in 2015. However, whilst these figures seem impressive, indeed some might say ambitious, they need to be considered in some detail to see the picture in relation to the use of nanotechnology for food and drink packaging. There are more statistics on the nanotechnology market in relation to packaging from other organisations, which confirm the potential of the technology (Chapter 9).

What is without question is that there is a belief that the implementation of nanotechnologies in consumer goods markets could grow rapidly. However, future developments and acceptance will depend on the outcome of current legislative and regulatory discussions in the US and Europe. There is currently no framework in place to assess the toxicity of nanomaterials and this is therefore a cause for concern in both the US and Europe. What is clear is that many large companies have adopted a watching brief and 'below the radar' involvement in nanotechnology R&D. Issues regarding migration of chemicals, supply and consumer safety need to be resolved. Leading companies, including Bayer, Kraft and Kodak, have carried out studies to investigate the use of nanoparticles to develop food packaging applications. UK applications over the next five to ten years will occur, but these will be dependent on regulatory issues, including recycling, disposal and reuse, safety issues and cost factors.

1.11.2 Westminster Food and Nutrition Forum

At the Westminster Food and Nutrition Forum[20] in London in July 2010, Kathy Groves of Leatherhead Food Research, UK, said: 'The emergence of new nanotechnologies in food packaging could help preserve the freshness and durability of products, so helping to cut waste.' This would support the view that in food and beverage packaging applications nanotechnology can assist by reducing spoilage in sectors such as bakery and meat products and in the protection of carbonated drinks and bottled water.

What is without doubt is that the consumer needs to be involved at all stages of the nanotechnology debate, as consumer acceptance is vital for the success of the technology. If the public can be convinced that nanotechnology used in food and drink applications can help reduce waste, this will be a major step to widespread commercial success.

It is true that the European market is currently small in comparison to the US and Japan, in terms of market value and current applications for nanofood technology. The whole nanotechnology field will be dealt with in detail in Chapter 9.

1.12 Consumer technologies, on-line retailing and social networking

1.12.1 Overview and the rise of Twitter

Although some in the US and elsewhere in the world would possibly differ, there are many who would agree with Eric Schmidt, Executive Chairman of Google, that the UK is still one of the world's leaders in technology, innovation and inventions. Schmidt, in an article in *The Daily Telegraph*, London of 16 May 2011, claimed that the UK is leading the world in e-commerce. This, he said, is a testament to the UK's, 'entrepreneurialism and willingness to embrace new technology'.

This opinion, which some, particularly in the US and the Far East would challenge, is supported by a survey in July 2011, on how the UK is using and adapting to Twitter, the social media that has transformed how information is spread and shared throughout the world. According to the survey by The Group,[21] which tracks how businesses use social media, by the end of June 2011 56% of FTSE100 companies had an active Twitter account, compared with 40% at the end of December 2010. Updated figures in 2012 indicated this number had risen to 61%. However, also according to the survey, only 12% of blue-chip companies had installed corporate blogs. Corporate businesses seem reluctant to engage in the 'blogasphere', but perhaps this will change in time. What the survey demonstrates is that companies are now realising Twitter is a medium that cannot be ignored. Indeed Twitter itself said in May 2012 that it now had more than 10 million users in the UK and 140 million global users in total. The UK is believed to be the fourth largest user country in the world, with the first three being the US, Brazil and Japan.

1.12.2 On-line shopping

Innovations in the FMCG market take many forms. We have experienced a massive rise in on-line shopping, with the floatation of the Ocado on-line retail business in the UK in June 2010 as an example of this important trend. Indeed, the Institute of Grocery Distribution[22] said that the UK on-line grocery market was worth £4.8 billion in 2010 and is forecast to rise to £11.2 billion by 2016. Waitrose Supermarkets have launched a web delivery service and Tesco and Sainsbury's are growing their on-line services in the UK. Others may follow in the not too distant future. Indeed Amazon, the international on-line retailer, sells grocery and food products from its web site on a global basis and the trend generally shows no sign of abating.

1.12.3 Facebook

Where these on-line developments and others, such as the phenomenally successful Facebook that had the largest ever internet company Initial Public Offering (IPO) of shares in May 2012, will leave the packaging market is not totally clear. However, an example of how Facebook could help is illustrated by a campaign launched by Friends of Glass,[23] a European consumer forum that supports and promotes the right of consumers to be able to choose food and drink products in glass packaging. It launched a social media campaign on Facebook in September 2011, introducing a competition that aims to reward people who recycle glass and to encourage them to recycle even more and in better ways. Called 'Pass the Bottle', the idea is that the 6000 Friends of Glass Facebook fans are asked to recycle a virtual glass bottle and invite new friends to do the same.

1.12.4 A multi-channel world

Retailers are finding the benefit of talking direct to customers by Twitter and other social media methods. More new product launches, more on-line packaging promotions and, via the use of technology, greater access to customer data and information are all part of the modern consumer world. Of course, traditional forms of shopping will continue, but more and more packaging brand owners and, as a result, packaging producers will embrace the multi-channel world. With the time constraints and hectic lifestyles of today, consumers will continue to make more of their purchases on-line, and on-line consumer technologies will have a large part to play in future strategies and innovations of the packaging supply chain.

1.13 References

1 Government Review of Waste Policy in England (2011) Available from: www.defra.gov.uk/. . ./files/pb13540-waste-policy-review110614.pdf
2 Courtauld Commitment Phase 2 (2012) 'Wrap'. Available from: www.wrap.org.uk/retail_supply_chain/. . ./courtauld_commitment
3 Wrap-Grant for Greenstar/Biffa (2012) 'Mixed plastic packaging plant'. Available from: www.wrap.org.uk/recycling. . .by. . ./plastic/new_mixed_plastics.html
4 British Retail Consortium (BRC) (2011) 'Recycling for thin plastics/cereal packaging liners/plastic bags'. Available from: www.brc.org.uk/brc_news_details.asp?id=1931
5 Global Packaging Project (2011) 'The Consumer Goods Forum'. Available from: www.theconsumergoodsforum.com/ [Accessed October 2012].
6 Europen (2011) 'European Organisation for Packaging and the Environment'. Available from: www.europen.be/ [Accessed October 2012].
7 Swedish Institute for Food and Biotechnology (SIK) (2011) Available from: www.sp.se/en/units/sik/sidor/default.aspx [Accessed October 2012].
8 Smithers Pira (2011) 'Functional Additives and Barrier Coatings Market Research'. Available from: www.smitherspira.com/the-future-of-functional-additives-and-barrier-coatings
9 Smithers Pira (2011) 'The Future of Active and Modified Atmosphere Packaging'. Available from: www.smitherspira.com/Future-of-Active-and-Modified-atmosphere-packaging
10 PCI Consulting (2011) 'European Flexible Packaging Market 2011'. Available from www.pcifilms.com>. . .>FlexiblePackagingMarketReports
11 Allied Development Corps (2011) 'Stand-up pouches market'. Available from: www.allied-dev.com/studies/stand-up_pouches_details.aspx [Accessed October 2012].
12 Freedonia Group (2009) 'Demand for Smart and Intelligent Packaging'. Available from: www.freedoniagroup.com/brochure/25xx/2515smwe.pdf [Accessed October 2012].
13 World Trade Organisation (2011) 'Food-borne diseases per year'. Available from: www.who.int/mediacentre/factsheets/fs237/en/ [Accessed October 2012].
14 CERES Logistics (2011) 'Percentage of recycled material in corrugated cartons in UK'. Available from: www.cereslogistics.co.uk/docs/home/index_home.html
15 Confederation of Paper Industries, UK (2011) Available from: www.paper.org.uk [Accessed October 2012].
16 OECD (2011) 'Report on international trade which is counterfeited'. Available from: www.oecd.org/dataoecd/13/12/38707619.pdf [Accessed October 2012].
17 International Anti-Counterfeiting Coalition (IACC) (2011) Available from: www.iacc.org/ [Accessed October 2012].
18 BCC Research (2011) 'Nanotechnology market'. Available from: www.bccresearch.com/report/NAN031D.html [Accessed October 2012].
19 Helmut Kaiser Consultancy (2011) 'Nanofood market'. Available from: www.hkc22.com/nanofood.html [Accessed October 2012].

20 Westminster Food and Nutrition Forum (2010) London, July 2010. Available from: www.westminsterforumprojects.co.uk/forums/publications.php?. . .wes
21 The Group (2011) 'FTSE 100 On-line Audience Report'. Available from: www.the-group.net/blog/index.asp?blogid=539 [Accessed October 2012].
22 Institute of Grocery Distribution (2011) 'UK on-line grocery market'. Available from: www.igd.com>. . .>Consumers&shoppers>Factsheets
23 Friends of Glass (2011) 'Pass the Bottle', European Consumer Forum. Available from: www.glassonweb.com>News

2

Modified atmosphere packaging and other active packaging systems for food, beverages and other fast-moving consumer goods

A. Emblem, London College of Fashion, UK

DOI: 10.1533/9780857098979.22

Abstract: Modified atmosphere packaging (MAP) is the fastest growing method of food preservation and is widely used for a variety of fresh and dry foods. This chapter examines the key principles of MAP, such as how it works, along with the packaging material options and the demands of the packaging operation.

Key words: modified atmosphere packaging (MAP), food spoilage, gases, vacuum packaging.

2.1 Introduction

For the purposes of this chapter, the term 'active packaging' is used to describe a packaging system that changes the condition of the product in some way. When exploring reasons why this may be of interest to manufacturers, retailers and consumers, undoubtedly one of the primary motivators, especially for perishable foodstuffs, is to delay spoilage and thus extend shelf life. Modified atmosphere packaging (MAP) is now an established active method of extending shelf life, and is the main subject of this chapter. However, there are other reasons for wanting to change a product's condition and packaging systems, such as those that can bring about product heating or cooling. These and other developments are covered in Chapters 1 and 9. The whole area of preservation systems, including oxygen scavenging and antimicrobial systems, is discussed in Chapter 4.

2.2 Development of modified atmosphere packaging (MAP)

Modified atmosphere packaging (MAP) is a packaging system that involves changing the gaseous atmosphere surrounding a food product inside a pack, and employing packaging materials and formats with an appropriate level of gas barrier to maintain the changed atmosphere at an acceptable level for preservation of the food. Changing the gaseous atmosphere may mean removing air completely, that is by vacuum packaging, or replacing air with other gases. While the latter is more common, vacuum packaging has a significant role to play in food preservation and is included in this chapter.

2.2.1 Market development

MAP is now a well-established method of food preservation, widely available in Europe since the 1980s, largely due to the retailer Marks and Spencer's introduction of MAP for cuts of fresh meat in the UK in the 1970s. Since then, its use has been extended to fish, fresh produce, pasta, pizza, other baked goods, and dry products such as nuts and snacks, and it is believed that MAP is the fastest growing method of food preservation at the expense of more traditional methods such as retorting and freezing. The UK remains one of the most prominent markets, along with France, and usage is increasing in other European countries, especially Eastern Europe. The relatively short distances and well-established distribution systems in Europe have favoured the use of MAP, while it has been slower to become established in the United States, reportedly due to fragmented and long-distance distribution chains. Globally, MAP is expected to show significant growth in the current decade (*marketsandmarkets.com* 2011).

2.2.2 Key reasons for the development of MAP

MAP is an attractive method of food preservation, because it provides extended shelf life with minimal/no deterioration in taste, texture or nutritional value of the food. This is in contrast to methods such as freezing, canning (i.e. heating) and the use of chemical preservatives, all of which have the potential to significantly change some aspect of the product. For example, frozen soft fruit will generally collapse and lose its structure on defrosting, canned vegetables often show colour changes and chemical preservatives such as salt and dilute ethanoic acid (vinegar) undoubtedly change the taste experience for the consumer. In contrast, MAP can result in none of these negative effects, delivering to the consumer food that remains in its original 'fresh' condition, with minimal use of preservatives. Today's busy lifestyles mean that consumers want pre-packed, easy and convenient to use products, but do not want to compromise on quality, and market studies consistently show that both freshness and 'preservative free' are highly valued attributes. MAP is well placed to deliver these desirable attributes.

Consumers also value the opportunity to see the product before making the purchasing decision, and this is readily achievable with transparent flexible/ semi-flexible packs, which offer all-round visibility, in contrast, for example, to canned foods. MAP can offer a clear view of the product as it is, ready for eating or cooking, allowing the consumer to inspect features such as colour and consistency. Another consumer benefit is that sliced goods, for example cooked meats and bacon, are easy to separate, whereas when such products are preserved by freezing, the entire pack usually has to be thawed out before slices can be removed.

These consumer-oriented benefits mentioned so far apply equally well to institutional users such as hospitals and school meals, and to the catering sector, thus further extending the appeal and popularity of MAP. The growth of in-store meat and fish counters has led to a demand for MAP semi-bulk packs, where a centralised butchery/fish preparation company carries out all of the cleaning, cutting and trimming operations and delivers packs, of say 12 pork chops or salmon fillets, to the supermarket. These are opened as required and displayed on the fresh meat or fish counter and sold to the consumer in the desired quantity. Only when the pack contents are sold, does the sales assistant open another pack, thus maintaining freshness and significantly reducing the opportunity for waste.

It is MAP's potential for waste reduction of highly perishable foods such as meat, fish and fresh produce, which appeals to product manufacturers and retailers. The extended shelf life available, which can be anything from 50 to 400% (Davies, quoted in Gould 1995, p. 305) depending on the product, means that such products can be distributed over longer distances and in larger quantities than when not using MAP. These benefits provide a brand owner with opportunities to penetrate more distant markets and can result in reduced distribution costs.

2.2.3 Negative aspects of MAP

When using MAP, successful product preservation relies first on choosing the correct gases for the given product (to be discussed later) and obviously this will necessitate careful trials to establish product safety. Selection of the correct level of barrier in the packaging is equally important, as is the security and integrity of the pack seal; a poor seal means loss of the protective atmosphere and hence loss of preservation. Thus, making an effective seal on the packaging line is key and this requires careful monitoring of sealing conditions, such as time, temperature, pressure and cleanliness of seal area, plus a reliable method of checking seal integrity, preferably a non-destructive on-line method in which 100% of production is checked.

As well as food-quality gases, MAP requires specialised packaging lines, staffed by highly trained operators who understand the critical nature of the processes being carried out and are capable of monitoring machine and personnel performance to ensure that agreed quality standards are met. Each of these requirements brings investment costs in equipment and people and when added to the on-going cost of specialised packaging materials, it should be clear that MAP is not a low-cost packaging solution and companies need to carefully assess that the realisable benefits outweigh the total cost.

A further potential disadvantage of MAP is that, except for vacuum packs, all MAP packs take up more volume than the product itself, due to the need for the protective gaseous atmosphere. This is a disadvantage for distribution costs, to offset against the benefits mentioned above, and may also have a negative impact on consumers' perception of the product's value for money or even on the legality of the offering on display. Care must be taken at the product and packaging development stage to minimise pack dimensions in relation to product dimensions; of course, this is a legal and commonsense requirement for all packed products, but it is perhaps a more pertinent consideration when using MAP.

Finally, the use of MAP to preserve perishable food products does not remove the need to maintain the products at reduced temperatures during all stages of the supply chain, both before and after packing. This means paying careful attention to temperature and hygiene conditions in the food preparation area and on the packaging line, and using the correct temperature in all warehousing, distribution vehicles and retail display equipment. Also, it is essential to provide clear instruction to the purchaser on how to store the product in the correct conditions at home prior to use.

2.3 Principles and applications of modified atmosphere packaging

This section provides only sufficient information to understand how MAP works; it is not a source of detailed information on food science, such as would be needed by product and processes development personnel responsible for bringing a product to the market.

As already stated, the main purpose of MAP is the extension of shelf life and the first requirement for success is to understand the spoilage mechanism of the product, such as what factors will cause the product to become unacceptable. Possible solutions to avoid/minimise spoilage can then be investigated, and in MAP this means making an appropriate choice of modifying gas and of barrier packaging.

2.3.1 Basics of food spoilage

Foods deteriorate by three main processes: microbiological, chemical and physical. The main organisms causing microbiological deterioration in foods are bacteria, moulds and yeasts, with bacteria being the most prevalent. The extent of the spoilage depends on factors such as the microorganism load (type and number) present in the product, the pH (acidity) and moisture content of the product and the range of temperatures to which it is subjected throughout its life. MAP has a role to play in manipulating two of these factors, pH and moisture content; however, with regard to microorganism load, it is important to understand that

MAP cannot reverse microbiological spoilage and if the product is of poor quality prior to packing, it will be of poor (probably worse) quality when the consumer opens the pack. Also, as has already been mentioned, MAP does not remove the need for maintaining food products at the optimum temperature to minimise microbial growth.

It is essential to know which particular microorganisms are present when considering an MAP solution to microbiological spoilage. Aerobic organisms such as pseudomonas can only survive in the presence of oxygen and thus removal of oxygen, for example by vacuum packaging, will inhibit growth. Unfortunately many pathogenic organisms, such as *Salmonella*, *Clostridium botulinum*, *Lysteria* and *E. coli* types are anaerobic, that is, they will survive without oxygen and thus vacuum packaging provides insufficient defence.

Chemical deterioration includes oxidative rancidity, which is a particular issue for fat-containing foods such as bacon, meats, nuts and potato snacks. Removal of oxygen, with or without replacing with gases such as carbon dioxide and nitrogen, will prevent/delay this type of spoilage. Other examples of chemical deterioration, which can be addressed by MAP, are enzyme changes that can result in undesirable effects such as breakdown and liquefying in fish, fruit and vegetables; and non-enzymatic browning due to the Maillard reaction, which produces complex compounds that are brown or black in colour and change the product's flavour.

A major cause of physical deterioration in foodstuffs is a change in moisture content. This can be moisture gain, causing products such as nuts and snacks to lose their crispness, and fresh produce such as salad leaves to become soggy, which in turn will lead to microbiological deterioration. Moisture loss can be just as critical, for example bread and bakery goods can dry out, as can fresh fruit and vegetables. Again, MAP can be used to delay/prevent these examples of deterioration.

2.3.2 Gases used in MAP

For more detailed information on this aspect of MAP, see Sivertsvik *et al.* (2002). This text has been used as reference material in preparing this section.

Once the spoilage mechanism of a food product is understood, it is necessary to consider the most appropriate gas or, more commonly, gas mixture to address this. The main gases used in MAP are nitrogen, carbon dioxide and oxygen; there is also some use of carbon monoxide, sulphur dioxide and the noble gas argon. Nitrogen is inert and where spoilage is mostly due to product respiration, for example in fruit and vegetables, gas mixtures high in nitrogen are used to minimise the respiration rate. Where spoilage is via oxidative rancidity, nitrogen is a good choice to displace the oxygen levels in the pack. It is almost insoluble in water or fat, which means that it is not absorbed into the product and thus MAP packs containing high nitrogen level are not prone to collapse in storage due to a reduction in gas concentration in the pack.

Carbon dioxide is the most widely used MAP gas, and it has particular uses in the preservation of non-respiring products, for example meat, fish, dairy products and bakery goods, often in combination with nitrogen and oxygen. Unlike nitrogen, it is not inert and plays a more active role in food preservation. It is soluble in both water and fat in the product and inhibits bacterial and fungal growth, due to a combination of different factors, including a reduction in pH and retarding of enzyme reactions. The actual mechanisms of preservation using carbon dioxide are complex and beyond the scope of this chapter. The disadvantage of carbon dioxide's solubility is that this leads to reduction in pressure inside the sealed pack, resulting in packs collapsing. To help to counteract this, it is common to use a high gas to product ratio of two to three times, that is, the volume of gas is two to three times the volume taken up by the product. Using a mixture of carbon dioxide and nitrogen will also help to counteract pack collapse. Another disadvantage of carbon dioxide's solubility is the formation of a liquid exudate from the product; again, the addition of nitrogen will help to minimise this effect, while placing the product on an absorbent pad in the pack will reduce the amount of free liquid that is visible.

Where oxygen is included in MAP gas mixtures, this is normally at low levels, so that it does not cause unacceptable spoilage due to oxidative rancidity or the growth of aerobic microorganisms. However, especially for respiring products, aerobic conditions are more desirable than anaerobic, which can result in fermentation and off odours and flavours, and this is achieved using levels of 2–3% oxygen. At higher levels, for example 80%, it has been found to significantly extend the shelf life of lettuce (Day, 1998). Perhaps the most common use of high levels of oxygen in MAP is in non-respiring products, specifically red meat, where the oxygen is used to maintain the bright red colour associated with freshness and good quality in meats such as beef, lamb and venison. This red colour is due to the dominance of oxygenated oxymyoglobin in the muscle, rather than myoglobin, which is purple and metmyoglobin which is brown. Gas mixtures of 70 to 80% oxygen and 20–30% carbon dioxide maintain the desirable red colour (due to the oxygen levels), inhibit the growth of pseudomonas spoilage bacteria (due to the carbon dioxide) and double the shelf life of chilled meat at 4°C from 2–4 days to 5–8 days. Pseudomonas bacteria are aerobic and thus removing the oxygen would further extend the shelf life, but of course it would also result in an undesirable colour. The Norwegian meat industry has addressed this by the development of an MAP system for red meats, which uses carbon monoxide rather than oxygen to deliver the desirable cherry red colour. Typical mixtures are 60–70% carbon dioxide (i.e. high levels for effective inhibition of spoilage organisms), 30–40% nitrogen as a filler gas to limit pack collapse and less than 0.5% carbon monoxide. Shelf life at 4°C is extended to 11–21 days and the low levels of carbon monoxide are not considered to be a health concern (Opinion of the Scientific Committee on Food, 2001).

Whichever gas or gas mixture is chosen, obviously it must always be clean and suitable for direct use with food. Also, it should be noted that in the countries of the European Union, Directive 95/2/EC lays down rules for the labelling of MAP products, including the need to add E numbers for the gases used, for example nitrogen E941, oxygen E948 and carbon dioxide E 290.

2.3.3 Vacuum packaging

Vacuum packaging refers to the technique of removing air from a pack prior to sealing and it predates the use of gases as a means of food preservation. Its principal purpose is to remove oxygen by pulling the packaging material into intimate contact with the product. It works particularly well for frozen poultry such as turkeys, where the exclusion of air helps to reduce freezer burn, and for fatty fish such as salmon. Hand and semi-automatically operated vacuum packaging chamber machines are available, offering a relatively low cost option (vs. MAP gas packaging), for small- and medium-sized companies such as farms selling their own fresh meat and bacon.

A significant advantage of vacuum packaging is that the pack volume is virtually the same as the product volume, with no 'empty' space inside the pack. However, this also means that products such as joints of meat are of unequal size, making them difficult to display. Dry goods such as pasta (provided the product can withstand the force of the packaging material being pulled around it without breaking up) lend themselves well to vacuum packaging. Products such as ground coffee and dried yeast can be packed in regular brick-shaped packs, thus minimising storage and distribution costs.

Moderate vacuum packaging (MVP) is a variation of traditional vacuum packaging, used for respiring products such as prepared fruit and vegetables. The product is packed in a rigid airtight container or a pouch and is surrounded by normal air, but at a reduced pressure (around one-third of normal atmospheric pressure). This slows down the metabolism of the product and the growth of spoilage organisms (Laurila and Ahvenainen, 2002).

2.4 Choosing the packaging materials and formats

Having selected the most appropriate gas or gas mixture to deliver the desired extension to product shelf life, clearly it is essential that the pack is capable of maintaining the correct atmosphere, from the point of production to the point of consumption, and thus a rigorous approach to pack selection is required.

As with any pack development, choosing the most appropriate pack must start with a consideration of the specific requirements of the product, which in this case includes the atmosphere immediately surrounding the product once it is packed, be that gaseous or a vacuum. This means that the gas and moisture vapour barriers of potential packaging materials are of crucial importance and this is a good point at which to start identifying suitable options.

Only glass, steel and aluminium foil above around 20 microns in thickness can be regarded as absolute barriers, such that gases will not pass through these materials, while all polymeric materials have some degree of permeability. See Table 2.1 for typical values of moisture vapour and gas transmission rates, along with other useful properties such as transparency and usage temperatures. The data shows the amount of vapour or gas that will pass through a given thickness and area of material, over a given period of time, and thus the lower the quoted

Table 2.1 Typical values for barrier and other properties

Material	Moisture permeability g/sqm/day 25 μ 38/90	Oxygen permeability cc/sqm/day 25 μ	Max use temp °C	Transparency	SG
Low density polyethylene (LDPE)	15–20	6500–8500	65	Translucent	0.91–0.92
High density polyethylene (HDPE)	5–10	1600–2000	115	Translucent	0.94–0.95
Cast polypropylene (CPP)	10–12	3500–4500	120	Yes – film	0.88–0.92
Oriented polypropylene (OPP)	5–10	1500–2000		Yes – film	0.88–0.92
Polyvinyl chloride unplasticised (UPVC)	30–40	150–300	70	Yes	1.23
Polystyrene (PS)	70–150	4500–6000	65	Yes	1.07
Polyethylene terephthalate (PET)	15–20	100–150	220	Yes – APET	1.36
Polyvinylidene chloride (PVDC)	0.6–1.0	2–4		Translucent	1.64
Polyamide – Nylon	300–400	50–80	220	Yes – cast film	1.13
Ethylene vinyl alcohol (EVOH)	20–100	0.4–1.5		Translucent	1.19
Acrylonitrile butadiene styrene (ABS)	40–80	800–1100	110	Yes	1.07
Aluminium foil 9 microns	0.3				

Note: Barrier properties for plastic films apply to 25 micron film.

figure, the better the barrier. Plastic packaging manufacturers can supply such data; when using it to make comparisons, take care to check that the units of measurement and the test conditions are consistent. Note also that the data given applies to pristine materials as produced by the manufacturers, and the processes of forming, product filling and sealing of packs on the packaging line, as well as the physical hazards faced by packed products during storage, distribution and retail display, all have the potential to introduce folds and creases that will adversely affect the material's barrier to moisture and gases.

The vast majority of packaging materials used in MAP are flexible films, and these are seldom single layer materials. Co-extrusions of different polymer layers or laminates of films, aluminium foil and paper are widely used, opening up a wide range of options. Given the range of barrier properties available, it is, of course, essential to know what degree of barrier is needed for a given product, and this will depend on the product itself, the storage temperature and the shelf life required. For example, products such as fresh meat will require high barrier packaging, while respiring products such as fresh produce are often packed in permeable films, to allow water vapour to escape. Short shelf-life products stored under refrigerated conditions will require lower barriers than long shelf-life products stored at ambient temperature.

Although barrier is an important property, it is not the only requirement, and the following should also be considered:

- Is product visibility important? If so, obviously this precludes the use of opaque layers such as aluminium foil or metallised films. It may also mean that films with an anti-fogging coating may be required, so that product visibility is not obscured by condensation.
- How is the pack formed on the filling machine? The main options are vertical and horizontal form, fill and seal, and tray thermoforming. Thermoforming invariably results in thinning of the packaging materials and consideration must be given to the acceptable level of thinning that can be allowed. All machine operations will involve sealing of thermoplastic film/sheet materials, the softening points and other sealing characteristics, which must be compatible with the machine capabilities.
- What is the temperature of the product at the point of filling, that is, what degree of heat resistance is required?

Of course, consumer needs are essential considerations in any pack development, for example:

- Where and how is the packed product going to be used? Is it destined to be used outside of the home, for example prepared salads for the lunch market?
- How is the consumer meant to open the pack, access the product, reclose if relevant?
- Where and how is it going to be stored?
- Is the product meant to be heated in the pack? If so, by what method, for example immersing in boiling water, or placing in a microwave or a

conventional oven? Again, the chosen method will determine the heat resistance required of the packaging materials.

- What information has to be provided? Is there adequate space available on the pack?
- Other considerations are the economics of the finished pack, as well as the environmental acceptability, and compliance with all relevant legislation (for which specialist advice must be sought).

The answers to all of these questions will inform choice of pack format and materials and provide a starting point to what may be suitable.

Clearly, there is no 'one-size-fits-all' solution and once potentially-suitable materials have been identified, carefully constructed and managed storage trials are essential. Such trials should be carried out using conditions as close as possible to the real-life conditions to which the pack and product will be subjected. Trial packs should be made up on the packaging line it is intended to use, so that machine handling and sealing conditions are properly evaluated. Seal integrity throughout the life of the packed product is critical and it would be prudent to carry out simulated handling trials, mimicking the likely hazards to which packs will be subjected, such as shock, vibration and compression, as well as fluctuations in temperature. Test protocols should be drawn up, stating test conditions, how packs will be evaluated after testing, according to what criteria. Product safety must be evaluated, using chemical analysis and checking microbiological condition, as well as organoleptic testing. When all testing is completed to satisfaction, the data gathered should be used in compiling packaging, finished product and process specifications.

2.5 Packaging operations and quality aspects

The importance of seal integrity in MAP cannot be over-emphasised and once the optimum sealing conditions have been established during the development phase, steps must be taken to ensure that these conditions are rigorously complied with during the packaging process. Packaging machinery must be validated as being capable of consistently maintaining the required temperature, pressure and dwell time, and these should be checked regularly for compliance with the agreed specifications. Correct alignment of sealing jaws is essential, as well as controls in place to ensure that sealing surfaces are not contaminated prior to sealing. Usage of the correct amount of modifying gas(es) is equally important, and all materials used must be traceable to their original supply.

Both US and EU legislation require that a Hazard Analysis and Critical Control Point (HACCP) process must be carried out for MAP operations, to identify and manage risks to human health. The entire supply chain should be included when identifying risks and setting appropriate critical control points, from selection and supply of product, modifying gas and packaging, through to the packer/filler operation, storage, distribution, display and sale to the consumer.

All of these requirements mean that MAP operations demand a high level of skill, with an effective training programme in place, and refresher training delivered as and when needed. Machine operators need to have a thorough understanding of the processes for which they are responsible, including the on-line and off-line quality checks that will be required. Packs must be checked for seal quality and the use of non-destructive 100% on-line systems is becoming common, as well as statistical sampling for more detailed testing such as headspace analysis and microbial load.

2.6 Future trends

As stated, MAP is believed to be the fastest growing method of food preservation and this trend is expected to continue. There is no doubt that MAP offers significant benefits in terms of shelf life and thus global distribution with minimal waste. However, the cost of MAP may prove to be a barrier to the realisation of its full potential. It is not only the cost of high-quality barrier films that has to be considered but, probably more significantly, the operational costs associated with maintaining product quality and complying with food safety legislation. European legislation, in particular, continues to impose more and more demands and whilst no one would want to see a lowering of standards to the point where consumer health is adversely affected, it is up to industrial bodies to promote a 'common sense' approach and not allow excessive legislative demands to obstruct market growth.

2.7 Sources of further information

Several sources have been consulted in the preparation of this chapter, including the following textbooks and articles:

Books

AHVENAINEN R. (ed.) (2003) *Novel Food Packaging Techniques*, Cambridge, UK, Woodhead Publishing Ltd.

GOULD G. W. (ed.) 1995. *New Methods of Food Preservation*, London, Blackie.

GUYNOT M., MARIN S., SANCHIS V. and RAMOS A. J. (2003) 'Modified atmosphere packaging for prevention of mold spoilage of bakery products', *Journal of Food Protection*, 66(10), 1864–72.

KILCAST D. and SUBRAMANIAM P. (eds.) (2000) *The Stability and Shelf Life of Food*, Cambridge, UK, Woodhead Publishing Ltd.

OHLSSON T. and BENGTSSON N. (eds.) (2002) *Minimal Processing Technologies in the Food Industry*, Cambridge, UK, Woodhead Publishing Ltd.

ROONEY M. L. (ed.) (1995) *Active Food Packaging*, London, Blackie.

Web articles

JOBLING J. (2001) 'Modified atmosphere packaging; not as simple as it seems', *Good Fruit and Vegetables*, 11(5). Available at: http://www.postharvest.com.au/MAP_article.PDF [Accessed October 2012].

MIR N. and BEAUDRY R. (2002) 'Modified atmosphere packaging of fruit and vegetables'. Available at: www.ba.ars.usda.gov/hb66/01map.pdf

Other sources

General information on MAP:

- http://www.foodpackages.net/freepress/index.php?lan=enModified Atmosphere: the handbook in English.

MAP gases – selection and supply:

- http://www.modifiedatmospherepackaging.com/~/media/map/Pictures/MAP_Poster.ashx [Accessed October 2012].
- http://www.boconline.co.uk/products/products_by_industry/food_and_beverage/food_applications/modified_atmoshere_pack/index.asp [Accessed October 2012].
- http://www.hilineindustries.com/productDetail.asp?uID=31 [Accessed October 2012].

Quality aspects of MAP:

- http://www.pbi-dansensor.com/Products.aspx [Accessed October 2012] (headspace analysis, leak detection, etc.).

Legislation, safety and training

http://www.food.gov.uk/multimedia/pdfs/publication/vacpacguide.pdf [Accessed October 2012].

http://www.food.gov.uk/multimedia/pdfs/publication/vacpack0708.pdf [Accessed October 2012].

http://vacuumpackingtraining.food.gov.uk/introduction/ [Accessed October 2012].

http://ec.europa.eu/food/fs/sc/scf/out112_en.pdf [Accessed October 2012]. Opinion of the Scientific Committee on Food on the use of carbon monoxide in MAP for fresh meat.

2.8 References

DAVIES A. R. (1995) in Gould G. W. (ed.), *New Methods of Food Preservation*, London, Blackie, 305.

DAY B. P. F. (1998) Quoted in Ahvenainen R. (ed.) (2003) *Novel Food Packaging Techniques*, Cambridge, UK, Woodhead Publishing, Ltd.

LAURILA E. and AHVENAINEN R. (2002) 'Minimal processing in practice: fresh fruits and vegetables'. In Ohlsson T. and Bengtsson N. (eds), *Minimal Processing Technologies in the Food Industry*, Cambridge, UK, Woodhead Publishing Ltd., 231.

Opinion of the Scientific Committee on Food, adopted 13 December, 2001.

SIVERTSVIK M. *et al.* (2002) 'Modified atmosphere packaging'. In Ohlsson T. and Bengtsson N. (eds), *Minimal Processing Technologies in the Food Industry*, Cambridge, UK, Woodhead Publishing Ltd.

3

Augmenting and securing the consumer brand experience through smart and intelligent packaging for food, beverages and other fast-moving consumer goods

J. Plimmer, Product & Image Security Foundation, UK

DOI: 10.1533/9780857098979.35

Abstract: Whilst the terms 'smart' and 'intelligent' are often interchangeable, the world of true intelligence in packaging is still a long way off. However, there are growing numbers of highly innovative technologies currently in use in the packaging and labelling sector that are often euphemistically termed smart or intelligent by their manufacturers. These technologies provide access to information that relates to the status of a product and informs on its condition and provenance in a way that directly engages with both consumers and internal staff within the supply chain. It is this process of direct interaction that makes such technology so interesting and relevant in today's fast-moving packaging marketplace.

Key words: intelligent, smart, authenticate, radio frequency identification (RFID), Near Field Communications (NFC), hologram, tamper evidence, track and trace, quick response (QR) code, digital watermarking, augmented reality.

3.1 Introduction

The main purpose of packaging is to adequately protect its contents and contain them on their journey to market and final point of consumption or use. Packaging also needs to adequately identify itself and draw attention to its presence in order to attract attention on a retail shelf or through other forms of display. This identification function is also necessary in order to provide pertinent information to the consumer about what it is and who made it (Branding), together with a variety of other data, some of which is obligatory and some of which is optionally provided or assists the brand owner manage the distribution process.

3.1.1 The attributes that make packaging intelligent or smart

For packaging or labelling to be termed smart or intelligent, it needs to include special attributes or functions, each of which are designed to be used in isolation or combination to assist in some way with identifying the product with more clarity, authentication, or by autonomously tracking the product through the distribution cycle, track and trace. In this way, packaging can be manufactured to interact with people and events in which it comes into contact.

Additional attributes such as those that assist with product status, for example delivering a confirmation of security on first opening and verifying that a product has not been exposed to unacceptable variations in temperature, can also be described as smart. Taken a step further, it is also possible to identify whether a product has been contaminated by microbial growth or has past its best consumption date.

In this chapter we will look at the various forms of technology that have been developed to assist in delivering smart and intelligent features to packaging, how this functions and where it can be used to secure products against threats such as counterfeiting, theft and refilling. We will also explore where such technology can help and inform users if products are fresh and ready for use, or are out of date and have been exposed to unsafe conditions before purchase or consumption.

3.1.2 Differences between smart and intelligent analysed

To simplify and explain each technology and how it enhances the packaging role, it is necessary to place the function it provides in context with each of its counterparts (Fig. 3.1).

It can be observed in the figure that the delivery of information to the user can be delivered through the traditional routes of human sense, such as sight and touch, and also through the use of artificial electronic sensing devices such as bar code scanners.

The traditional practice has always been to provide specialist reading devices, such as bar code scanners within the closed supply chain, because the information they provide is also essential to the distribution function. Until now, such 'reading' devices have not been readily available for public use and consumers have been unable to easily access all the data available on a product, which can assist them to make a well informed choice when considering their purchase.

This tradition is now being challenged as smart phones become ubiquitous and many now offer the ability to read radio frequency identification (RFID) tags, bar codes and embedded watermarks through 'no cost' applications that can be loaded onto the phone in a matter of seconds by their user. The future of such technology and its impact on users will be discussed later in this chapter.

3.1.3 Intelligently managing the identification of a product

Since a primary function of intelligent packaging is about managing the identification of a product, it follows that the use of a unique number in this role

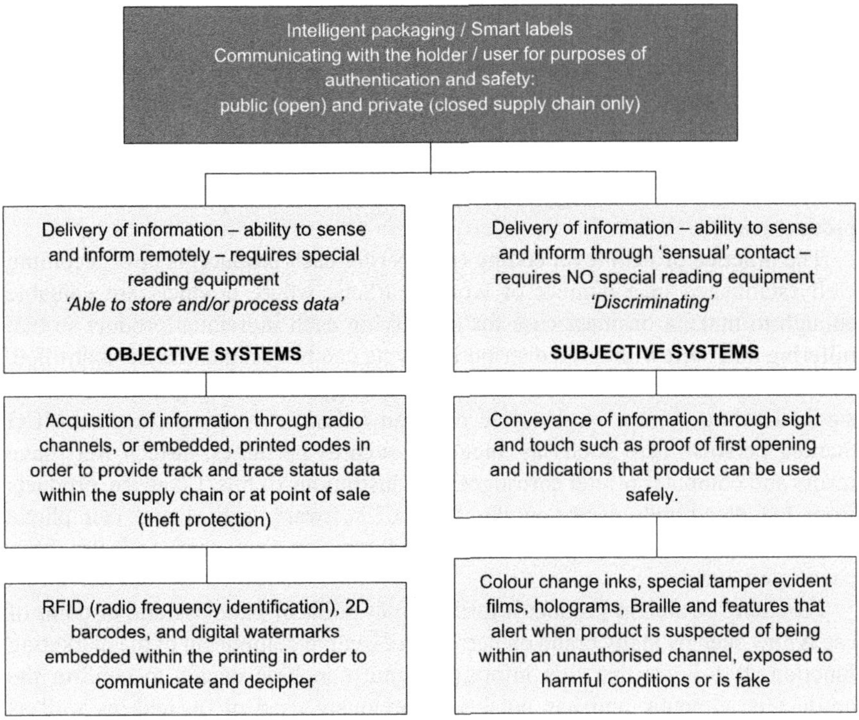

Fig. 3.1 The various functions and delivery mechanisms available to the intelligent and smart packaging user.

would be beneficial in much the same way as a bank account number or passport number is used to identify an individual. For a number of years, it has become established practice to apply sequential numbers to product packing, in an attempt to identify each pack so that it can be traced individually through the supply chain right through to point of consumer purchase.

Such numbers can be easily found on wine bottles, specialist food packaging and consumer goods such as personal care items and electronic products. Numbers provide the attribute of traceability and can be combined with other useful information such as date codes, batch and lot references and identify where a product was assembled or produced. The numbers by themselves do not make products smart, but when such codes are combined and written in a form that can be recognised automatically by readers linked to intelligent databases, items become 'visible' to the computer controlled distribution system and are therefore more easily tracked and accurately identified within the distribution process. It becomes easier to control stock, audit goods in warehouses and identify where each product is sold, thereby providing valuable information to the retailer and to the brand owner that assists with their marketing management.

3.2 Integrating 'intelligence' into labels and packaging

We are all familiar with bar codes and can recognise the universal product classification (UPC) code when we see it in use on packs. These codes identify a product by type, but not each product individually. Bar codes are also applied to secondary packaging and pallets, as a method of identifying consignments in bulk. However, until recently, it has been unusual to see unique numbers on every product pack.

The practice of item level coding (numbering each product) is now becoming well established in a number of world markets, where products are valuable enough to make a business case for identifying each individual product so that pilfering and theft within the distribution cycle can be recognised and controlled. Products at most risk are wines and spirits and high-value food products such as smoked salmon, extra virgin olive oils and balsamic vinegars. In the FMCG market, personal care goods in categories such as perfumes, hi-tech wet shave razors and computer printer cartridges are illustrations of this. Likewise, products from the electronics sector in the form of software, games and cell phone accessories also require a high degree of protection from theft and also from counterfeit attack.

Providing sufficient product information at point of purchase and at point of use is now seen by many brand owners as an essential component of the marketing function. It follows that this information must include details to confirm the product is authentic and has not been previously used or opened, as well as statutory data such as ingredients, safety notices and instructions regarding use. Established methods of communicating these facts have involved smart printing technology and this subject forms the structure of the second part of this chapter.

Delivering essential product information through intelligent packaging functions is a recent trend and a number of advanced technologies now exist that provide consumers with the ability to 'interact' with products and further enhance the experience by gaining a 'history' of each purchase. A method that is becoming more popular for tracking, tracing and authenticating products, and thereby revealing their history and pedigree through their labelling and packaging, is radio frequency identification (RFID).

3.2.1 Intelligent packaging and radio frequency identification (RFID)

It is unrealistic in this chapter to discuss the physics behind RFID, but the technology is well proven and has been widely adopted as a secure carrier of information within the closed supply chain. The intention within this section is to set the scene as to how RFID can now assist with communicating product safety, and secure source of origin information, to those points within the supply chain that act as guardians to protect against infiltration by counterfeit or unauthorised items, and to the end user in public 'open' systems.

RFID has typically been used in the supply chain for 20 years or so as a mechanism to track and trace products at pallet level. As such, it is used to transfer information relating to each batch of product and provides data that is useful in identifying the transaction (pallet number, distribution reference, product type, volume, etc.)

In many ways RFID can be viewed as a 'radio' type of bar code. Data can be acquired at a greater distance using radio signals than can be achieved with bar code readers. A further attribute of the technology is that identification information carried by a number of products can be collected with one scan.

Each electronic code is carried in a label or 'tag', which is fixed to a product either as a self-adhesive label or hang tag, or as part of its packaging. Tags carry a silicon chip and an antenna. The chip is encoded with data and the antenna provides distance communication and also powers the tag when an incoming signal is detected. This signal prompts the release of data from within the tag back to the reader, which provides a handy communication channel to a computerised, networked, stock control and distribution system.

Tags in this format are referred to as 'passive' in order to differentiate them from active tags that carry a power source in the form of a battery. Active tags are more often used for applications requiring longer range reading, such as asset tracking and vehicle toll collection, the latter being reusable and the former being disposable.

As previously mentioned, the technology is well established as a form of electronic identification and various frequencies have been allocated internationally to ensure that radio identification is able to operate without interfering with other users who are utilising radio transmission and reception for other purposes (Table 3.1).

Table 3.1 An illustration of the different frequencies used in RFID and their most appropriate applications

Band	Frequency	Read range	Application
Low Frequency (LF)	100–500 khz	Up to 20 inches (50 cm) Sometimes over-dependent on power available	Access control, animal identification, vehicle key locks
High Frequency (HF)	13.56 khz	Up to 36 inches (1 m)	Access control, smart cards and tickets, item level tagging (brand protection) libraries and electronic article surveillance – EAS anti-theft systems
Ultra High Frequency (UHF)	866–956 MHz	Up to and sometimes over 20 feet (6 m)	Supply chain use – pallet tracking, baggage handling at airports, toll collection
Microwave	2.45 GHz	3–10 ft (1–3 m)	Item tracking and toll collection

It can be observed in the table that the most appropriate form of tag used in protecting and tracking products in intelligent packaging applications falls into the high frequency (HF) format. This is the most popular format used in packaging and labelling applications and a whole variety of RFID tags are available in different shapes and sizes to suit every application (Fig. 3.2). Previously, RFID has been seen as too expensive for item level use, but economies of scale and new manufacturing techniques are now able to deliver tags for application to consumer goods, high-value food and drink and electronic products at more affordable prices.

Basic tags take the form of silicon chips and antennae and become electronic number plates. As such they only carry a reference to the UPC and a sequential number in order to securely record the number of products produced. In this format, tags are 'read only' and the reference number encoded in the tag is only used to track its progress through the supply chain.

Further enhancements can be made to the functionality of a tag by adding extra fields that offer further attributes, such as the ability to write a variable location to the tag or provide the tag with anti-theft functions, such as Electronic Article Surveillance (EAS).

3.2.2 Adding anti-theft functionality to product packaging

RFID has been used for several decades now as an anti-theft device and we are all familiar with the EAS gates we see at the entrances and exits of retail stores. In these systems, EAS tags act as an alarm mechanism and this is set off if the tag is not disabled at the point of sale. This system has traditionally been kept separate from other RFID applications, but it is now possible to incorporate EAS functionality into the track and trace function of a more sophisticated RFID tag.

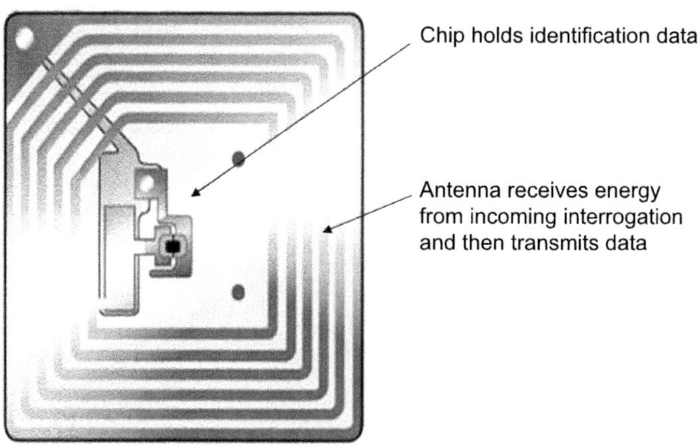

Fig. 3.2 Illustration of a RFID tag in label format.

RFID tags are capable of recording information on their way through the supply chain and also provide the capability to disable a function remotely when necessary. Therefore tags that incorporate the EAS function can be deactivated when a product is legitimately sold, but also protect the product from theft on its way through the supply chain. Adding EAS functionality to tags is seen as a method of improving the business case for RFID and it follows that if further useful tasks can be incorporated into such tags, then the business case would be improved further.

3.2.3 Introducing RFID to brand protection

Brand protection can be considered as the management of systems and technologies that protect the product from threats such as counterfeiting, theft, refilling and other activities that have a negative effect on the brand. RFID has up until recently been considered to be too expensive for brand protection applications at item level, other than those that involve luxury goods and high-value prescription drugs.

The inability of the public to read RFID tags has also been an important limiting factor, because buyers are unable to check packs for such intelligent tags since suitable reading equipment has only hitherto been available to supply chain operatives and retailers. Furthermore, security has also been viewed as an inhibitor because of fears that tags could be easily copied (cloned) or removed from authentic packaging with a view to recycling them on fake products.

Both cloning and recycling attacks have been adequately addressed over the last few years through a number of initiatives that have been developed to solve these concerns. First, the introduction of non-cloneable RFID tags from a number of sources has led to the development of Physically Uncloneable Functions (PUFs) within the silicon chips embedded in the tag. Since each chip has minute differences within its structure, which are formed when it is fabricated, it is now possible to identify these and store them within the chip as a unique material signature. This means that each chip is self-checking and this checking mechanism is itself protected through encryption keys that are only known to the user.

Second, the risk of recycling authentic RFID tags by removing these from discarded packaging and applying them to counterfeit products has been countered through the addition of anti-tamper mechanisms to the self-adhesive label that fixes the tag to the product pack (Fig. 3.3). This is achieved through adding a perforation that breaks when a pack is opened and destroys part of the tag circuit antennae, which makes the RFID tag inoperable.

A further benefit of supplying a straightforward disabling mechanism (Fig. 3.3) is to provide a means by which tags can easily be disarmed after product purchase, thus providing privacy from eavesdropping by other retailers during the shopping trip. This concern has often been voiced on the behalf of shoppers by advocates of privacy. To ensure that such tamper proof tags are not prematurely disabled, in order to circumvent EAS systems, it is possible to design the disabling mechanism in such a way that even if it is activated by shoplifters it still functions (in EAS mode) at the exit gate.

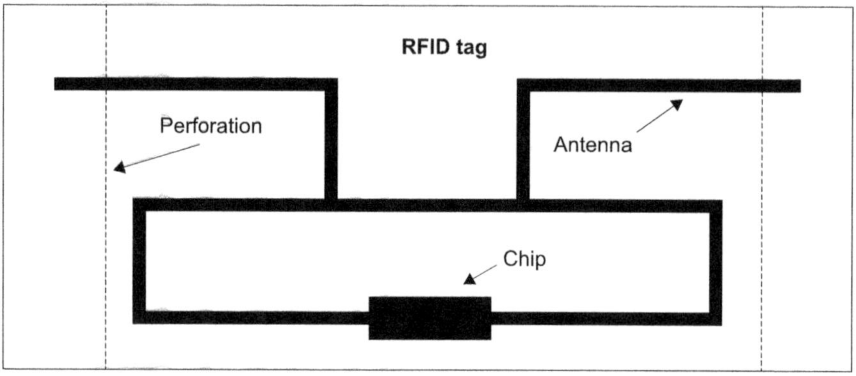

Fig. 3.3 Adding tamper evident functions to an RFID label to ensure tags cannot be reused.

3.2.4 RFID moves into the public domain

As can be seen from the above paragraphs, RFID has up until recently only been available as a tool to enhance internal brand owner and retail supply chain logistics. This landscape is changing rapidly and the years 2006 to 2011 saw RFID in a 'near field' form become an important player in banking, transport and access control.

Whilst these applications have nothing in common with packaging and brand protection, they have acted as a proving ground for tags and reading systems that utilise the facility of Near Field Communications (NFC) readers, embedded in many of today's smart phones. NFC is a protocol that allows RFID-enabled items to be read in close proximity to the reader. Reading distances are often no more than 4 cm, and at these ranges it is not possible for other 'tags' in the vicinity to interfere with the recognition of a chip in a NFC card used for payment, or for ticket validation on transport. Since RFID is a contactless form of data interchange, the technology is naturally referred to as 'contactless payment'.

This trend has driven NFC into the smart phone market, with RFID functionality being embedded in phones, allowing them to be used in the same way as contactless payment cards. Part of this functionality also allows the phone to act as a contactless reading device and now places RFID reading technology firmly in the hands of the general public. Smart phones work in conjunction with 'apps', or applications, which are sophisticated mini-programs designed to provide the user with entertainment, lifestyle advice, help with shopping and other useful tools such as mobile banking and, of course, interaction with brand owners through social networking sites such as Facebook.

By including NFC-enabled tags in their packaging and labels, brand owners are able to provide their customers with valuable extra information about their purchases, which builds brand loyalty and enhances the 'brand experience' over and beyond that possible through communicating via the 'printed' word on the surfaces of a product's packaging. In this way, brand owners are able to offer extra value and thereby be in a better position to positively influence the 'sale'.

An illustration of this is the ability of a brand owner in the wines and spirits sector to include a NFC tag on the bottle – usually as part of the neck or main body label. Such a tag carries a code that includes product details, a web site address (URL or uniform resource locator) and a unique serial number that can be used to identify each product securely.

Customers are encouraged through advertising and promotional material provided with the purchase, to scan the tag after downloading an app. This process takes them directly to a dedicated web site where content is downloaded in the form of video, text and graphics that is both entertaining and at the same time educational. Downloaded content may be instructions and options on the various cocktail recipes that are possible using a particular brand of spirits, or flavour combinations that work best with a regional wine or liqueur.

Through on-line video streaming of information, provided by the intelligent tag link, brand owners are able to provide a much richer experience for their customers, and encourage further purchases and suggest choices of other products within their range that may be of interest to the user.

By combining RFID with packaging and labelling in this way, the brand owner is able to expand the envelope of information provided by the tag from within the supply chain right out to point of use. During the data acquisition process, at the final customer boundary, the brand owner is able to allow the extraction of important information relating to the authenticity of the product, the region of purchase and encourage further dialogue through their web site.

3.2.5 Where RFID fits with packaging and labelling conversion

There are a number of methods available to the printer and converter of packaging and labels, which allow the easy addition of NFC/RFID tags to their manufacturing process. Tags are readily available in a wide variety of sizes and designs to suit every type of product packaging from:

- Brand identification labels.
- Content labels.
- Anti-tamper seals and closures.
- Instruction and promotional labels.
- Neck labels on bottles.
- Caps on bottles and jars.
- Primary and outer packaging, such as rigid boxes and cartons as well as cases.
- Blister and clam shell packs.
- Swing tickets and tags.

It is, of course, necessary for the converter to consult and liaise extensively with the brand owner to choose the correct tag for their application, as well as deliberating on the data to be carried and its structure. Initially it will be useful for printers and converters first entering this field to request advice from the major suppliers of self-adhesive label-stock and/or packaging board, since most now supply RFID and NFC tags in the range of materials they supply to the trade. Tags are most often supplied in continuous rolls of self-adhesive labels and are applied during the conversion and printing process.

3.3 Intelligent packaging: integration with cell phones, camera phones and smart phones

It is a fairly well-established practice in some consumer good market sectors for manufacturers to include promotional codes on every pack. These codes are designed to support marketing campaigns and offer customers a chance to win a prize, or receive other inducements such as vouchers in an attempt to encourage purchase.

These promotional codes take the form of alpha-numeric strings of data that can be uploaded using a cell phone key pad and 'texted' to a dedicated server, or be entered via computer keyboard and internet connection through a web browser on a PC. Codes are added to packaging using inkjet or laser ablation technology and this form of encoding takes place in real time during the final packaging process, or in the case of bottling, at the time of label application.

To be cost-effective, such coding systems must be capable of operating at the same speed as the packaging line and at present only inkjet and laser ablation technology can deliver these rates of application. Laser ablation is effected by vaporising part of the printing ink decorating the pack by using a precisely positioned and focused fast pulsing laser beam to 'burn' away the ink and form characters in the white paper or board below. Inkjet inks are available in a number of fast dry indelible options and can be applied to every type of packaging material, including glass and plastic. The process involves injecting droplets of ink at high speed to create characters on a 'matrix' basis.

The introduction of digital colour print systems into the labelling and packaging markets also means that print suppliers that use such resources are also able to apply unique numerical codes synchronised to the background print process. As this coding practice has become established, many brand owners have recognised that numbering each and every product with a unique reference is a useful way of controlling stock, identifying pilferage and providing a route for customer interaction over and above the normal helpline contact now offered on many products.

In addition to being useful as a promotional tool, codes can also be easily modified to include checking procedures that allow the brand owner and the user to verify the uniqueness of the code by using a cell phone or internet browser, thereby providing counterfeit detection right through the supply chain and at point of sale for customers that may be worried about product authenticity. To prevent duplication attacks of the codes by counterfeiters, it is necessary to randomise the code and also to employ basic cryptography as a means of checking the code, either on line or through a dedicated smart phone app.

Cryptography is a mathematical method of checking the code for authenticity and if the cryptographic key used to create the code is kept secret, then attempts to copy or generate passable imitations are nullified.

3.3.1 The use of 2D and QR codes in product protection and promotion

Since the introduction of the camera phone and now the smart phone, the development of 'quick response' (QR) codes has driven the market for interactive and intelligent content forwards at great speed over the last three years. QR codes provide a near instantaneous method of obtaining promotional and/or authentication content. When used in conjunction with an app on a smart phone, or through the multimedia messaging service (MMS) available on other non-smart handsets, a 2D bar code provides an immediate 'clickable' link to brand owner-provided internet content. Such phones use their camera to directly link via the 2D bar code and provide promotional and/or authentication messages in real time, as well as the quick facility to 'interact' with the brand by sharing video or other multi-media content.

A variety of QR reading apps[1] are available free for download and work with Apple's i-Phone, Symbian and Android operating systems. Some apps, such as ScanLife,[2] can read UPC, EAN and Datamatrix codes too and in these applications they provide a facility for reading product codes and aid purchasing decisions by allowing the user to consult price comparison web sites so that they can be sure they are getting the best deal, as well as an authentic product.

Digital print systems are able to combine content-rich variable bar codes for authentication purposes during the label and packaging manufacturing process. Those printers and converters without this digital resource are able to print the static background matter using conventional processes such as flexographic, gravure or litho and incorporate the bar code through on-line inkjet personalisation systems or dedicated off-line processes that may be linked to cutting, folding and creasing operations.

The majority of QR codes now seen on packaging relate to static codes for promotional use. These are combined with the printer's artwork during origination. However, there is a steady growth in the application of QR and other matrix serialised codes for product protection; these are becoming noticeable in the auto-part aftermarket and in the protection of regional fine wines.

3.3.2 The growing use of 'randomisation' linked to bar coding as an intelligent brand protection tool

In some markets where counterfeiting is seen as high risk, such as luxury goods, or where health and safety are at stake in the case of counterfeit liquor, it may be necessary to increase the level of security by combining the bar code with an artificial randomisation feature that adds a unique material biometric to the label or pack being protected. Randomisation is an artificial feature that is added at the print production stage and consists of an unsystematic array of dots, lines or in one case bubbles, encased in a plastic overlay. The process involves introducing an artificial 'chaos structure' that is linked to the bar code and this chaos structure is stored in a database linked to a web server through the bar code.

Imagine a small number of matchsticks held in the hand over a table. The matches are tossed into the air so that they land in formation on the table surface and thereby create a pattern of overlaying sticks. Do this a number of times and you will see that every time a different (random) pattern is created. You will need to repeat this process an infinite number of times before you exactly recreate a duplicate pattern. Increasing the number of sticks increases the chances of accidental replication exponentially.

Now imagine a printing process that imitates this procedure through the random distribution of lines within the framework of say, a square inch. The resulting code structure will create in effect billions of unique combinations of lines each dissecting each other.

Each discrete self-checking code is created in the form of a label that combines the variable random series of intersecting lines with a 2D bar code. This label becomes a secure 'number plate' that can be used to verify the authenticity of a product using the camera on a smart phone for members of the public, or a proprietary reader in a closed track and trace security system, such as is found in the automotive replacement parts supply chain.

Such codes are in effect impossible to copy, since the sequence is left to chance and is totally random. Only pre-enrolled codes are stored in the database and since these codes are loaded by, or on behalf of a brand owner, they can only be checked through scanning a 2D bar code and linking through a secure web site address also embedded in the bar code. Even though this technology is newly developed, a number of apps are already available to support brand owners wishing to use the system to empower customers to verify the source of their products and ensure that they are within their sell by date.

Print-based security coding systems, such as those outlined above, are invasive of the 'real estate' on each pack or label and have to compete with brand information, content listings, safety instructions and sell by dates, etc., all of which need to be present for reference on the packaging. RFID and NFC tags used for brand protection and supply chain management can be invisibly incorporated within the packaging. Therefore it is often necessary to strike a balance between the high cost of RFID and its invisibility on the one hand, and lower cost serialisation and its invasive encroachment of valued space on the other.

It should be noted that all the technologies discussed in this section require a form of notification to alert their presence. The use of symbols, such as those that now indicate NFC/RFID connectivity (an 'N' in white on a blue square background and a 'radio waves' signal logo, respectively) are becoming more widespread. QR codes are now so widely established that there is little need other than a text reminder to scan these for more content delivery.

Those technologies that depend upon augmented reality (AR) and digital watermarking (discussed herein) also require a visual confirmation of their availability; otherwise they will remain hidden from the user. These can be viewed by visiting either the Apple 'App Store' or Google's Marketplace and searching for 'augmented reality' and/or 'digital watermarking'.

3.3.3 Future trends: authentication through digital watermarking, the link to augmented reality and the power of social networking

Embedding covert digital watermarks within the print that is present on the surface of a label or pack is seen as a novel method of removing the disadvantage of space invasion, which 2D bar codes and other visible systems of security need to operate authentication systems. Digital watermarking is not new. The technique has been around for at least two decades within the high security print sector. It is also commonly used to protect the IP on photographs used by picture agencies and in the video industry.

Only recently has digital watermarking been seen to offer distinct advantages to the packaging and labels sector. The main reason for this is the development and availability of apps for the smart phone that recognise digital watermarks and are able to convert these marks into commands that are used to link the phone to multi-media internet content. Digital watermarking for secure brand protection works only through digitally produced print created on equipment, such as the HP Indigo or Xeikon. Alternatively, it can also be produced by conventional label or packaging machinery fitted with a suitable digital (variable print) station.

'Watermarks' are placed within the printing by altering the shape and size of the pixels used to create the print. This process is controlled by proprietary software that inserts the variable mark into the print during ripping. Since digital printing offers the facility for every sheet produced by the press to be 'individual', it provides an ideal facility for deploying a different covert variable code within every sheet. Watermarks can also be encrypted to protect content or made available for public checking through a number of apps used by smart phones.

This facility is very new and it is expected to also generate interest in using smart phones to 'browse' extra content and promotional media that brand owners are not able to place on labels or packs because of the limitations of space mentioned earlier. Using their camera, in conjunction with apps such as Digimarc Discover,[3] owners of smart phones are provided with an instant link to web content supplied by the brand owner. Embedded security code within the watermark can provide authentication detail and customers are often encouraged to comment on their purchase through social media sites such as Twitter and Facebook.

This process of enhancing the 'brand experience' is called Augmented Reality (AR) as it mixes the information available in the real world with that provided in a 'virtual' setting. It is this process, linked with social networking, which offers the most opportunity for brand protection in the future, since end users will be able to alert each other to fakes through their own networks and more extensively through products such as Twitter.

Digital watermarking is not the only process that provides AR with the ability to process what is 'seen' through the camera lens of a smart phone and converting this code into a link. Other techniques exist in parallel, such as the ability of apps to recognise objects, logos and other images associated with a product. In this instance, the ability of the app to recognise content does not depend upon variable print, since no sequence code is needed because the technology is only used for promotional purposes, not brand protection. In these cases, conventional printing methods can be used since the app is programmed to recognise logos and other features within the printed page, such as an advertisement.[4]

One further development in this sector offers the promise of capturing covert codes embedded in labels and packs through the screen of a smart phone. Such smart phone screens operate using touch, and a new code, 'touch code', offers the potential for users to interact with print without having to switch on their cameras. Touch code requires a special extra layer of print to be placed on, or close to, the surface of a label, hang tag or pack. It is this layer of print that reacts with the capacitive screen, in conjunction with an app, to load the promotional or authentication message that originates from the brand owner's web site or cloud hosted web server.[5] Capacitive touchscreen displays rely on the electrical properties of the human body to detect when and where on a display the user is touching. Because of this, capacitive displays can be controlled with very light touches of a finger. Examples of devices with capacitive touchscreens are the Apple iPhone, the HTC Desire and the Samsung Galaxy S.

3.4 Smart labels and their uses in authentication and in reporting on product safety

Smart labels and packs have the ability to 'inform' users on the status of a product, often before the packaging is opened. They function by communicating through the physical senses available to users (usually sight or touch), rather than through electronically processed messages that are generated by the electronic technologies covered earlier in this chapter.

Previously it could be said that 'intelligent' labels and packs supply unambiguous information to the user. Judgement is made by a user based on information processed autonomously and derived from a bar code, RFID tag or embedded watermark. These systems can be considered 'objective' since they confirm a product's status independently. However, smart authentication solutions on labels and packs require a physical judgement to be made by the user to confirm they are real. This judgement will be reached through a process of knowledge acquired through experience, or through immediately recognisable signs and signals that indicate whether a product has been opened previously and poses a risk, or carries an authentic indication of provenance – or not.

Such systems are considered to be 'subjective', since it requires a judgement be made as to their status and the way they react to an external stimulus such as tilting, rubbing or changing colour under intense light.

It is recognised practice, when dealing with authentication solutions in this field, to classify devices according to their hierarchy. This process is best described as ranking systems as to whether they can be recognised quickly through the senses of touch and sight (quickly react visually to an external stimulus). Illustrations of this activity are the use of colour-changing inks and the 'feel' of raised print on banknotes to confirm they are genuine. Such systems are officially referred to as 'primary', since they are the initial methodologies established for certification of authenticity at first hand. They are also called 'overt' systems. Should these primary systems fail to provide an acceptable indication of authenticity, then it is necessary to create another secondary or 'covert' level of tests that can be applied to establish provenance.

This covert, or secret level, is only available to trained officials and inspection teams within the supply chain. It requires a tool such as a magnifying glass to observe microscopic printing or etching placed within a certain position on the pack. Other systems deployed depend on the use of polarisation filters, scanners and laser reading devices, which can be used to isolate visual and chemical signatures that are present in the printing ink or substrate used to create the product label or packaging. Since such covert systems are often kept confidential for purposes of safekeeping, we will concentrate on those that purely provide overt responses and that can be used by the public as well as within internal supply chains to monitor the status of products.

3.4.1 The application of inks in secure labelling and packaging environments

The most widely used devices in brand-protected packaging include those that deliver a visual signature of conformity and that are difficult for counterfeiters to imitate. These devices are delivered in the form of colour-change inks and reflective foils, often embossed with security designs or holographic images.

Inks are a useful tool in the printer's anti-counterfeiting tool box, since they can be applied using existing printing and conversion equipment. Inks that are used in brand protection applications require a high degree of exclusivity since, if they are widely available, they lose their 'secure' status.

It is also important to recognise that such technologies require careful quality control, in order to ensure that their effects are uniform and can be monitored against agreed standards – otherwise wide-ranging variations in their visual appearance would nullify their benefit as a recognition tool. Therefore it is essential to ensure that secure inks and any other anti-counterfeiting element included in the final label or pack is auditable. Traceability is significant here, since if it is possible for bona-fide security devices to 'leak' from the system, they cease to be 'secure'. Some of the most secure inks used in brand protection provide a distinctive colour switch and are referred to as optically variable. These inks can also be seen on currency and they offer a quick method of confirmation when they are tilted in the field of view. Colour changes vary from green/gold through to red/purple and blue/green, with many different combinations in between.

Inks in this form are often highly viscous in order to carry the pigments that provide their colour change characteristics. Therefore it is only possible to print such inks using screen printing techniques, either flat or rotary, depending upon the equipment available. There are some variations to this rule and both intaglio and gravure processes are able to deliver these effects too, but intaglio is a process that is only used in the currency and high security print markets and not in the packaging and label printing industry.

Colour-change inks can also be delivered in a format that reacts to heat and cold as well as friction – when rubbed with a fingernail for instance. Colour change is also possible through illumination using high intensity light, such as that available from a laser pen or from a high intensity photographic quality flash light. Alternatively, ultra violet (UV) and infra red (IR) wavelengths can deliver a wide range of effects that range from invisible to visible and distinctive colour change effects when exposed to the correct excitation source. The light from an excitation source (laser pen, UV or IR lamp), providing the addition of sufficient energy to an electron, atom, atomic nucleus, or molecule to raise it from its lowest energy level ground state to a higher energy level, leads to a significant colour change.

Photochromic inks offer a distinctive colour change when exposed to high intensity light, and certain phosphors (within the ink) can also deliver a unique colour change when they are exposed to light from a laser pen. These inks, together with UV inks, can be applied using litho, flexo and letterpress printing methods.

3.4.2 The application of inks for reporting on product safety

Colour change is also an effective method of delivering warnings that perishable products, and those affected by excesses of heat and cold, are capable of displaying their 'status'. This is necessary to identify and remove them from display or from the supply chain, or alert the buyer that they have become contaminated, over-reached their useful life, or been exposed to water.

Other 'smart' characteristics, such as ripeness indicators and packaging inserts that extend shelf life through delaying the process of decay, are also available to the packaging supplier. Some of these technologies include oxygen scavenging labels, anti-microbials, ethylene scavengers, moisture absorbing labels and ripeness indicators. These topics are dealt with in other chapters in this book (Chapter 10).

However, there are a number of printed 'smart' technologies that can be used to alert users that products have been exposed to water, too much heat or cold and maybe an unhealthy influx of microbial contamination because the packaging has been damaged in transit. Water contamination is a common occurrence in the FMCG electronics sector. Products such as cell phones, i-pods and such can accidentally be dropped into water during use. This invalidates their guarantee, but repairers may not be aware when a product is returned for service that water ingress is the cause of the malfunction. Often it is in the interest of the user not to disclose that a broken cell phone has been dropped into a drink or a sink full of water by mistake.

Water contamination indicators are supplied in label form and irreversibly change colour alerting to the fact that water damage has occurred. This technology is useful to both the user and the original equipment supplier, since it prevents misunderstandings occurring when malfunctioning products are returned for repair.

Extreme heat and cold alerting systems are also available in ink format and can be printed using conventional printing systems. Once a set of parameters has been established, both thresholds can be monitored for unwanted or unhealthy exposure. For instance, extreme heat in the form of extra hot drinks being consumed can be alerted with a red/orange indicating label that flashes up as soon as an agreed unhealthy temperature is reached. This warning disappears when a safe temperature for consumption is reached. Conversely, colour-enabled labels can also inform when a safe high temperature has been reached in cooking food in a microwave.

Colour change is also used in an informative way for time temperature labels that are used to record the number of days a product has been stored in a fridge, or that it has not been exposed to higher than safe temperatures in distribution (Fig. 3.4). Time recording labels function in accordance with a 'wicking' effect that slowly releases a coloured dye into a clear, viewable narrow pipe that is graded in hours or days, indicating how long the label has been in use since activation.

Finally, in the product safety arsenal, there are labels that can advise that products have been exposed to unnecessary or damaging 'shock' in the form of dropping or tilting, through the use of stickers that include fragile filaments that fracture when exposed to excess 'g' forces, or offer a visual confirmation that tilting has occurred through a one-way mechanism similar to a spirit level.

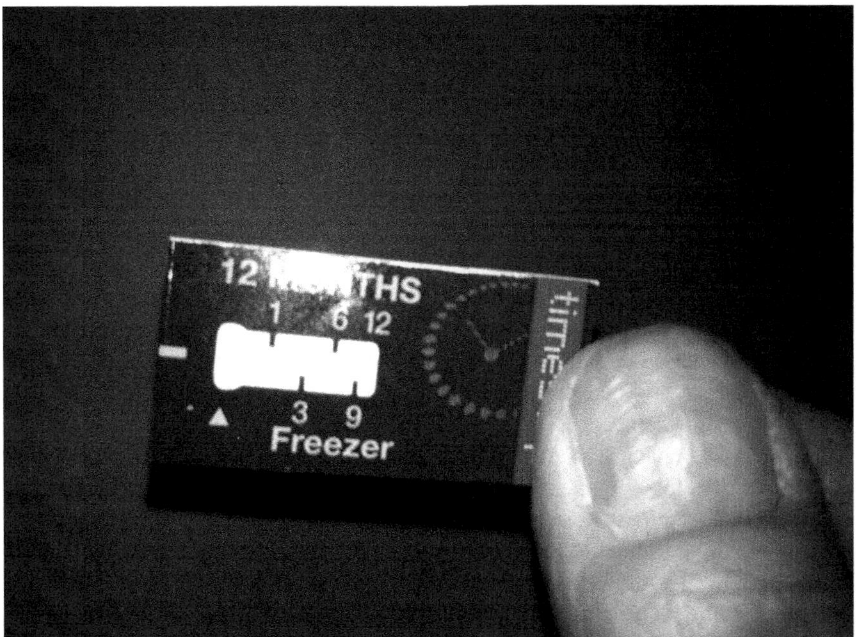

Fig. 3.4 The white space inside the time temperature label shows a colour gradation as the label ages after activation. This supplies the user with a record of how long a product has been in open storage or outside a safe temperature range.

3.4.3 The use of foils in counterfeit deterrence devices

We have seen how colour is used in an effective way to engage and deliver confirmation of authenticity, and also consumer alerts and safety warnings if perishable or fragile products have become unusable. Kinetic colour, that is, bright colour alerts delivered through movement, can also be useful in areas where high-cost consumer products are at risk from counterfeit attacks.

Refracted light provides an ideal medium for attracting attention. Utilising this property in security holograms has become a common sight on banknotes and now in the high street on consumer products where diffractive optical devices (dovids) combine complex embossed films with self-adhesive or heat transfer adhesives for inclusion in foil stamping and labelling applications. These deliver recognisable security designs in order to provide customers with an assurance of provenance. They can also be combined with shrink sleeves and tear tapes to provide tamper evidence protection, which will be covered later in this chapter.

Wines and spirits are vulnerable to counterfeiting and refilling attacks and high-quality holographic foils are being used to protect these products wherever they are on open sale. Probably one of the most noticeable product protection holograms can now be observed on all bottles of Spanish Rioja. Wine from this region has become a target for fakes and customer confidence has been ensured through the use of these devices, together with a serial number on every bottle, to provide a quick check for consumers and supply an audit trail for distributors and retailers.

Holograms can also be seen protecting expensive brands of tea, coffee, balsamic vinegars, virgin oils and luxury foodstuffs such as caviar and truffle oil; indeed, anywhere where scarcity and proprietary methods of production add value to a brand that is at risk of compromise through copy attack. Security holograms have also been included in a number of other non-food sectors, such as household electronics, printer consumables and multi-media products.

There is a wide range of holographic effects available that include 2D, 3D, 3D with animation and more complex designs that combine movement with colour change and alternate negative/positive images when the device is tilted. One major attribute of the technology is that it may be used as a promotional as well as a security device. Low security holographic embossed images are often used to draw attention to products on the shelf through their ability to reflect light in such a way that it draws the eye.

Used in personal care products such as toothpaste, mouthwashes and shampoo, holographic-embossed papers and boards offer an inexpensive method of brand promotion that simultaneously offers counterfeit deterrence – albeit at a very low level. Even such mundane products as these require attention to security, since counterfeiters are drawn by the difference between manufacturing costs and selling price, which is large in this sector. It should not be forgotten that manufacturing costs are driven up by the high levels of advertising spend, which those supplying fakes do not have to fund.

A further attribute of these light-refracting foils is that they can be applied as a hot or cold foil transfer, purchased as ready-made base metallised material and also be used as capping, lidding or tamper evident sleeves. Recently their incorporation into printed metal packaging, as seen in the decoration and protection of products such as outer tubes for whisky bottles, automotive engine oils and as containers for high priced luxury goods, indicates their flexibility as security devices across all forms of packaging.

Films and foils in undecorated format also provide a degree of protection from counterfeiting. A number of metallisation and de-metallisation processes exist for producing optically variable foils, but most of these are so well-established as decorative devices that they have lost their appeal as anti-counterfeit tools.

When packaging and labelling is utilised for physical and secure product protection, it must offer both deterrent and detection properties that are difficult to replicate. It is this degree of distinction that makes bona fide products easily identifiable – but with unique characteristics that are hard to emulate. To be effective, anti-counterfeiting devices do not need to be impossible to copy – just impractical to copy.

Those manufacturing methods that require high investment and skill, such as holographic foil production, or those that depend upon secrecy, either though novel IP or system protection such as encryption methodologies, often make the best choices for inclusion in brand protection programs.

3.4.4 Delivering 'smart' tamper evidence combined with authentication

As mentioned above, space is always at a premium on labels and packaging for fast-moving consumer products. Branding space is most important, since this is what sells the product. However, space for instructions, tables of contents, product bar codes, sell by and use by dates, illustrations of the product in use, etc. all vie for position on the pack.

Compromises often have to be reached between the communication of essential practicalities in everyday use and promotion of the brand and other associated products. In effect, brand protection devices offer another level of competition for valuable space on every pack and may often compete (and distract) from the important branding insignia.

Clever designers will be able to overcome this paradox by incorporating anti-counterfeiting devices with the materials used to construct the pack or label, or utilise the anti-counterfeiting feature as a brand enhancement by combining a colour-change ink or hologram with the brand insignia for instance.

It is also worth considering including a security feature within a tamper evident seal or closure. This either removes the pressure from such devices vying for space on the main pack or complements other security devices that may be incorporated on the pack or on secondary packaging. There is a wide range of choice available when consideration is given to a tamper evident seal. Whilst its primary function is to secure the 'first opening' and deliver a message to the consumer that the contents of a pack are pristine and have not been interfered with, the seal also delivers a number of other messages. On a subliminal plane it communicates that a brand owner is taking care to ensure their product reaches their customer intact and in the best possible condition. It also demonstrates that a pack (or bottle) has not been refilled after its first use, and then put back into circulation.

Whilst refilling attacks used to only be a concern in undeveloped countries, instances of illegal recycling of previously used packaging filled with counterfeit product occur regularly now within what we term the 'developed world'. Instances of illegally refilled wine and spirit bottles have been regularly occurring in the UK and mainland Europe over the last few years. These events have been widening in scope and geographic area too. Remember it is the taxation cost that attracts the criminals. Every bottle they can refill delivers a premium way and above the cost of the contents. In order to effectively address these attacks – and threats of attack – brand owners need to act responsibly when it comes to sealing products at risk from such dangerous activity.

Tamper evident tear tapes used in conjunction with suitable wrapping films or combined with opening devices on shrink sleeves are useful tools to indicate proof of first opening. Such tear tapes often include an authentication feature in the form of a holographic metallised finish, so they protect against counterfeit attack and offer a secure confirmation of first opening.

Alternatively, tamper evidence can be introduced in self-adhesive label form, either through the use of paper or frangible filmic materials as a printing substrate. A variety of special adhesives are also available to prevent and deter tamper evident seal removal and replacement. These adhesives are tinted so that any attempt at removal leaves a coloured residue behind on the pack that provides an alert. Further security can be built into the adhesive coating by printing messages such as 'opened' or 'VOID' in silicone ink on the back of the label material. This has the effect of transferring the adhesive easily from the back of the label onto the packing material on which the seal is fixed. It is the transfer of this coloured adhesive from one surface to the other that develops the 'opened' message. In most tamper evident systems the adhesive is killed on first opening and cannot be re-fixed in place.

Through the clever use of holographically embossed metallised films as a label substrate in such security sealing systems, tamper evident adhesives can be designed to destroy the metallised film. This process offers even higher barriers against tamper attacks, especially when combined with serialised bar codes that offer a method of track and trace too.

The economies available from combining authentication, track and trace and tamper evident devices where practical, are reinforced by the advantages of ease of use and the addition of dexterity required by anyone wishing to compromise the system.

3.4.5 Future trends in smart labelling

The fundamental difference between smart labels and intelligent labels is that the former depend upon the human senses of sight and touch, and the ability of the public at primary authentication level to make subjective judgements based on the authentication features they are presented with. Keeping these security technologies constantly beyond the reach of counterfeiters requires a high degree of R&D in order to keep pace with the ability of fakers to compromise and copy security devices with new novel alternatives as they become available.

An illustration of this point is the substitution of optically variable varnishes and paints used in ladies make-up and automotive finishes for similar optically variable inks used in security packaging and labelling. It is therefore necessary to be constantly developing new combinations of colour-change features and ensure that they are exclusive to the brand protection field.

Some interesting developments are now coming to the fore in the field of photonic colour. This technology offers enhanced optical effects beyond the scope of those achieved with established optically variable pigments. Photonic colours can be changed by a variety of stimuli, through passive detection or on command. This means that users will be able to add electronic and chemical stimulation of colour effects to those usually observed using passive optical methods such as unaided visual sight.

New stunning colour-change effects have been developed from honeycomb nanostructures that reflect wavelengths perceived as colour by the eye. Such nanostructures are delivered as a film that contains a colloidal structure, or are created within materials such as paper by using microscopic engraving techniques to create nano-sized holes that reflect and/or refract alternative narrow wavebands of light at various viewing angles. These are genuine iridescent effects and, since they do not depend upon pigments or dyes, they deliver brighter colour in a form similar to that of gemstone opals.

One of the so far unquantifiable attributes of photonic crystal technology is that it can also act as a 'nose' and change colour when atmospheric contact is made with water molecules or certain elements present in the gases given off by foodstuffs. It is still too early to say where this technology will lead, but some of the early developers are predicting that it may find a role in quality control. If that is the case, it will need to be combined with labelling and packing in order to function effectively.

Other optical devices that are predicted to find wider use in brand protection circles are photopolymer holograms. These were developed several decades ago, but failed to find traction because they were expensive to produce. Meanwhile, the ubiquitous embossed metallised hologram has so many sources of manufacture now that such devices are now seen as losing their security value in all but very complex forms.

Since scarcity and difficulty in manufacture are both attributes of photopolymer holograms, they are beginning to appeal to brand owners since they offer 360-degree viewing of easily recognisable objects that require less attention to detail during authentication. This in turn means that they are more practical when used as primary devices by those untrained in specialist knowledge.

Finally, in tamper evidence, the advances in shape memory polymers (SMP) are also worth noting. Such devices offer dual authentication and tamper detection characteristics. Since SMPs display 'smart material' characteristics, they respond to external stimuli such as heat and change form and shape to deliver messages previously embossed in the film. It is this process that can be described as 'memory', since the material 'remembers' its original shape when exposed to the correct conditions – in this case heat.

Messages are delivered in the form of embossed text that is released irreversibly. For instance, the film can be embossed with a 3D text message that changes to 'OK' or 'Genuine Original' from a more generic message such as 'check this out'. The film can be made into self-adhesive labels and the covert part of the message is delivered only seconds after activation. Whilst SMPs have been available for nearly 30 years, it is only a recent series of modifications that have made them suitable for tamper detection and brand protection applications.

3.5 Conclusions

Counterfeiting and the other various forms of product abuse covered in this chapter are not new threats. Indeed, illegal replicas and fake products have provided a serious challenge to brand owners for half a century or more. Before the onset of globalisation, when brands were more local or regional, fake reproductions still existed, but these were by and large contained within the community where they originated and were easier to control.

The upsurge in offshore manufacturing, together with the delegation of production and raw material outsourcing by global brands, can be viewed as one reason for the advance of counterfeiting, which now affects a considerable proportion of world trade in some sectors.

Global brands often operate only as IP holders and designers these days and it is inevitable that criminals will continue to be attracted by the high mark-ups that are made in most market sectors.

Fakers pay no taxes, have no responsibilities for the conditions in which their employees labour, and often benefit from retailing their products through the 'no cost' internet, so it is important that effective brand protection procedures counteract the problem. It is the responsibility of those employed within the packaging industry as a whole to ensure that they continue to develop and deploy suitable brand protection devices that are attractive and cost-effective to use, so that those tasked with fighting the problem can continue to receive the tools for the job.

The counterfeit deterrence technologies and techniques available to printers of security packaging and labels are extremely diverse and often require highly specialised skills and methods of quality control in order to reproduce them consistently. It is also necessary to ensure a high degree of site and material security too. For reasons of security, the methods and applications of such secure technologies are only discussed at a general level within this chapter.

3.6 References

1. *mobile-bar codes.com* (2012) 'QR-Code Readers'. Available from: http://www.mobile-bar codes.com/qr-code-software/ [Accessed 14 September 2012].

2. SCANLIFE (2012) 'QR Codes that Work better, Faster and Smarter'. Available from: http://www.scanlife.com/en [Accessed October 2012].
3. DIGIMARC CORPORATION (2012) 'Digimarc® Discover'. Available from: www.digimarc.com/discover [Accessed October 2012].
4. AURASMA (2012) 'The World's First Digital Browser'. Available from: http://www.aurasma.com/ [Accessed October 2012].
5. PRINTECHNOLOGICS (2012) 'TouchCode'. Available from: www.touchcode.de [Accessed October 2012].

4

Developments in plastic materials and recycling systems for packaging food, beverages and other fast-moving consumer goods

T. A. Cooper, ARGO Group International, USA

DOI: 10.1533/9780857098979.58

Abstract: This chapter describes the present status of plastic packaging materials and technology employed in the food, beverage and other fast-moving consumer goods industries. This includes current types of packaging polymers used and advances and future trends in plastic packaging materials; water-soluble packaging; high performance barrier films; packaging coatings and additives, including clarifiers and scavenger systems; rigid, thin-wall and flexible packaging; pouch packaging; packaging sustainability; and current mechanical recycling systems.

Key words: plastic packaging, barrier packaging, clarifiers, water-soluble packaging, plastic packaging recycling, rigid, thin-wall, flexible and pouch packaging.

4.1 Introduction

Packaging is a major use for plastic materials, which is constantly increasing. It consumes about 100 million tonnes per annum worldwide, or about a third of all plastics (37% in Europe), and is the largest single end-use for plastics globally. The value of plastics used for packaging worldwide considerably exceeds that for metals, glass, paper, paperboard and other materials, and in industrialized countries, plastics are used for packaging about half of all products.

This chapter will describe the current use of plastic packaging materials in the food, beverages and other fast-moving consumer goods industries; discuss some advances and future trends in plastic packaging materials, including rigid and flexible packaging, high-performance barrier films and packaging coatings and additives; and outline current recycling systems. Biobased and biodegradable plastics will be discussed in Chapter 5.

4.2 Major types of petrochemical-based plastic materials used for packaging food, beverages and other fast-moving consumer goods

Most plastic packaging is currently petrochemical-based. About 4% of all petroleum used is converted into 240 million tonnes of plastics annually, of which packaging represents about 40%. Many types of commodity and specialty polymers have been used in packaging and some of the major ones are described below. Plastic packaging is currently dominated by polyolefins, such as polyethylene (PE), polypropylene (PP) and polystyrene (PS), and polyesters, predominantly polyethylene terephthalate (PET). Advantages of plastic packaging are that it is lightweight, can be designed and tailored for a particular product and is relatively cheap, so that even in food packaging it constitutes less than 10% of the final product price.[1,2]

However, conventional plastic packaging also has associated problems of disposal, littering (including ocean pollution), reuse and recycling, because of the very slow rate of environmental degradation and the lack of collection and recycling infrastructure in many countries. Even in Europe, only about 10% of the 50 million tonnes of plastics used annually is reused, mechanically recycled or incinerated to produce energy. Only PET and high-density polyethylene (HDPE) bottles have high recycling rates over 40%, and less than 10% of most other packaging is recycled, largely because of problems of collection, sorting and recycling mixed plastics.

4.2.1 Polyethylene

Three major types of PE are used in packaging: high-density polyethylene (HDPE); linear low-density polyethylene (LLDPE); and low-density polyethylene (LDPE).

High-density polyethylene (HDPE)

High-density polyethylene (HDPE) is the second-most widely used packaging plastic after PET. It comprises essentially linear hydrocarbon chains and is made by low-pressure polymerization using Ziegler-Natta, metallocene or single-site catalysts. Because there is very little branching, the chains pack tightly and develop a large fraction of ordered crystalline regions (lamellae) on cooling from the melt. Consequently, HDPE polymers are highly crystalline (65–90%) translucent or opaque materials, with a high density (0.940–0.965), high modulus (~1200 MPa) and tensile strength, higher melting points than other PE types, and moderate impact strength (notched Izod ca. 16 kJ/m^2). The molecular weight of packaging grades is generally around 100 000 to 250 000.

A variant of this is medium-density polyethylene (MDPE) (density 0.926–0.940), which has somewhat more branching than HDPE due to the incorporation of very small amounts of an α-olefin comonomer. MDPE has better impact and stress crack resistance than HDPE.

Processing and other mechanical properties of HPDE largely depend on the polymer structure (molecular weight and molecular weight distribution) and comonomer content, which varies with the catalyst systems used. Broader MWD materials are more suitable for film blowing and casting and blow molding; narrow MWD materials are better for injection molding to minimize internal stresses and distortions, improve solvent and chemical resistance and reduce environmental stress cracking.

HDPE is used in many consumer goods packaging applications, particularly for bottles and industrial drums made by extrusion blow-molding, rigid packaging such as clamshells and other injection-molded containers, blown and cast films for bags and pouches, and caps and closures. To avoid environmental stress cracking for applications requiring chemical or solvent resistance, such as packaging for oils, personal care products, cosmetics, detergents, bleaches, cleaning products and household chemicals, high molecular weight grades or copolymer grades with a reduced level of crystallinity must be used.[3]

Nampak recently introduced an ultra-lightweight HDPE milk bottle design containing 15% recycled materials, which is 16% lighter on average than the standard designs, yet exceeds current performance specifications. This is achieved by avoiding forcing the material as far into the bottle corners and repositioning the handle.[4]

Linear low-density polyethylene (LLDPE)

Linear low-density polyethylene (LLDPE) is polymerized by similar processes to HDPE, but controlled short-chain branching is introduced by incorporating small amounts (generally 1–10%) of α-olefin comonomers such as 1-butene, 1-hexene or 1-octene. LLDPE properties markedly depend on the comonomer type and content, as well as molecular weight and density. Increasing the comonomer content decreases the density (range 0.915–0.940) and crystallinity, thereby decreasing the heat resistance, tensile strength and modulus, creep resistance, hardness, impact strength, chemical and solvent resistance, and barrier properties towards oxygen and water vapor compared with HDPE but increasing its clarity.

Major LLDPE applications are films and sheets (including agricultural films, stretch wrap, bubble wrap and multilayer films) because of its greater transparency than HDPE and higher tensile and tear strength, elongation and puncture resistance than low density polyethylene (LDPE). However, it is not as easy to process as LDPE and is often used in blends with LDPE, where the LLDPE provides strength and the LDPE provides heat sealability and processability. The world LLDPE market in 2009 was nearly US$ 24 billion.[5]

Molecular weight, molecular weight distribution, chain branching, and melting behavior are crucial factors in the sealability and hot tack properties of metallocene PEs.

The development of metallocene and other single-site catalysts has produced new ranges of high-comonomer level LLDPE resins, such as hexene-(Exxon Exact) and octene-(Dow Affinity) types with very low density and narrow molecular weight distribution, which are used as flexible packaging seal layers, since they provide a lower seal initiation temperature than conventional PEs and better hot tack, which determines the process cycle time.

Low-density polyethylene (LDPE)
Low-density polyethylene (LDPE) is made by a different process from HDPE or LLDPE, involving high-pressure free radical polymerization of ethylene. It is highly branched and contains long-chain branching. The chains cannot pack so tightly as with HDPE, leading to lower crystallinity (maximum ~45%), and hence a lower melting point, lower density (0.91–0.94) and tensile strength, but also improved flexibility, ductility, transparency and gloss.

Major applications are flexible blown films, bags and pouches, rigid containers, extrusion blow-molded bottles, and caps and closures for personal care packaging and cosmetics. The global market was about US$ 22 million in 2009.[6]

4.2.2 Polypropylene (PP)

Polypropylene (PP) is a semi-crystalline polymer made by propylene homopolymerization or copolymerization with a small amount of ethylene comonomer, either in liquid propylene or gas-phase reaction. It is widely used because of its low density (lower than HDPE and LLDPE) combined with low price, and improvements over the past 20 years have provided increased melt strength for film and foam applications and higher heat distortion temperatures moving it into styrenics markets. However, the price situation has become more volatile with increased availability of cheap natural gas in North America, which is likely to reduce ethylene, and hence PE prices compared to PP. The world market for PP was 45.1 million tonnes (~US$ 65 billion) in 2008.[7]

PP has three stereochemical forms, isotactic, syndiotactic and atactic, but only isotactic PP, made by Ziegler-Natta catalysis, is currently used in packaging. Both homopolymers and random copolymers are used, where the comonomer is generally ethylene, to reduce the high crystallinity found with homopolymer PP. The copolymers have higher clarity and flexibility than the homopolymer PP and a lower melting point. Impact modified versions are also available, where the PP is blended with an elastomeric phase (generally ethylene-propylene rubber).

Because of the extensive varieties of commercial PP polymer types, which can be designed to provide a wide range of properties, and their easy processability by all the standard thermoplastic methods, PP is used in an extensive range of packaging applications, including flexible films, rigid containers, trays, bottles, tubs, tubes, caps and closures for products such as foodstuffs, beverages, personal care, cosmetics and household products. For example, high molecular weight types are used for extrusion blow-molded bottles, jars and other containers and for extruding sheet for thermoforming pots, tubs and trays. Lower molecular weight (high melt-flow) grades are used for injection- and compression-molding of pots, tubs, cups and other containers, caps and closures, and food service items. Because PP is both stiffer and harder than HDPE, it does not provide an elastic seal when used in caps and closures and a soft liner (generally ethylene-vinyl acetate copolymer) must be incorporated. Very high melt flow PP grades are used for thin-wall molding.

Polypropylene homopolymer
Polypropylene homopolymers are characterized by high crystallinity (although lower than HDPE). They have a higher melting point, temperature resistance and modulus, and lower density than HDPE or LLDPE, but with the drawback of low impact strength. PP homopolymer gives a high water-vapor barrier, but a poor oxygen and carbon dioxide barrier. The molecular weight of packaging grade PPs is typically 200 000 to 600 000, and both molecular weight and molecular weight distribution can be controlled by the catalyst systems used. However, PP is more susceptible than PE to oxidative degradation, leading to chain scission and molecular weight reduction, so that antioxidants are often added, as well as antistatic agents.

PP homopolymer has good melt flow properties over a wide flow rate range and processes well as a thermoplastic. Films can be made by both film blowing and casting and can be oriented to improve their optical and tensile properties. Oriented PP films are useful for stretch applications. Nucleation using specially designed clarifying additives can reduce the crystallite size to allow excellent clarity to be obtained (Section 4.5).

Because of its high heat resistance (mp 160–170°C), PP can be used for autoclavable containers and products which are sterilized after packaging, and in hot-fill or dishwasher applications, but is not suitable by itself for packaging oxygen-sensitive foods or beverages and must be sealed with aluminum foil for dairy product tubs. Like HDPE, it is often used for caps and closures, even when the container itself is made of another material such as PET (for carbonated drinks) or glass. However, unlike HDPE, it exhibits an unusual 'living hinge' property, which makes it useful for many types of closures and lids.

Polypropylene copolymer
Random polypropolyene (PP) copolymers contain low levels (generally 1.5–7%) of ethylene or other olefin co-monomers. They have a lower degree of crystallinity than PP homopolymers or rubber-modified impact homopolymers, are lower melting (incorporation of 7% ethylene reduces the melting point to 152°C), and easy to process as thermoplastics. Copolymers have lower modulus and better flexibility, clarity and impact resistance (particularly at low temperatures) than unmodified homopolymer, and are also good barriers to moisture vapor but not to oxygen or carbon dioxide. The random copolymers exhibit good resistance to acids, alkalis, alcohol and aliphatic hydrocarbons but not to aromatic hydrocarbons. Non-oriented copolymer films are also soft and easy to heat seal.

PP copolymers are predominantly processed by extrusion, injection molding and blow molding. Applications include packaging for baked goods and produce, clothing and medical packaging. They can also be used where good transparency and gloss are required for bottles and packaging for personal care products, cosmetics and food, although HDPE and transparent polymers such as PET are more often used for bottles.

Impact-modified polypropylene
Impact-modified polypropylene is generally a two-phase rubber-modified highly crystalline PP containing small proportions of ethylene-propylene rubber. It has good temperature resistance and relatively high modulus together with high impact strength, particularly at low temperatures (below freezing), and processes well. Like other PP types, it is a good water-vapor barrier but a poor oxygen and carbon dioxide barrier, and because of its two-phase structure is opaque. The ethylene-propylene rubber modifiers can be added by compounding, but are now increasingly polymerized directly during the manufacture to produce the rubber as finely dispersed domains *in situ*.

Biaxially-oriented polypropylene (BOPP)
Because of their high crystallinity and rapid rate of crystallization, both homopolymer and copolymer PP films can be oriented by standard methods, such as tenter frames, to give increased strength, stiffness, clarity, gloss and barrier properties towards water vapor, oxygen and carbon dioxide compared with unoriented films, but with loss in heat sealability since the oriented films distort. Because of their unusual property combinations such as good shrinkage, stiffness, transparency, sealability, twist retention and barrier properties, they are widely used for high-barrier modified atmosphere packaging (MAP), for snack foods, fresh produce and confectionery, and flexible packaging applications such as shrink wrap.

BOPP films continue to be developed using coatings to improve their properties. A new example is ExxonMobil high-barrier Bicor MB866 OPP film for modified-atmosphere long shelf-life packaging for oxygen-sensitive dry products such as fruits, nuts, cereals, snacks, roasted coffees and pet foods. This film has a polyvinyl alcohol (PVOH) coating on one side and an acrylic coating on the other and the PVOH coating is adhesive laminated to a sealant film, such as blown PE or cast PE. This gives an oxygen permeation level below 1 cc/m^2/day, similar to metallized polyester and below other barrier films, including polyvinylidene chloride- or ethylene vinyl alcohol copolymer-coated or co-extruded films. The Bicor films can also be used for stand-up packaging.[8] MAP is described in detail in Chapter 2.

4.2.3 Polystyrene (PS)

Polystyrene (PS) used for packaging is an atactic amorphous polymer made by free-radical polymerization of styrene in the liquid phase or in suspension. PS polymers generally show poor barrier properties to moisture vapor, oxygen and carbon dioxide. Despite this, they are often used for packaging oxygen-insensitive and short shelf-life food products, although special food-grade materials must be used, in which unreacted styrene monomer has been removed to very low levels.[9,10]

General purpose (crystal) polystyrene (GPPS)

Unmodified homopolymer (crystal or general purpose) PS is highly transparent and has good tensile strength, modulus and dimensional stability, but extremely poor impact properties, because the bulkiness of the phenyl ring substituents resists chain rotation. Density is about 1.05 g/cm^3 and the glass transition temperature (Tg) is about 100°C.

GPPS can be easily processed by injection molding, extrusion and thermoforming. Many grades are available with differing molecular weights; high MW versions generally have high heat resistance and strength, while lower molecular-weight, higher melt-flow grades are used for molding thin-wall containers. Some types offer resistance to environmental stress cracking with household chemicals and oils, but PS is generally more chemically reactive than PE or PP.

Packaging applications include CD and DVD 'jewel' cases and food service items. Oriented polystyrene (OPS) is produced by stretching extruded PS film, which reduces haze and increases modulus. It is often used as transparent packaging, since it is less expensive than very clear plastics such as PET and has less haze than PP. Its major disadvantage is brittleness, and it cracks and tears easily.

High-impact polystyrene (HIPS)

High-impact polystyrene (HIPS) is produced by polymerizing the styrene in the presence of butadiene rubber, which is incorporated partially as a rubber blend and partially as a grafted copolymer. HIPS has higher impact strength but lower heat distortion temperature than crystal PS. Because of the two-phase structure, it is generally opaque and many grades are available with different rubber contents, phase structures, molecular weights and tensile, modulus and impact properties. However, PP has been increasingly moving into HIPS uses over the past two decades, particularly for dairy products and other food packaging.

SBS block copolymer thermoplastic elastomers

Block copolymers of styrene and butadiene (SBS), and their hydrogenated forms (SEBS), are made by anionic polymerization. They have a two-phase structure of styrene hard segments (25–30%) and polybutadiene soft rubbery segments joining the hard segments together and the phases are small enough to minimize visible light scattering and retain transparency. The PS segments (segment molecular weight 10 000–20 000) associate to form glassy domains, which act as virtual crosslinks at ambient temperature, but which soften at higher temperatures to allow flow and processing as a thermoplastic. These materials exhibit flexibility and impact resistance, making them suitable for thermoformed transparent disposable packaging such as deli trays and cosmetics packaging, and they can also be blow-molded and used in flexible packaging as part of multilayer structures.

Polystyrene foams

Polystyrene (PS) foams are also widely used, both extruded foams in which a blowing agent is injected into the melt and expanded PS bead foams made by impregnating suspension polymers with blowing agents such as pentane. Extruded foams are mainly used for applications such as insulation, while closed cell bead foams are extensively employed for protective packaging and cushioning materials (plastic peanuts) and food service items such as hot-cups.

For the manufacture of bead foams, styrene monomer is polymerized in suspension to form small beads of diameter about 0.2 to 3 mm, with a weight average molecular weight of 160 000 to 260 000. The beads are then impregnated with 4 to 7% hydrocarbon blowing agent, such as butane or pentane under pressure and dried. To produce the molded expanded PS product, the beads are then pre-expanded by controlled treatment with steam to form 'pre-puff', allowed to come to equilibrium with ambient air and then placed in a closed mold and steam heated again to make the final fused structure in the mold shape. The density of the final foam is controlled by the pre-expansion process.[11,12]

There have been increasing attempts to find alternatives to PS foams over the past few years, particularly for takeout food services (clamshells, hot-cups) and 'peanuts', and they are continually attacked by environmental organizations, even though data compiled by the Plastics Foodservice Packaging Group of the American Chemistry Council show that less than 1% by weight of US solid waste is PS.[13] Bans on PS foam packaging in restaurants have been enacted in many locations, particularly in California, and sometimes legislation requires the use of biodegradable packaging.

4.2.4 Polyesters

Polyesters are produced by the reaction of a diol with a dibasic acid, and copolyesters are made using mixtures of diols, dibasic acids or both. Both semi-crystalline and amorphous types are employed in packaging.[14–16]

Polyethylene terephthalate (PET)

Polyethylene terephthalate (PET) is the most common thermoplastic used in packaging. It was initially developed in the 1940s for synthetic fibers, which are still the dominant application. High molecular weight polymers became available in the 1970s to produce a transparent, lighter bottle material to replace glass. Like glass, PET is strongly resistant to microbiological attack and does not react with food or biodegrade. It is now ubiquitous in bottle manufacture, and a report by IHS indicates that 31% of PET will be used for rigid packaging, including bottles (in 2012, ~18 million tonnes/annum), 4% for films, 3% for other niche applications, and the remainder for fibers.[17]

PET is also expected to increase its share of total packaging markets, because of the development of better gas barriers and ultraviolet (UV) light protection, which extends the shelf life of the packaged products, as well as improved blow-molding technology and hot-filling processes. It is also showing greater use in beer packaging, resulting from advantages in product quality and transport costs and legal constraints on the use of glass outdoors and in sports venues, etc. Other developments include continued light-weighting and growing emphasis on sustainability and recycling. PET recycling is a key issue, owing to growing consumer environmental concerns, legislation and government targets. Recycled PET (rPET) growth is also being driven by companies seeking environmental, economic and marketing advantages from its use. However, there is still some uncertainty over how widespread rPET use will be in bottle production, which will depend on the relative costs of rPET and virgin PET and the recycle availability and quality. PET recycling is discussed further in Section 4.11.

PET is a high molecular weight, semi-crystalline (up to 60%) polymer formed by the copolymerization of terephthalic acid or dimethyl terephthalate with ethylene glycol. It can be processed and converted by injection molding, extrusion and injection stretch blow molding. Major factors determining its properties are the molecular weight, degree of crystallinity and Tg. The initial melt-phase polymerization produces a low molecular weight polymer, which is then further polymerized by heating in the solid state (solid-stating) to increase the molecular weight and cause crystallization. The PET pellets generally reach about 60% crystallinity, which then disappears in the production of the amorphous preform in blow molding and returns to about 20% in the final bottle due to strain-crystallization in the blow-molding process. Very small crystallites are formed, which allow the molded product to be highly transparent and have high gloss, improve the mechanical and gas barrier properties, and provide good creep resistance under pressure (which is important for carbonated soft drink bottles) and good impact resistance.

The crystallinity of PET is strongly influenced by the processing and cooling conditions. Depending on the heat history, PET can be made either amorphous (APET) and transparent by rapid cooling or by incorporation of comonomers (see below) or semi-crystalline by slow cooling, heat setting or annealing, or by the incorporation of nucleating agents to both increase crystallization rate and maximize the degree of crystallization and crystallite size (CPET). Consequently, PET can appear transparent if the crystallite phases are small compared to the wavelength of visible light (particle size <500 nm) or translucent, opaque or white with larger crystallites as in CPET. Heating above the Tg allows the crystallites to grow and the degree of crystallization to increase up to a limit of about 60% for packaging films and sheet.

APET combines structural strength with glass-like clarity, is crack resistant and does not stress-whiten. It can be easily thermoformed, tolerates temperatures from –40°C to +70°C and can be sterilized with ethylene oxide and gamma radiation, so that uses include disposable food and medication packaging, trays and blister packaging. CPET suffers less deformation under stress than

APET, particularly at high temperatures, but is considerably more brittle at low temperatures. However, thermoformed CPET trays can withstand both freezing and oven-baking temperatures and can be used for frozen foods, ready meals, etc.

PET copolymers

Many PET copolymers containing additional monomers over the ethylene glycol and terephthalic acid comprising PET are also commercially available and may have advantageous properties for specific applications, although they are more expensive than standard PET grades. They are produced by incorporating an additional glycol or diacid (or both) into the molecular chain and generally reduce the polymer crystallinity but increase its melt strength. For example, cyclohexane dimethanol (CHDM) can be used to replace part (generally 20–30%) of the ethylene glycol in the polymer backbone to make glycol-modified PET copolymers (PETG). Because of its larger size than ethylene glycol (it has six more carbon atoms), it inhibits crystallization and the copolymer is totally amorphous with a distinct Tg but no crystalline melting point. This provides high transparency and gloss and low haze, but poorer heat resistance and barrier properties than PET.

The copolymers also exhibit good chemical and scratch resistance, good mechanical properties such as high stiffness, hardness, ductility and toughness (even at low temperatures), higher melt strength and easy processing. However, like PET itself, which has a heat distortion temperature of 65–70°C, PETG does not have very high temperature resistance. Unlike PET, PETG does not require injection-stretch blow-molding and can be extrusion blow-molded into bottles and extruded into sheet and film, tubes and profiles, thermoformed and injection molded. Processing is characterized by fast throughput and lower energy requirements as no annealing is needed. Major uses are bottles, jars and other containers for household and food products, caps and cases, cosmetics and personal care packaging, and blister packaging. Since PETG can be sterilized with ethylene oxide or gamma radiation, it is very useful for medical device packaging.

Tritan™ copolymer was introduced in 2007 by Eastman Chemical Co. and incorporates tetramethyl cyclobutanediol (TMCD) as a comonomer.[18,19] It has high Tg (110°C), high clarity and high impact resistance and is sterilizable with ethylene oxide or gamma rays. It is finding uses in water bottles, containers, serving trays, blender bowls, tumblers and other kitchen products to replace polycarbonate, since it is bisphenol-A (BPA) free. Thus Rubbermaid has converted all components previously made of polycarbonate to Tritan in 174 food-handling products worldwide. Tritan resists heating, cooling, scratching, chipping and detergents, and withstands 1500 commercial dishwasher cycles without hazing or stress cracking, while retaining its toughness and shatter-proof properties. It can be used for large containers with deep-draw shapes, often using polycarbonate molds.[20] Tritan is also a good candidate for replacing glass in drinking vessels in bars, nightclubs, sporting events and festivals rather than polycarbonate, which crazes over multiple dishwasher cycles.[21]

Other polyester copolymers are made by modifying the polyester acid component, most commonly replacing part of the (linear 1,4-linkage) terephthalic acid with isophthalic acid (1,3-linkage) or ortho-phthalic acid (1,2-linkage). This creates a kink in the PET chain and inhibits crystallization.

These types of amorphous or copolymer PET can be advantageous for packaging applications such as thermoformed trays or blisters. Conversely, crystallization may be important in applications where mechanical strength and dimensional stability are required. The incorporation of small amounts of CHDM, diethylene glycol or isophthalic acid comonomers can be useful for PET bottles to slow down crystallization but not eliminate it, thereby enabling injection-stretch blow-molded bottles to be made with enough crystallinity to provide sufficient aroma and carbon dioxide barrier for carbonated beverages.

Blends of PET with other polyesters, such as polyarylates, have also been used for packaging requiring higher temperature resistance (i.e. cosmetics packaging containers, which can often be left in automobiles in hot sunshine). Blends with polyethylene naphthalate (PEN) have also been employed to improve the PET barrier properties. However, all of these systems contain partially a blend and partially new copolyesters formed by transesterification. PET and PEN homopolymers are immiscible so that blending necessitates special methods to cause transesterification and make the material homogeneous. PEN is highly crystalline, depending on its heat history, and has many desirable attributes and enhanced properties over PET, such as four to five times the water vapor and oxygen barrier, higher tensile strength and stiffness, chemical, hydrolysis and UV resistance, and shorter processing cycle times. However, it is much more expensive than PET, so that its use in packaging is now limited despite its earlier promise.

Oriented PET film and sheet

Biaxially-oriented PET film (e.g. Mylar) is used in many applications because of its high mechanical strength. It can be metalized or aluminized to reduce its gas permeability and is stable under high temperature and high humidity conditions. This type of film has replaced PVdC and aluminum foil in barrier laminate structures such as packaging dry foods, snack food, toiletries and lidding for soup containers, and medical applications The film is heat sealable and can also be used for extending the shelf life of fresh or minimally processed foods.

PET processing[22]

PET processing must be performed so as to avoid degradation of the polymer arising from hydrolysis, thermal breakdown and oxidation. Such degradation can lead to reduced molecular weight, discoloration, cross-linking (gels and fish-eyes) and production of acetaldehyde, which is a taste problem for bottled water, even in low concentrations such as 10 to 20 ppb. Since PET absorbs moisture, to avoid hydrolysis and molecular weight loss on heating, it must be dried before processing to less than 50 ppm, and the drying temperature must also be less than 160°C. Use of copolymers, such as those incorporating cyclohexanedimethanol or isophthalic acid, can also help avoid or eliminate degradation and acetaldehyde formation by lowering the processing temperature, pressure, shear conditions and residence times required. Stabilizers, generally phosphite antioxidants, can also be used.

Another major challenge in PET processing is its low melt strength, which prevents extrusion blow-molding of the normal crystalline grades, so that increased molecular weight polymers or copolymers must be used for extrusion blow-molding. Consequently, injection-stretch blow-molding must be employed for standard PET grades to make bottles.

There have been considerable developments in both products and processing for PET bottles and other containers over the past ten years. To prevent deformation of oriented PET when it is subjected to high temperatures, heat-set technology has been developed for bottles, jars and films, and PET containers can now be used for hot-filling systems, typically around 85°C, to provide sterilization and for pasteurization systems at 60°C.[23] PET barrier systems are now also being used for beer bottles and allow carbon dioxide levels to be maintained and oxygen permeation prevented, to provide safe containers for sporting and other events. A major example is the recyclable green PET beer bottle developed by Heineken for the 2012 London Olympics to mimic its standard green glass bottle and incorporating the Constar Oxbar oxygen scavenging and barrier layer system.[24] Barrier and scavenger systems are discussed further in Sections 4.3 and 4.4 below, respectively.

There has also been considerable advance in container design to allow light-weighting. In 1980, a typical 2-liter PET bottle weighed 68 g but today it weighs only 47 g. Some water bottles are now not even strong enough to be stacked unless they are pressurized with nitrogen. PET is also now being used for larger containers such as 40 liter kegs for beer and hard cider[25] and for bottles for wine and other liquors, and this is expected to increase with oxygen barrier improvements. PET is also used for packaging food, including edible oils, syrups and spices, and for toiletries and personal care products. CPET is used for microwavable containers for frozen foods, and biaxially-oriented PET is used in meat and cheese packaging and laminated packaging for snack foods. PET pouches are used for boil-in-bag frozen foods and sterilizable pouches for medical uses, while PET extrusion coatings on paperboard are used to make ovenable board for frozen food applications.

4.2.5 Polycarbonate (PC)

Polycarbonate (PC) is an amorphous polymer produced by melt condensation of bisphenol A (4,4′-(propane-2,2-diyl)diphenol) with diphenylcarbonate. It is highly transparent, hard, rigid and impact resistant, and has high enough temperature resistance to allow autoclave sterilization. However, it has mediocre gas and moisture barrier properties, which have limited its use in food and beverage packaging. It also exhibits poor scratch resistance unless treated with special coatings and undergoes environmental stress cracking with many chemicals, oils and solvents, particularly ketones, so that containers must be annealed to eliminate molding stresses. PC is readily processable by extrusion blow molding, injection molding and thermoforming and has been used for packaging applications such as large water containers, beverage bottles and baby bottles, as well as medical packaging applications.

Because of concerns about the toxicity of residual BPA, which can leach out from the plastic, its use for baby bottles and other food packaging has been limited or banned in some countries, and it is being replaced by Tritan or other amorphous copolyesters, which do not contain BPA. BPA is now banned in baby bottles in the US,[26] but on 30 March 2012, the US Food and Drug Administration rejected a petition from environmentalists to ban it from all food and drink packaging, including plastic bottles and canned food, saying there is insufficient evidence to justify new restrictions, although studies will continue.[27–29] This topic is covered further in Chapters 1 and 11.

4.2.6 Polyamides (nylons)

Polyamides are linear thermoplastic condensation polymers containing recurring amide groups made by the reaction of diamines with dicarboxylic acids or by the condensation of amino-acids, which have both amine and acid functionalities in the same molecule. Polyamide properties are largely controlled by the amide groups, which provide strong intermolecular hydrogen bonds.

PA-6 and PA-6,6

The major packaging types are PA-6 and PA-6,6. PA-6, or polycaprolactam, is made by the ring-opening polymerization of caprolactam and has 6 carbon atoms in its repeat unit. PA-6,6 is made by the condensation polymerization of adipic acid and hexamethylenediamine and has 12 carbon atoms in its repeat unit. The strong intermolecular hydrogen bonding results in unusual property combinations, including high melting point and heat resistance, good mechanical properties over a wide temperature range (strength, flexibility, puncture resistance and toughness), transparency, chemical resistance, and high barrier to oxygen, carbon dioxide and many solvents and aromas, but not to moisture vapor. For most packaging applications, PA-6 and PA-6,6 are made into flexible films or used as components in co-extruded or laminated multilayer structures with PE, PP or ethylene-vinyl alcohol (EVOH) for packaging oxygen-sensitive foods including fruit juices, meats, fish, cereals, cheese and other dairy products, and prepared meals such as TV dinners and ready meals. PA-6 and PA-6,6 are thermoformable and are used in multilayer sheet structures because of their high deep draw depth and ability to maintain barrier layer thickness in the corners of the thermoformed product. An exterior polyamide layer gives high heat resistance to the multilayer film structure, allowing high sealing line speed and good seal quality.[30]

Standard extrusion processes and equipment can be used. However, polyamides are extremely hydrophilic and can absorb 6 to 8% of water by weight, which must be removed by drying before processing to a level of about 0.1 wt%. Blown and cast films can be produced, as well as biaxially-oriented films and extrusion coatings. Melt temperatures of 250 to 270°C are typical for PA-6,6 and about 20 to 30°C lower for PA-6. Since the resins are semi-crystalline, rapid cooling gives lower crystallinity and more transparent, flexible, thermoformable films.

Polyamide films will re-absorb moisture after processing. Since the water acts as a plasticizer, this reduces film strength and stiffness and increases flexibility, elongation at break, drawability, puncture resistance, impact strength and thermoformability. Consequently the films should only be used or converted after conditioning with water or allowing them to reach equilibrium with ambient humidity to produce high-quality dimensionally stable film. The polyamides can crystallize to very high levels, depending on the cooling rate during film production: rapidly quenched films are nearly amorphous, while most slowly cooled film is highly crystalline. Amorphous or partially crystallized film will then post-crystallize in the solid-state and shrink when it absorbs ambient humidity, which causes the Tg to fall below the ambient temperature, or is reheated above the Tg (55°C), so that it can be cut to the desired width only after conditioning. This post-crystallization does not affect the advantageous properties of rapidly cooled film, including high transparency, high flexibility, high puncture and impact resistance and good thermoformability.

The oxygen transmission rate of PA-6 is increased when it absorbs water in high humidity environments. While it is 100 times lower than polyolefins, it is still 10 to 100 times higher than high-barrier EVOH, although EVOH is also sensitive to moisture, so that in high humidity environments the EVOH barrier properties can be degraded enough to attain parity with barrier polyamides such as PA-MXD6 (a copolymer of m-xylylene diamine and adipic acid, discussed further in Section 4.3.3 below). The polyamides exhibit only poor barriers towards highly polar molecules such as water or lower alcohols, but exhibit high barrier properties to low-polarity substances such as hydrocarbons and aroma chemicals from foods such as limonene, camphor, vanillin and menthol. Nanocomposites based on PA-MXD6 and PA-6 are also being developed for barrier packaging (Section 4.3.3).

Oriented polyamide films

Biaxially-oriented polyamide (BOPA) films exhibit extremely high strength, stiffness and puncture resistance, low elongation at break and good printability. They can be metalized or coated with inorganic barrier layers such as 'SiO_x'. Orientation of cast polyamide film increases the oxygen barrier by 30%.

4.2.7 Polyvinyl chloride (PVC)

Polyvinyl chloride (PVC) is a thermoplastic homopolymer of vinyl chloride. The polymers employed for packaging are mostly made by suspension polymerization (80%), although some are by emulsion or solution polymerization. PVC is largely amorphous, with only a low level of crystallinity. However, there are strong interchain interactions because of the high carbon–chlorine bond polarity, and the unplasticized polymers are rigid at ambient temperature. Packaging uses include rigid packaging for bottles, blister packs and clamshells and films for flexible packaging.

Unmodified PVC is difficult to process because its decomposition temperature (starting around 100°C) and melting temperature (Tg 82°C) are very close and decomposition produces hydrogen chloride, which is extremely corrosive. Consequently, stabilizers must be added to rigid PVC and plasticizers to make flexible PVC, and the PVC properties are strongly influenced by the amounts and types of additives used. A wide range of properties can be obtained, from rigid containers and blow-molded bottles to soft, flexible films, depending on the plasticizers used. PVC polymers are highly miscible with a wide range of plasticizer types, many suitable for food contact, and impact modifiers can also be added.

Formulated PVC can be processed by extrusion, film blowing, injection molding and blow molding. Bottles have been used for edible oils, liquor, household products and toiletries. Flexible films are tough and resilient and have been widely used for food wraps, including fresh fruit and vegetables, chicken in chilled trays and especially fresh red meat. The oxygen barrier properties of the plasticized film can be adjusted to a level allowing the meat to retain its red color and fresh appearance. Other applications are medical device packaging and blister packaging. For more information on MAP, see Chapter 2.

According to the 2011 report by GBI Research, worldwide PVC use for packaging in 2011 was 3.7 million tonnes (out of a total PVC demand of 32.3 million tonnes).[31,32] However, the use of PVC for packaging has been constrained over the past five years because of environmental and toxicological issues, high density and higher costs than PET and PP, and general advances in competing materials. Many markets once dominated by PVC, such as food and beverage bottles, have now largely transferred to PET, although PVC is still used in pharmaceuticals and blister packaging. Companies such as P&G are moving to eliminate PVC from their packaging.[33] Also, new barrier MAP systems have been introduced to compete with the EPS tray/PVC film technology that has previously dominated case-ready meat packaging.

Current developments in PVC technology by major companies such as DuPont, Dow and Teknor Apex, and many plasticizer manufacturers, are promoting the use of environmentally friendlier and toxicologically less suspect PVC formulations incorporating biobased plasticizers or high molecular weight phthalates, which are not considered to be estrogen mimics, to replace materials such as bis-(2-ethylhexyl) phthalate, dibutyl phthalate, di-isobutyl phthalate and benzyl butyl phthalate, which are now classified as 'Substances of Very High Concern' for REACH in Europe and for which there is a sunset date of 21 February 2015 under their recommendation for authorization under Annex XIV.[34] In addition, new stabilization systems have been developed that are less toxic than the cadmium- and lead-based stabilizers previously used, such as calcium-, zinc- and barium-based stabilizers. Many of these developments are being driven by the VinylPlus initiative in Europe[35,36] and may stop or even reverse this decline in PVC use for packaging.

4.2.8 Ethylene-vinyl acetate copolymers (EVA)

Ethylene-vinyl acetate (EVA) copolymers are random linear copolymers produced by the copolymerization of ethylene and vinyl acetate (VA), either in a high-temperature and high-pressure free radical process or, for high vinyl acetate materials, in emulsion. The properties are largely dependent on the molecular weight and the vinyl acetate content. While commercially available materials range from about 5% to 50% VA, those containing 5–20% are mostly used for packaging applications, particularly food packaging. The relatively bulky VA comonomer reduces the crystallinity relative to PE until, at 50% VA, the EVA copolymers are totally amorphous, and hence increases the flexibility, stress-crack resistance and toughness, particularly at low temperatures; increases the clarity and gloss; extends the heat seal temperature range, adhesion and hot tack properties; and provides better barrier properties. Likewise, increasing the molecular weight increases the heat seal strength and hot tack, flexibility and toughness. Because of their ease of processing and high adhesion, EVA polymers are frequently employed for extrusion coating and as co-extruded heat seal layers in multilayer constructions such as cheese wrap with other polymers such as PET, cellophane and biaxially-oriented PP films.

4.2.9 Ionomers

Ionomers are based on acid copolymers, such as ethylene-acrylic acid and ethylene-methacrylic acid, partially neutralized with cations such as sodium (Na^+), zinc (Zn^{2+}) or lithium (Li^+). This produces highly polar plastics with higher melt strength, outstanding toughness especially at low temperatures (better than LDPE) and better transparency than the un-neutralized acid copolymer. The ionic groups provide random interchain ionic cross-links, decreasing the crystallinity and yielding properties more like high molecular weight materials, while still allowing normal thermoplastic processing in conventional equipment designed for PE resins, such as extrusion, co-extrusion, blown and cast film, and extrusion coating, since the ionic bonds are more easily broken than covalent bonds. Processing temperatures are generally 175–290°C and the ionic bonds also cause high elongational viscosity leading to excellent pinhole resistance. Barrier properties are fairly modest, but they can be combined with other polymers, such as high density polyethylene or polyvinylidene chloride or metallic foils, to provide multilayer structures with good barriers.

Other advantageous properties of ionomers for packaging are their sealing performance, formability, oil and grease resistance and high hot-draw strength, which permit higher packaging line speeds and fewer package failures. A wide range of grades with different property combinations can be tailor-made by adjusting the acid comonomer content, molecular weight, molecular weight distribution, degree of neutralization, type of neutralization ion used and the incorporation of additives. In general, the sodium ion types show better optical properties, hot tack and oil resistance, while the zinc ionomers are more resistant to water and show better adhesion properties in extrusion coating and co-extrusion applications. Some commercial products are DuPont Surlyn® (random ethylene-methacrylic acid copolymer generally containing <15 mole% MMA), Dow Amplify and Honeywell Aclar resins.

Ionomers are often used for packaging applications requiring easy formability, high toughness and good appearance. The largest single use is for food packaging films, since they are highly resistant to oils, fats and other chemicals, provide high seal reliability over a wide temperature range and adhere well to aluminum foil. They are also widely employed as heat-seal layers in multilayer films for meat, fish, cheese and snack food packaging, and in structures such as Tetra Pak juice boxes.[37,38]

4.2.10 Polyvinyl alcohol (PVOH)

Polyvinyl alcohol (PVOH) is a water-soluble polymer, which is structurally an addition polymer of vinyl alcohol. However, since vinyl alcohol is not stable but tautomerizes to acetaldehyde, PVOH is produced by the hydrolysis of polyvinyl acetate made by the free radical polymerization of vinyl acetate. The properties of PVOH are largely governed by its molecular weight and branching structure, which are determined by the polyvinyl acetate used as the starting polymer, and by the degree of hydrolysis of the polyvinyl acetate, which governs the proportion of the acetate groups that remain unhydrolyzed to the alcohol in the PVOH polymer. PVOH polymers with a low degree of hydrolysis (generally 88% or greater) show a low degree of crystallinity and are soluble in cold water. As the degree of hydrolysis is increased, the degree of crystallinity also increases and the dissolution temperature required rises, so that at 97% hydrolysis, a temperature of about 70°C is required for fast dissolution in water. Thus PVOH polymers are actually copolymers of vinyl alcohol with small proportions of vinyl acetate comonomer, and their structures can be tuned to obtain the requisite solubility, physical, thermal and barrier properties for specific applications. In addition to their crystalline properties, PVOH polymers also show strong interchain interactions from hydrogen bonding by the highly polar hydroxyl groups.[39–41]

PVOH is the principal polymer used for water-soluble packaging. It is generally produced as a film, either by casting on rolls from aqueous solution and then drying, or by extrusion film blowing or casting. PVOH films can also be used in multilayer co-extrusions or laminates. For many years it was considered that PVOH could not be processed as a thermoplastic, since its melting point (180–240°C, depending on its degree of hydrolysis) is generally higher than the decomposition temperature, particularly for high hydrolysis (high crystallinity) grades, and the polymers are both heat and shear sensitive and degrade by dehydration, oxidation and crosslinking. However, over the past 15 years, specialized processing technology and formulations containing plasticizers such as glycerol, other polyhydric alcohols, glycols or polyglycols, have been developed to allow thermoplastic processing of PVOH compounds, including extrusion, film blowing and casting, injection molding and blow molding.

PVOH films show high tensile and tear strength, flexibility and puncture resistance, and can be made pinhole free. They have contact clarity and are inherently antistatic, can be sealed by heat or solvent welding with water and are non-toxic. However, their properties will vary with ambient humidity, since PVOH absorbs moisture from the atmosphere to reach an equilibrium water content of about 3–5%. The films, particularly the high hydrolysis types, show high barrier to oxygen, carbon dioxide, organoleptics and hydrocarbons, but not to water vapor. However, the barrier properties deteriorate in high humidity environments, so that they are extremely good for the dry film (better than ethylene-vinyl alcohol (EVOH) copolymers), but fall off rapidly with increasing water content. Depending on the other formulating ingredients, PVOH films have also been approved for food contact applications.

PVOH is also inherently biodegradable, particularly in solution, and degradation is most rapid for the low hydrolysis grades in compost and soil, but has a reverse trend in aqueous solution.[42] It is attacked by at least 20 different genera of bacteria, yeasts and molds, which occur in a variety of disposal environments such as activated sludge, compost, facultative ponds, landfills, anaerobic digesters and septic systems and in natural soil and aquatic environments. The degradation mechanism is unusual, since it occurs randomly along the PVOH polymer chains by attack on vicinal hydroxyl groups to form β-diketones, which are then further oxidized to form ketones and carboxylic acids. Biodegradation is discussed in more detail in Chapter 5.

Major packaging uses for PVOH are for applications requiring pre-measured and unit dosing and water-soluble delivery systems. Examples are hospital infected laundry bags (the bag dissolves in the washing machine cycle) and a variety of sachets and pouches for dishwasher and laundry detergents, pesticides, herbicides, fertilizers, dyes, pigments and cement additives, to avoid pouring and measurement in use and to provide user convenience. As a recent example, Procter & Gamble, working with MonoSol LLC, have recently introduced premeasured unit dose cold water soluble liquid Tide PODS, which have a 3-chamber construction for a 3-in-1 concentrated laundry detergent that cleans, removes stains and brightens. The PVOH packaging dissolves completely, even in a cold-water wash. These PODS reduce total packaging material use by 11% per wash load, compared to liquid laundry detergent bottles.[43]

4.3 Barrier polymers and technology

There have been tremendous advances in barrier polymers and technology over the past two decades. The world transparent barrier film market value in 2012 has been estimated at \$6.7 billion and is expected to grow at 4.2% per annum to reach \$8.2 billion in 2017, with the electronics sector showing the fastest growth.[44]

In the highly competitive food and beverage market, packaging producers have developed an extensive range of barrier and scavenger materials, which allow manufacturers to extend shelf life and preserve the taste, color, odor and overall quality of packaged products. For foodstuffs and beverages, the oxidation of oils, fats and other food constituents can produce off-flavors, discoloration and loss of nutritive value, so that protection against oxygen is vital and a continuing industry goal is to produce barrier systems, which will protect oxygen-sensitive products by limiting oxygen ingress. In other products, such as carbonated soft drinks, it is necessary to prevent carbon dioxide loss while still retaining recyclability of the package. The permeation rate of these gases limits the shelf life of beverages and foodstuffs packed in plastics such as PET. Other demanding uses include plastic bottles for beer, where both carbon dioxide egress and oxygen ingress must be controlled, and teas. Such applications for the most part use multilayer barrier PET constructions although some monolayer or coated systems are also employed.

Major types of barrier polymer are discussed below, as well as several coating technologies. The rise in demand for ready meals, driven by such trends as single-person households and families where all adults work, together with the trend towards sustainability (single serve packaging can provide large reductions in food waste) has led to an increasing need for barrier systems. Barrier bottle systems are also of high importance because of the need to obviate potential problems which could be caused with the PET and HDPE bottle recycling streams. Many new systems have been introduced, including the use of new barrier resins, monolayer and multilayer systems, surface coatings and oxygen scavengers, and this development is still continuing. Because of the wide variety of packaging applications where barrier properties are needed, no single system has yet been found to meet all requirements.[1,2,45,46]

4.3.1 Ethylene-vinyl alcohol (EVOH)

Ethylene-vinyl alcohol (EVOH) copolymers are widely used as barrier polymer layers in multilayer packaging structures, both in films and molded products. The two major producers are Kuraray (EVAL resins) and Nippon Gohsei (Soarnol resins). EVOH polymers are highly crystalline random copolymers of ethylene and vinyl alcohol groups where, as in the case with PVOH (Section 4.2.10), hydrogen bonding by the highly polar hydroxyl groups both increases the intermolecular interactions and the hydrophilicity. Since vinyl alcohol does not exist as a discrete polymerizable monomer, EVOH polymers are made by polymerizing ethylene with vinyl acetate and then hydrolyzing the vinyl acetate to vinyl alcohol.

EVOH polymers provide extremely good barriers towards oxygen, carbon dioxide and organoleptics (aromas, flavors and fragrances) as well as excellent solvent, oil and organic vapor resistance towards low polarity materials such as hydrocarbons, although the resistance to alcohols is much lower since these may be absorbed in the EVOH. The EVOH barrier, physical and processing properties all depend on the copolymer composition. Higher vinyl alcohol levels give better barrier towards oxygen and carbon dioxide and worse towards moisture vapor, but also induce greater propensity for moisture absorption, which conversely decreases the gas barrier properties. Because of the hydroxyl groups in the polymer structure, the EVOH polymers are hygroscopic and special precautions must be taken with applications involving high humidities in processing or use, although the barrier property deterioration with moisture absorption is reversible.

The higher the ethylene content in EVOH, the lower the density, melting point and Tg and the poorer the barrier to oxygen and carbon dioxide. The melt flow properties also depend on the ethylene content while thermoplastic processability improves at higher ethylene levels. Packaging grades generally have ethylene contents of 27–48 mole %, and can be processed by film blowing or casting, sheet extrusion and thermoforming, and injection- and blow-molding, and usually constitute part of a multilayer structure. They can be co-processed with a wide variety of other packaging polymers, such as PE, PP, PS, PET and polyamides. As thin a layer of EVOH as possible, consistent with the required barrier properties, is used since these polymers are expensive. Co-extrusion, lamination or co-molding between layers of high moisture-barrier polymers will reduce the sensitivity of the EVOH to moisture and also provide a moisture barrier for food and beverage packaging. Organoleptic barriers have become more important with the growth of convenience stores where many types of food and other goods such as detergents are displayed and stored in close proximity and cross-contamination of taste and odor components must be avoided.

EVOH polymers are suitable as barrier layers for packaging foods, beverages, household chemicals, cosmetics and medical packaging, where oxygen must be kept out or carbon dioxide kept in. They are used in a wide range of flexible applications using co-extruded films, retort pouches, and bag-in-box applications for foods and wine, and also in many rigid applications including trays, cups and lightweight high-barrier lidded containers, and for co-blow molded bottles, made using PP, HDPE and PET for foods such as soups, mayonnaise, salad dressings, tomato ketchup, juices, carbonated soft drinks and beer. EVOH is also used for extrusion coating for paperboard for juice packaging to prevent absorption of non-polar flavor components by PE. However, problems can arise when such multilayer films and bottles are recycled, particularly for PET bottles where an EVOH layer can lead to transesterification during the recycling process, causing deterioration of the recycled PET.[47–49]

4.3.2 Polyvinylidene chloride (PVdC)

Polyvinylidene chloride (PVdC) is a semi-crystalline copolymer generally containing 70 to 94% vinylidene chloride, with other monomers such as vinyl chloride, acrylic esters and vinyl nitrile, generally in amounts of 6 to 30%. The homopolymer decomposes below its melting point and these comonomers allow processability as a thermoplastic by decreasing the melting point to 140–175°C. PVdC is used in relatively small amounts as a packaging barrier layer both in films and in rigid packaging, although in most rigid applications it has now been supplanted by ethylene-vinyl alcohol and barrier polyamides. Unlike EVOH, it has excellent moisture barrier properties as well as being an oxygen and carbon dioxide barrier. It is a good organoleptic barrier and has excellent resistance to oils, fats and chemicals. It can be processed by extrusion, co-extrusion, film blowing and casting, lamination and injection molding, although care must be taken to avoid degradation and the production of 'black specks'. As with PVC, PVdC also has a disadvantage of high density and has come under attack for environmental issues, so that other barrier polymers have tended to supplant it in packaging.

The major packaging applications of PVdC are as barrier layers in food, pharmaceutical and cosmetics packaging; as a monolayer film in household wrap; as a component in shrinkable multilayer films, most commonly co-extrusions with polyolefins, for packaging meat, cheese and other moisture- or oxygen-sensitive foods; and as barrier layers in thermoformed containers. PVdC is also used as coatings for paper and paperboard, to give grease, water and oxygen barrier properties, and as barrier coatings for cellophane, BOPP and PET.

4.3.3 Barrier polyamides

The major polyamide used for barrier applications is PA-MXD6, which is a copolymer of m-xylylene diamine with adipic acid manufactured by Mitsubishi Gas Chemical. It is semi-crystalline and has good transparency and high gas barrier properties. It experiences lower water absorption than conventional polyamides such as PA-6 and PA-6,6 and exhibits higher tensile properties, modulus and Tg (85°C). However, it is still hygroscopic and moisture absorption degrades the gas barrier properties under high humidity. The moisture sensitivity is intermediate between PVdC and EVOH.

MXD6 polymers have a high melt temperature (237°C) and a wide temperature processing range, so that they can be used with PE, PP, other polyamides and PET to produce multilayer flexible films or rigid packaging by extrusion, thermoforming, injection-molding and blow-molding. MXD6 can also be biaxially-oriented by stretching to make high-barrier multilayer films, and is retortable.[50]

Mitsubishi Gas Chemical and Nanocor, a nanoclay supplier, have developed a nanocomposite/PA-MXD6 blend called Imperm for multilayer PET bottles, which shows high oxygen, carbon dioxide and moisture barrier properties based on the combined barrier contributions of the MXD6 and the nanoclays. The barrier properties are two to four times greater than attained by the use of MXD6 alone, depending on the temperature and humidity, while still retaining high clarity and allowing light-weighting by the use of thinner bottles than single-layer PET bottles while still retaining equivalent barrier protection. The nanoclay also provides mechanical reinforcement for the matrix PA-MXD6 polymer, improves high temperature properties, increases the crystallization rate and reduces moisture sensitivity, which is much lower than experienced with EVOH. Target markets are beer, juices, salad dressings and carbonated soft drinks. These materials have also been used for thermoformed sheet for multi-layer deli containers and for flexible multi-layer polyolefin film for snack food packaging.[51,52] The use of nanotechnology is discussed in more detail in Section 4.3.9 and Chapter 9.

PA-6 compositions containing low levels of nanoclays are also under development for barrier packaging applications and can produce film and sheet with a four-fold oxygen barrier improvement over unfilled PA-6, together with a 35°C increase in heat distortion temperature and increases of 30% and 50% in tensile strength and flex modulus, respectively. Honeywell Specialty Polymers has introduced a family of barrier PA-6 nanocomposites containing nanoclays, called Aegis. These provide barrier layers with good adhesion to PET and high delamination resistance for packaging applications, such as multilayer PET bottles, and work well in co-injection stretch blow-molding. Because the nanocomposites have increased heat resistance, modulus and strength over the base PA-6 resin, they may enable all-polymer packaging materials to be produced to replace present foil packaging.

The Aegis family includes the Aegis OXCE grade, which also incorporates a proprietary oxygen scavenger system containing a polydiene dispersed in the polyamide nanocomposite for high-clarity beer bottles and other applications where very high oxygen barrier and recyclability is required. The nanoclay barrier prevents oxygen egress while the oxygen scavenger absorbs any oxygen in the head space and bottle contents. It also gives a high barrier against carbon dioxide. This system has been used in South Korea for the Hite Brewery Hite Pitcher beer bottle, which is reported to provide a six month shelf life similar to glass bottles. The Aegis nanocomposites used without the oxygen scavenger can also replace EVOH in flexible packaging (blown and cast films), since they are have high puncture resistance and are higher barrier than EVOH, allowing for light-weighting, and are less expensive. Potential packaging applications include foods, cosmetics, medical products and electronic equipment These nanocomposites can also be employed as a PA-6 replacement to improve barrier properties and reduce moisture uptake in paperboard coatings.[53]

4.3.4 Polyglycolic acid (PGA)

Polyglycolic acid (PGA) is a highly transparent, semi-crystalline aliphatic polyester formed by polycondensation of glycolic acid and is the simplest polyester structure possible. While it has been used for many years as a bioabsorbable suture, there has been limited production because the water liberated by the esterification process tends to hydrolyze the polyester back into short chains. However, a new process, which produces high molecular weight polyglycolic acid via the formation of a glycolide dimer that avoids the above hydrolysis problem and can be used on a large scale, has now been developed by Kureha and a 4000 tonnes/yr plant built in Belle, West Virginia.

PGA has higher density (1.50 amorphous, 1.70 crystalline) than other polyesters, but exhibits high mechanical strength and modulus. Melting temperature is 220°C (close to PET, which facilitates the co-injection of bottle preforms), while the Tg is 40°C and the heat distortion temperature is 168°C. It can be extruded into film and sheet, and injection- and- blow-molded and co-molded. It can also be oriented to dramatically improve its physical properties.

PGA shows excellent barrier properties against oxygen, carbon dioxide, aroma chemicals and moisture vapor and high resistance to many common chemicals and solvents. Its oxygen and moisture barrier properties are both better than EVOH, PA-MXD6 and PVdC. Because of its moisture resistance, its barrier properties against oxygen and carbon dioxide are not greatly affected by relative humidity, unlike EVOH and MXD6. It also shows better barrier behavior against flavor and aroma components such as d-limonene, L-menthol and vanillin than either EVOH or PVdC. PGA exhibits 100 times the carbon dioxide barrier of PET, so that incorporation of only 2% PGA as a barrier layer in PET carbonated soft drink bottles allows total bottle weight to be reduced by 35%, while maintaining an equivalent carbon dioxide barrier and shelf-life performance. No tie layers are required in multilayer systems with PET and unlike other barrier layers, it causes no problems with PET bottle recycling since it hydrolyzes and dissolves in the alkaline wash step and meets the Association of Postconsumer Plastic Recyclers (APR) Critical Guidance Protocols for Recycling.[54] While it is not yet biobased, PGA is biodegradable and compostable and meets EN13432 and ASTM D-6400 certifications (Chapter 5).[55,56]

4.3.5 Starch-based barrier systems

Starch is an excellent barrier for oxygen and carbon dioxide, but a poor barrier for water vapor and its properties are sensitive to moisture. It can be used for gas barrier layers between protective water barrier polymer layers in multilayer structures. Starch is further discussed in Chapter 5.

A recent development is Plantic Technologies' ecoPlastic™ multilayer ultra-high barrier systems for food packaging. These use more than 90% renewable, GM-free hydroxypropylated high amylose starch formulations as the barrier layer together with moisture-resistant PE or PP outer layers. While the starch-based barrier material has higher permeation to carbon dioxide and oxygen than ethylene-vinyl alcohol, it is also much less expensive, so that greater thicknesses can be used to provide equivalent or higher barriers than EVOH-based multilayer systems. It also forms excellent taint and odor barriers.

The ecoPlastic™ materials are supplied as thermoformable sheet, generally with the structure PE/starch-barrier/PE for form, fill and seal applications, and PE/starch-barrier/PP for preformed trays, where the PE layer is at the top to allow adhesion of the lidding film and gas flushing for MAP. The sheet can be thermoformed on existing equipment, with no change in processing conditions. A 500 μ thick tray contains about a 400 μ starch-barrier layer thickness and provides a higher barrier and increased product shelf life than an EVOH-based tray containing about a 25 μ thickness of EVOH in a total thickness of 500 μ, while still being less expensive. The oxygen transmission rate for a 450 μ tray is less than one-tenth that for a 450 μ PET/EVOH/PE tray at 10°C, 50/90%RH, and requires a lower sealing temperature of 120–130°C compared to 135–175°C for the traditional tray. There is a shelf-life increase of 15% for ground beef and 20% for chicken over traditional systems, leading to significantly lower wastage. It is anticipated that lidding films and shrink films containing high amylose starch barrier layer formulations will also be introduced by the end of 2012. A 25 µ lidding film will contain about a 15 µ starch-barrier layer.[57,58]

In addition to the high barrier properties, these ecoPlastic™ systems show significant sustainability and environmental advantages over traditional systems. They have a high renewable content (>80%) for the overall structure. Lifecycle analysis (LCA) calculations indicate that 50 million trays, each weighing 23–25 g, use an aggregate of 1100 tonnes of starch-barrier formulation and save enough energy over traditional PET barrier trays to power 770 average homes for a year, as well as saving carbon dioxide equivalent to 463 100 trees or taking 1430 average cars off the road for a year. Plantic is also partnering with Braskem to introduce equivalent systems using Braskem biobased PE with renewable content over 90% without changing the other properties.[59]

4.3.6 Cellulose-based barrier systems

Like starch (Section 4.3.5), cellulose has good gas barrier properties against carbon dioxide and oxygen and provides a good barrier against microorganisms, flavors and aromas, but poor moisture barrier properties unless it is coated. Further information is provided in Chapter 5.

4.3.7 Protein-based barrier systems

There has been considerable development work over the past five years on protein-based bioplastics, including their use in packaging since they show high barrier properties towards oxygen and moisture.

A good example is the use of whey proteins, which are obtained in large quantities from cheese manufacture and have excellent barrier properties versus oxygen (almost as good as EVOH) and moisture barrier similar to PET. Thus, the Spanish-based Patronal de la Petita i Mitjana Empresa de Catalunya is coordinating the WheyLayer Consortium, consisting of 14 partners from seven EU countries, in the development of economically sustainable whey-based multilayer packaging materials. One of the partners, the Fraunhofer Institute, has made multilayer structures in the laboratory, which have high oxygen and moisture barrier properties, and adequate adhesion and mechanical properties but improved recyclability over multilayer systems containing EVOH or PVdC, since the whey protein can be removed enzymatically or chemically. The whey proteins also naturally contain antioxidant and anti-microbial compounds, which can extend the shelf life of food and beverages. These systems are now being scaled up.[60]

4.3.8 Barrier coating systems

There has been considerable development work on plasma-deposition barrier coating systems for packaging films and bottles, with the aim of providing substantial reductions in permeability while retaining clarity and lack of color without causing problems for recycling. Major approaches have involved the generation of silicon oxide or amorphous carbon barrier coating layers on the packaging material. The amorphous barrier layer also prevents migration of components from the packaging material, such as PET, to the package contents.

The Sidel Actis process produces a 0.1 μ deposit of amorphous carbon on the inside of the bottle employing an acetylene plasma produced with microwaves. It can be used with bottle sizes from 0.2 to 0.7 liters for carbonated soft drinks, beer, juices, sports drinks and tea. An improvement of up to 30-fold for the oxygen barrier and 7-fold for the carbon dioxide barrier over unmodified PET bottles is claimed.[61] The Mitsubishi Shoji process is also reported to produce a diamond-like coating.[62]

In other systems, such as the KHS Plasmax and Krones processes, a silicon oxide (SiO_x) barrier coating layer about 0.2 μ thick is produced on the inside of a blow-molded PET bottle, by pulsed low-pressure microwave induced plasma coating using hexamethyldisiloxane and oxygen. This extremely thin layer produces a substantial improvement in barrier properties, without reducing the bottle clarity or adversely affecting recycling.[63] The Toyo Seikan SIBARD process uses an SiO_x film coating on an organosilicon adhesive film to adhere it to the PET bottle wall.[64] Use of such processes can allow bottle thickness to be reduced, while still providing adequate barrier properties for many applications.

Plasma barrier coating techniques have also been adapted to improve the barrier properties of PET, o-PP, o-PA and polylactic acid (PLA) films with an SiO_x coating, without compromising their optical properties or the compostability of the PLA. The SiO_x coating imparts greatly improved barrier towards oxygen, carbon dioxide, moisture and organoleptics, which is not affected by temperature and humidity.[65] However, coated PLA films (Amcor Ceramis) show poorer adhesion than PET and require pretreatment with an oxygen plasma to obtain high bonding, and uncoated PET still shows better barrier than coated PLA or PP towards oxygen and carbon dioxide.

Ceramis-PLA film is finding uses in lidding films for thermoformed food trays and other rigid packaging and for flexible pouch laminates. The films can be used for MAP, and for aseptic, hot-fill, pasteurization and sterilization applications, are transparent to microwaves and unlike EVOH systems show no retort shock. They have high flex-crack resistance and are printable by conventional methods.[66] Polylactic acid is discussed in more detail in Chapter 5.

Other inexpensive barrier coatings based on minerals such as calcium carbonate and clays have been introduced for paperboard. For example, EarthCoating is a blend of LDPE with calcium carbonate and generates a moisture vapor transmission rate about 60% lower than for LDPE alone, yet with enhanced barrier versus oil, grease and fatty acids and comparable seal performance and is also FDA approved for direct food contact.[67]

4.3.9 Nanocomposite barrier systems

The incorporation of nanosized and sub-micron-sized additives is also under intense investigation for enhancement of barrier properties in both rigid and flexible packaging, including barrier bottles, while often allowing light-weighting. Because of the small particle sizes and high aspect ratios of the nanoclays and other mineral particles incorporated to form these nanocomposites, particularly montmorillonites and other stacked silicate structures, highly tortuous pathways are produced within the polymer structure for the permeation of the gases or water vapor, and large improvements in the barrier properties can be obtained by the incorporation of relatively small proportions of nanomaterials. The incorporation of nanoclays can also improve mechanical and thermal properties such as strength, stiffness, puncture resistance and heat resistance of the material and allow package light-weighting without adversely affecting the optical properties, since the particle sizes of the nanomaterials are generally much lower than the wavelength of visible light. Consequently, there is minimal light scattering and the transparency is not adversely affected, although there can be some increase in haze. Such improved systems incorporating nanoparticles are also seen as a recyclable potential for foil packaging replacement.[68] For more information on the use of nanotechnology, see Section 4.3.9 above and Chapter 9.

4.4 Scavenger systems

Scavenger systems for removal of oxygen and ethylene have been developed to increase shelf life of oxygen-sensitive materials and to delay ripening of packaged fruits and vegetables.[46,69,70]

4.4.1 Oxygen scavengers

Several active packaging technologies are being developed to help protect the packaged product from oxidation to improve shelf life and reduce spoilage and wastage. According to Albis Plastics, which manufactures Amosorb oxygen scavengers, the primary technology is oxygen scavengers that reduce the oxygen level inside the package, which have a global market value of around US$ 1150 million, followed by self-venting films at US$ 610 million, moisture scavengers at US$ 513 million and microwave susceptors at US$ 406 million. The primary market is meat packaging, followed by snack foods and ready meals. Further information on active packaging is given in Chapter 2.

Oxygen absorbers can be incorporated in packaging to remove oxygen already present in the package headspace, the interior package walls or the packaged product itself, and to continually remove oxygen permeating into the package on storage. To maximize their useful lifetime and effectiveness, the oxygen scavengers are generally used in combination with passive barrier packaging systems. Many types of oxygen scavenger have been developed, mainly for food and pharmaceutical applications.[71–73]

Early forms were sachets placed within the food package and most commonly contained iron powder (together with halide salts as a trigger) or ascorbic acid. An example is Mitsubishi Gas Chemical Ageless products, which can reduce oxygen levels in sealed containers to 0.1% or less.[74] However, because of concerns over users thinking the scavenger was a 'flavor additive' and consuming it along with the product, and to simplify packaging line operations and allow high line speeds, and because sachets do not work with liquid products, systems have been developed for both flexible and rigid packaging where the oxygen scavenger is incorporated in the packaging plastic or is itself a polymeric composite. For example, Toyo Seikan Oxyguard is based on iron salt additives, and this company also has a five-layer PET bottle structure, PET/barrier agent/PET/barrier agent/PET, where the barrier is its proprietary SIRIUS oxygen scavenger.[64]

Other systems use oxidizable polymers such as polyamides, polybutadienes and polymers with unsaturated side chains together with a transition metal salt catalyst. Thus Amosorb is based on a blend of a copolyester and polybutadiene incorporating an iron salt, which provides an active barrier compatible with PET with good delamination resistance. The original Amosorb was developed by BP Amoco, then sold to Ciba in 2001 and to ALBIS in 2009, but its use has been limited by its large particle size and cost. The next generation scavenger has a smaller particle size (so that it can be used in thinner films), higher absorption and better dispersion and is compliant with EU regulation 450/2009 on active and intelligent packaging. Oxbar, made by Constar, employs a blend of PET with an aromatic polyamide such as MXD6 incorporating a cobalt salt. This is co-molded between two PET layers as a barrier layer for PET bottles, such as beer bottles.[75] Cryovac OS 1000 and 2000 scavengers incorporate Chevron technology based on 1,2-polybutadiene. Yet other systems are based on photo-reducible compounds such as anthraquinone, which forms hydroquinones when exposed to UV that can then react with oxygen.

4.4.2 Ethylene scavengers

Ethylene is produced by all plants and induces fruit and vegetable ripening, softening and deterioration, even at low concentrations. This is detrimental to the shelf life and quality of many fruits and vegetables during storage and distribution. Consequently, ethylene scavengers (ethylene absorbers) are starting to be used by large supermarket chains such as Tesco and M&S for fresh fruit packaging to reduce spoilage and increase shelf life.[76] More information is given on these systems in Chapter 11. An inexpensive carbon nanotube sensor that detects ethylene has recently been developed at MIT and could be incorporated into packaging and shipping boxes to show ripeness of packaged fruit to help reduce wastage. More information is provided on nanotechnology in Chapter 9.

4.5 Nucleating and clarifying agents

Nucleating and clarifying additives are extensively used with PE, PP and PET. In polyolefins, they are generally used to control crystallization rate and reduce crystallite size to a range smaller than the wavelength of visible light to produce greater transparency, particularly in HDPE and PP. In PET, they are used to increase the degree of crystallization to make CPET, which is opaque.

Nucleating agents used for PE and PP have traditionally been based on organic acids such as adipic, benzoic and stearic acids and their metal salts or on sorbitol derivatives such as 1,3:24-bis(3,4-dimethylobenzylideno)sorbitol. Major recent advances in the field have alleviated previous problems with dispersion and taste generation. Nucleating clarifiers can now improve PP clarity to be comparable to glass, PET, PS and polycarbonate. However, it is more difficult to nucleate PE than PP, because of its much faster crystallization rate and less work has so far been done on it.

Major commercial producers are BASF (Irgaclear®, purchased as part of Ciba Specialty Chemicals and based on benzenetrisamides), Rika International (acetal sorbitols) and Milliken, which has the Millad and Hyperform ranges. Milliken Millad® NX8000, a sorbitol derivative introduced in 2007 for food and beverage packaging, provides up to 50% haze reduction in injection- and blow-molded PP over previous standard products, and has excellent dispersability allowing lower processing temperatures, high process stability, good organoleptic properties and has food contact approval.

A recent development for PE and PP is Milliken Hyperform HPN-20E, a high aspect ratio long, thin tabular-shaped organic salt with a median width of 2.8 μ. During melt processing, these long tabular structures orient in the machine direction to varying degrees, but to the greatest extent in processes with non-turbulent laminar flow. With PE, since the tabular structures are flat, the PE crystalline lamellae stack on top to form cuboid crystalline regions different from other nucleators that form spherulites (i.e. spherical crystallites). This controlled crystallization, altered crystal structure and decreased crystallite size with HPN-20E provides improved optical properties (contact clarity; haze and gloss). Furthermore, the specific crystal orientation leads to enhanced physical properties such as 5–20% increase in machine direction modulus with little loss of tear and impact properties, and anisotropic shrinkage useful in round and radial applications to give stronger parts. It also provides a more tortuous pathway for water vapor permeation and improves barrier properties by 25–50%. HPN-20E can be used in a wide variety of packaging thermoplastic processes including extrusion, blown film, injection molding and extrusion blow-molding.

Addition of HPN-20E to HDPE increases the DSC peak crystallization temperature from 118.5°C to 121.4°C, which can generate a productivity increase in processes such as injection-molding of up to 25%. For HDPE blown film, 25–50% improvement in water vapor, oxygen and solvent barrier properties can be obtained, suitable for meat, cheese, poultry, dry foods and cereal packaging. Likewise, LLDPE blown films show up to 50% increase in clarity without blending with LDPE and 15–35% improvements in barrier properties towards water vapor, oxygen and solvents. HPN-20E reduces haze from 18.4 to 8.22 for butene copolymers, 13.2 to 8.14 for hexene copolymers and 11.5 to 10.5 for octene copolymers. Major markets are LDPE/LLDPE blends for fruit and vegetable packaging. The use of HPN-20E in injection-molding gives easier processing and improved productivity, and the nucleation effect is better than that of most pigments, as shown by the higher crystallization temperature. For injection-molding of caps and closures, it provides a leveling effect resulting in more consistent processing and improved dimensional stability at increased productivity. For rigid tubes, for example caulk cartridges, there is significant liqht-weighting potential, since the water vapor transmission rate for a 1.5 mm thick industry standard cartridge is matched by a 1 mm thick cartridge using HPN-20E nucleant.

Milliken HPN-600ei is a nucleating agent for PP thermoforming that also gives excellent clarity, good mechanical properties and increased productivity. Use of this clarifier technology can produce high-clarity PP, which can compete with other highly transparent polymers such as PET.[77,78]

Because of its crystalline structure, PP is generally more difficult to thermoform than amorphous materials, due to its sharp melting point and low melt strength, so that the processing window is very narrow. However, PP can crystallize in two different crystal forms, the α- or monoclinic form that melts at 165°C in the syndiotactic homopolymer, and the β- or hexagonal form that melts at a temperature 12–14°C lower and also produces improvements in impact resistance, reduces the yield strength and changes the drawability. In a typical extruded or injection-molded material, the α-form crystals comprise more than 95% of the crystals and the nucleants/clarifiers described above retain this crystal-form distribution. However, proprietary β-nucleants have also been developed, such as Mayzo Corporation MPM-2000 and Rika International NJSTAR NU-100, to increase the proportion of β-crystallites. The different properties of the β-form crystals from the α-form are advantageous in PP thermoforming. As the proportion of β-form crystals increases, the melting range and hence the processing window increases due to the presence of both α- and β-form crystals (~146–163°C compared to 160–163°C for the α-crystalline PP), allowing a decrease in cycle time, and the increased ductility and lower tensile yield strength also facilitate thermoforming and give a more uniform wall-thickness, higher rigidity and top-load crush strength.[79,80]

PET exhibits slow crystallization from the melt but can be nucleated with heterogeneous particles such as talc or (more effectively) nanoclays, by carboxylic acid salts such as sodium stearate or by sodium ionomers. This increases the rate and uniformity of crystallization and facilitates the growth of small and uniform spherulitic crystals, which is advantageous since the formation of large crystals leads to brittleness.[81]

4.6 Antimicrobials additives and coatings

Antimicrobial agents are used to prevent attack on plastics by microorganisms such as fungi, molds and bacteria. While most plastics are not readily biodegradable, they are often compounded with other materials such as plasticizers, which are subject to microbiological attack. However, most packaging applications for fast-moving consumer goods do not have long enough shelf- and storage-lifetimes to make this a problem and antimicrobials, which are used in packaging, are usually present to protect the packaging contents.[69]

A common example is the use of metallic silver nanoparticles or silver ions to retard foodstuff spoilage. A recent development is GardIn Fresh's FreshExtend silver-coated nanoparticle additive, where an inorganic core such as calcium carbonate, barium sulfate or silica is coated with polymer particles coated with silver. This extends the shelf life of perishable products by inhibiting fungal, mold and bacterial growth, while also absorbing ethylene gas produced by the ripening of fruit or vegetables. These silver-coated nanoparticles are claimed to be appreciably less expensive than pure silver nanoparticles. Recent work has demonstrated that the antimicrobial agent is actually silver ions, not a nanoparticle-specific toxic effect.[82] More information on nanotechnology is presented in Chapter 9.

In a completely different recent approach, Oplon Pure Science has signed an US$ 8 million development agreement with PepsiCo for packaging which will significantly inhibit bacterial and other microorganism growth in foods so that, for example, milk could be left at room temperature for a month and fruit juice for two weeks. In this technology, a layer of anti-bacterial material is sprayed on the package interior to prolong product shelf life. The antimicrobial coatings are described as polyelectrolytes contained inside a polymer matrix, which create a surface electric field in contact with liquids that disrupts cell membranes of yeasts, molds and bacteria and even certain viruses.[83]

In many countries, antimicrobial additives are regulated as indirect food additives (e.g. FDA registration in the United States and EC registration in Europe).

4.7 Active and intelligent packaging

Active packaging relates to packages that do more than passively contain and protect food, while smart or intelligent packaging senses and informs. This is an extension of the traditional functions of packaging to provide protection and greater convenience, and allow extended food shelf- and storage-life. Thus future packaging will not just provide a barrier but also interact with the packaged materials.[84–86] Examples include: sensors where color changes indicate the freshness of seafood (COX Technologies' FreshTag) or show when beer is at the correct drinking temperature or when beverages are extremely hot; Photonic BioSystems oxygen-sensitive luminescent dye that can be incorporated inside sealed food and beverage packages and scanned to determine product quality; vegetable packaging that absorbs oxygen to prevent rotting if the temperature rises; milk carton linings incorporating an enzyme that destroys lactose for the lactose-intolerant; packaging wrap impregnated with carbon dioxide to prolong freshness; anti-microbial packaging materials to prevent food decay or inhibit mold growth; and package caps that release taste and aroma as a beverage is drunk.[87–89] Packaging sensor technology is also being developed by a collaboration of Bemis Company and Thinfilm Electronics.[90] For more detailed information on active and smart packaging technologies, see Chapters 3 and 10.

4.8 Rigid packaging

Shatterproof, lighter-weight rigid plastics are increasingly replacing traditional glass, metal and paperboard packaging. PET is now used for a third of world rigid packaging production, greater than PE. Recyclability has become an important factor and suppliers of all packaging types, both food and non-food, are looking to increase sustainability by using light-weighting, recycling and biodegradable and/or biobased materials. A key challenge is to develop greener packaging with properties equivalent to that made from conventional petrochemical-based plastics. Biobased materials are discussed in detail in Chapter 5.

4.8.1 Recent examples of sustainability improvements

There is a worldwide effort to develop packaging with improved sustainability and only a few examples can be given here.

The RPC Group has many rigid plastic packaging developments aimed at light-weighting, using post-consumer recyclate, and replacing glass. Its Jugit system, which incorporates a milk bag refill in a reusable container, uses 75% less packaging than a traditional HDPE bottle. The 300 ml PET bottle for Marks & Spencer's Essential Extracts personal care range incorporates 30% recycled content.

In the US, Ecologic Brands has developed the eco-bottle, which has a recyclable plastic pouch inside a molded fiber shell and uses up to 70% less plastic than a normal plastic bottle. The shell is also 100% recyclable and compostable. Petainer in the UK recently developed a refillable PET bottle incorporating more than 25% PET recyclate, with identical performance to a refillable bottle made of virgin material. Krones in Germany has likewise reduced Lidl's drink packaging PET bottle weight by almost 30%, while maintaining functionality and handling properties. Krones also claims its NitroHotfill 15.5 (a 15.5 g bottle for 500 ml) is the world's lightest hot-fill container. In 2011 Amcor reduced PET resin use by 10 million pounds using lighter-weight containers. The Amcor Big Mouth PET jar for hot-filled foods is 86% lighter than glass and 34% lighter than traditional heat-set PET. Amcor's new Active Hinge technology also reduces container weight by as much as 20% in 8, 10, 12 and 16 oz bottles. Nestlé Waters is progressively reducing its rigid plastic bottle weight and cut plastic use by 24% per liter of container volume between 2004 and 2009.

Klöckner Pentaplast has developed TerraPET rigid film with 30% renewable polymer content, which still has the same performance as traditional APET. Linpac recently launched its RfreshMB sustainable meat and fish packaging PET trays containing up to 95% food-grade post-consumer recyclate, but with the same barrier performance and rigidity as trays made from virgin material.[91]

4.8.2 Thermoformed packaging[92]

A new study by ReportsnReports projects the global thermoformed plastic market to increase at 4% CAGR from 7.0 billion lbs in 2012 to 8.5 billion lbs by 2017. Packaging accounts for about 70% of this, amounting to 4.8 billion lbs in 2011, with projected growth to nearly 6 billion lbs by 2017. Transparent packaging, such as clamshells and blister packaging, facilitates product sales since the purchaser can see the product.[93] Another report from *Plastics News* projects North American thermoforming growth of 6% through 2014, but with further consolidation in the packaging sector.[94] Thermoforming may replace injection molding in many packaging applications, because of its production speed and lower tooling costs. Focus on sustainability will grow with increased use of thin-walling and recycled and bio-based materials.

Thermoformed rigid packaging for retorting and 'retort shock'

Many retort food packages are multi-layer structures, incorporating an EVOH oxygen barrier between PP layers. These PP layers protect the EVOH from moisture and maintain package structural integrity when exposed to high temperatures and steam during retorting. However, while PP is a good moisture barrier at ambient temperatures, it allows excessive moisture permeation at high temperatures during retorting, which causes a significant, although temporary, reduction in the EVOH effectiveness (Section 4.3.1). Consequently, oxygen can enter the packaging for a period of time after retorting, which reduces shelf life. It takes 10 to 14 days to fully regain the EVOH layer oxygen barrier properties, although most of the oxygen enters in the first 100 to 200 hours after retorting.

New REVOH technology from Winpak Portion Packaging alleviates this problem of retort shock, by incorporating micron-sized mineral particles and other additives and specially formulated blends in the outer PP layers to improve their moisture barrier properties. This enhances the post-retorting oxygen barrier of thermoformed containers and reduces oxygen ingress 30 to 35% during the critical period immediately after retorting, thereby increasing shelf life of the package contents. These systems also provide an 8 to 15°C increase in heat distortion temperature for better microwaving performance, as well as good low-temperature impact resistance and about twice the thermal conductivity of standard containers, which improves the heating uniformity of the contents. Furthermore, because of the incorporated minerals, less petrochemically-derived PP is used. Less EVOH can also be used to give the same oxygen barrier as the conventional PP systems, thereby reducing package cost.[95]

4.8.3 Developments in thin-wall packaging

There has been rapid growth in thin-walled plastic packaging, mostly for injection molding and thermoforming applications, such as containers and lids for both food and non-food products, to reduce packaging weight and materials use and to provide new possibilities for design of complex or unusual package shapes. Due to the increasing demand for single servings and smaller packaging, and the ongoing requirements to lightweight and reduce packaging, this technology is being increasingly used for a wide variety of food packaging, including dairy product containers such as yogurt pots, butter and margarine tubs, frozen foods, ready meals, fruit and fruit juices, vegetables, other fresh produce, and meat packaging, as well as non-food packaging such as pharmaceuticals, cosmetics, paint and adhesives. AMI Consulting has estimated that over 2.8 million tonnes of polymer are used yearly for thin-wall packaging in Europe, one of the main drivers being consumer convenience. PP and PS are the two most important materials, followed by PET and some PVC. CPET is used for thin-wall packaging of microwaveable ready meals, while APET is preferred in many dairy products such as chilled desserts because of its transparency.[96,97]

New materials have recently been introduced for injection-molded thin-wall packaging. These include random PP copolymers such as the Borealis BorPure™ family, which has controlled molecular weight distribution and nucleants to provide fast crystallization, high transparency and good combinations of modulus, creep resistance, stackability and impact properties. These materials show improved organoleptic properties, which will not impart taste or odor to food products and their low melt viscosities allow processing at lower temperatures using less energy than competing materials. Several grades are available for specific applications such as deep freeze ice cream packaging and portion pack thin-wall food containers. Other examples are the Sabic Flowpact range, which includes two impact copolymers and a homopolymer.

Sabic PP Flowpact FPC 100 is an impact copolymer for packaging dairy products and yellow fats and contains both processing and antistatic additives. It exhibits high melt flow and crystallization rate, allowing reduced cycle times, high-speed running in multicavity molds and energy savings, as well as a good balance of impact strength and stiffness. In a third example, because of the increasing demand for high clarity packaging at modest cost, National Petrochemical Industrial Co (NATPET) has introduced two new high-clarity Teldene random copolymer PPs for thin-walled food and non-food packaging and injection-molded containers. Both incorporate Milliken's Millad NX8000 clarifier, which provides reduced processing temperatures, cycle times and process energy consumption. The molded containers have excellent optical properties, thought to result from greater solubility of the Millad NX8000 in PP compared to conventional clarifiers. Thus, at 1 mm thickness, haze is only 6% compared with 15–20% in grades incorporating conventional clarifiers when processed at low temperatures. These grades are aimed principally at markets in Europe, Turkey, the Middle East and Asia.

Compostable plastics are also available for thin-wall molding (Chapter 5). Thus, FKuR Kunststoff provides PLA-based and cellulose acetate-based materials for injection-molding and thermoforming. PLA has disadvantages of brittleness and noise, poor high temperature resistance and susceptibility to hydrolysis, but additive systems have now been developed to dramatically improve its properties. FKuR Biograde V0291 gives clear injection-moldings at wall thickness as low as 0.35 mm, while Bio Flex S 5630 can be thermoformed on standard equipment with similar cycle times to PP.

Both molders and packagers appreciate the improved appearance, processing improvements, economics and carbon footprint reduction obtained with these new materials for thin-wall packaging and further advantageous developments are anticipated.[97–99] Further information is presented in Chapters 1 and 11.

4.9 Flexible packaging

Flexible packaging can be defined as: any package, or part of a package, whose shape can be easily changed; packaging constructed from one or more plies of plastic film and sealed with heat and/or pressure; and films assembled to form a bag, pouch, cup, tray, wrap, label or combination of these. It is not normally considered to include retail and consumer plastic bags.[100–102]

Plies can be combined by processes such as co-extrusion and lamination and the layers can include outer 'bulk' layers, barrier layers, tie layers and seal layers. The major bulk resins used are LDPE, LLDPE, HDPE, PP, PS and PET, which all have the requisite thermoplastic processability, melt strength, low gel levels, physical, optical and thermal properties, compatibility with other resins and moderate cost. PE is expected to remain the most commonly used film because of its versatility, cost-effectiveness, ability to be down-gauged and good barrier properties.[103]

Flexible packaging has been one of the fastest growing packaging segments over the past decade and new product development is continuing. It offers significant performance advantages and also reductions in both material and transportation costs compared with rigid packaging and optimizes the transfer of packaging between the converter, packaging facility and end user. Not only does it use less space when empty than rigid packaging (unless this is highly stackable) but also allows packaging to be assembled from roll materials at the filling location, reducing or removing the need to transport ready-formed but empty packaging. The filled packs are also lighter and thinner and require fewer truck journeys and less fuel use. According to Flexible Packaging Europe, a 0.2 liter metalized flexible stand-up pouch requires half the number of standard truck loads than packaging in 0.2 liter glass bottles, due to the lower weight and less space between packages.[104]

Over the past decade, flexible packaging has dramatically developed from simple monolayer bags and wrappings to multilayer co-extruded, laminated and coated structures incorporating several different materials and coatings, including plastics, paperboard and metal foils, to provide high barrier properties and enhanced shelf life. Other features now available include zippers, spouts, peel-off lids, self-venting films (which allow foods to be steam-cooked inside the packaging), and microwavable pouches. Flexible packaging properties can also be tailored to meet specific property-combination requirements for packaging a wide range of products, and this allows sustainable, recyclable and biodegradable materials to be incorporated in the structures. There is increasing demand for food and convenience products using high-barrier materials for enhanced shelf life and for ecologically friendly packaging.

The most rapidly growing areas for converted flexible packaging are snack foods, beverages and meat products. Growth drivers are demographic trends such as empty-nest and single-person households, and families with all adults working, and the need for engineered films providing extended shelf life. These trends are generating increased demand for pouch-packaged food, smaller pack sizes and foods with shorter preparation time. Increasing consumer demand for healthy eating and on-the-go eating is also promoting single-serve packaging of beverages, snacks and baked goods. Such single-serve packages usually require more packaging material per unit weight of product than conventionally sized packages, but also generate less waste of the packaged material.[105]

Major companies in flexible packaging worldwide include:

- AEP Industries Inc.
- Amcor Limited.
- American Packaging Corporation.
- Bemis Company Inc.
- Clondalkin Group plc.
- Constantia Packaging AG.
- Exopack LLC.
- Huhtamäki Oyj.
- Mondi Group.
- Printpack Inc.
- Berry Plastics.
- Rexam plc.
- Sealed Air Corporation
- Sonoco Products Company
- United Flexibles GmbH.

More information on flexible packaging is given in Chapters 1 and 11.

4.9.1 Pouch packaging

According to recent data from the Mintel Group, pouch use for consumer products, including food, shampoo and pet food, is up by 37% since 2007, with especially strong growth in snacks. In 2011, 1210 new products were first introduced in pouch packaging, up from 885 in 2007, representing 3.3% of all new products. Key drivers include: significant material and cost savings and environmental friendliness compared to rigid containers; high growth in emerging markets; and continuing development of new applications, particularly for constructions such as integrated zippers and spouts and for wider product ranges including non-food use such as detergent refills.[106–108] More information on pouches is given in Chapters 1 and 11.

A continuing challenge with these pouch constructions is that they are generally made from 3-ply laminates, such as Alox coated barrier PET/oriented polyamide/cast PP, and are not recyclable. However, with the trend in consumer preferences towards environmentally-friendly and recyclable packaging, there is pressure for the development of sustainable alternatives to replace these current structures to allow markets to be retained or expanded. Sustainable plastic materials and structures with a significant reduction in environmental impact are needed that can withstand both retort conditions (up to 130°C) and microwaving, be easily openable and printable, provide a one year shelf life and meet food contact regulations.

Some examples of pouch packaging

As noted above, this is a rapidly growing field, so only a few examples can be discussed here.

Campbell's Soup and Heinz are both offering soup products in pouches in 2012, as well as sauces (Campbell's) and tomato ketchup (Heinz). Campbell's is targeting the new pouched products at trendy millennials for whom canned soups seem archaic but who are attracted by the modern aesthetics and appeal of pouches printed with elaborate graphics. Heinz aims to improve its price position relative to cheaper private labels, taking advantage of the 10–15% savings in packaging costs as well as the reduction in shipping costs.

Heinz Dip & Squeeze pouch packaging for ketchup was introduced in 2010 to supersede the ketchup sachet, which had been on the market for 42 years but which many consumers found difficult to open. Dip & Squeeze is a red bottle-shaped pouch made of polyester laminated to a high-barrier sealant web and has a laser score at the top for easy opening and squeezing. Alternatively, the lid can be peeled back from the opposite end for dipping rather than squeezing, and the package contains three times the amount of ketchup found in the traditional sachets. It is now available across the US and more than a billion have been sold. Heinz is also introducing a 10 oz size spouted stand-up ketchup pouch with a reclosable cap to sell for US$ 0.99 to complement its normal 20 oz bottle, which sells for US$ 1.99. The ketchup is dispensed by removing the cap and squeezing the flexible package – no more shaking the bottle!

Campbell's Go! Soup, introduced in the summer of 2012, is a premium line with trendy flavors like coconut curry with chicken and shiitake mushrooms and is packaged in a pouch with attractive graphics to attract a younger market. Campbell's is also launching a line of skillet sauces packaged in pouches later in 2012 for easy preparation of upscale dishes such as shrimp scampi or Thai green curry chicken. Both these products will be packaged in retortable stand-up pouches made from tailored, high-barrier laminated structures. The 14 oz microwavable Go! Soup pouch material is an all-plastic laminate and the 9 oz Skillet Sauce pouch is a plastic/aluminum laminate. These are single-use products so the pouches are not resealable. The sauce pouch is not designed to be microwaveable but to be emptied into a skillet for cooking with other ingredients.[106,109]

Kraft recently introduced the YES Pack – a squared-off pouch made of a high-strength flexible nylon-PE blend film, with dual handles and a rigid plastic mouth, holding one gallon of salad dressing for commercial operations. This is a squeezable package, so almost all of the contents can be removed without scraping the sides of a rigid package. Another recent trend is the use of stand-up pouch packaging for wine, such as the Astrapouch, which holds the equivalent of two 750 ml bottles. It employs a resealable, one-way plastic spout to keep opened wine fresh for up to a month when refrigerated.[110]

Stick packs are also increasingly being used for nutraceuticals and energy drinks. The stick packs were originally designed to facilitate pouring the contents into a water bottle (e.g. Crystal Light On the Go flavored drink powder stick packs to replace tubs). However, they can now be used for low-cost direct-to-mouth packaging for ‘active’ lifestyles. An example is encapsulated caffeine (Encaff Products Inc.), which can be poured directly onto the tongue without needing water (i.e. energy drinks without the water). This also opens up opportunities for nutraceuticals and pharmaceuticals to provide convenience and dosing accuracy. Another example is Good Times Beverages individual cocktail pouches for convenience in making single frozen cocktails. The pouches can be placed in the freezer, squeezed several times and the contents poured into a glass or drunk straight from the pouch.

Pouch packaging is also being designed for permanent, flexible, reusable packaging, which is washable and extremely portable, such as the Vapur-Ampac foldable, water bottle.

4.9.2 Stretch wrap

Stretch wraps are blown or cast films designed such that once the polymer molecules are stretched they will try to return to their original configuration and hold onto whatever contents they have been stretched around. The end of the film generally clings to itself to secure the package. The most common stretch wrap material is LLDPE. The hexene and octene copolymers provide the best stretch film characteristics, particularly elongation at break and puncture resistance. Other types of PE, EVA and PVC can also be used, and sometimes co-extrusions to produce different degrees of cling on each side of the film so that it adheres to itself when wound on a pallet but does not stick to adjacent pallets. Many such films exhibit about 500% stretch at break but are only stretched to about 100–300% in use to avoid necking. Once the film is stretched on, the load is kept tight by the elastic recovery of the film. Other important properties, apart from cling and break strength, are tear resistance, static dissipation and clarity.[111,112]

4.9.3 Shrink wrap

Like stretch wrap, shrink wrap uses the film memory effect to revert to smaller dimensions after stretching and deformation to provide a tight wrap around the packaged object. However, in the case of shrink wrap, the packaging is done loosely and it is then exposed to heat to shrink the film.

Major polymers used are PE, PP, PVC and (recently) PLA, often lightly cross-linked by electron beam irradiation. The polymer film is heated and then oriented, either uniaxially or biaxially, and the orientation frozen in by rapid cooling.

Subsequent heating of the film allows it to try to return to its original dimensions. Shrink wrap is used for packaging food, compact discs, DVDs, software, books, magazines and electrical goods. It is also used to provide tamper resistance. For shrink wrapping foodstuffs, co-extruded or laminated structures may be used to give required mechanical properties as well as shrink, such as sealability, toughness and slip, and barrier properties.[111–113]

4.10 Sustainable packaging

Sustainability has become an issue in many areas involving plastics over the past few years. However, this has been more noted in plastic packaging than other plastics applications, and started earlier, about two decades ago, because of the visibility of packaging materials to the consumer. Thus, there have been growing pressures from governments, regulators, non-government organizations, environmental groups, suppliers, customers and others to improve packaging sustainability by reducing its environmental and societal impacts, and for the development of 'green' products. Because of this, packaging, and particularly plastics packaging, is being subjected to intense scrutiny along the total supply chain, moving from raw materials suppliers to packaging producers to packer/fillers to retailers to consumers, from governmental and environmental organizations around the world. While there have been immense advances in this area, it has been a challenging process because of the complex interactions between products and their packaging and the multiple functions of packaging in the supply chain.[114]

Sustainable packaging has been defined by the Sustainable Packaging Coalition (August 2011) as:[115]

- Beneficial, safe and healthy for individuals and communities throughout its lifecycle.
- Sourced, manufactured, transported and recycled using renewable energy.
- Optimizing the use of renewable or recycled source materials.
- Manufactured using clean production technologies and best practices.
- Made from materials healthy throughout their lifecycle.
- Being physically designed to optimize materials and energy.
- Effectively recovered and utilized in biological and/or industrial closed loop cycles.

Many companies are moving towards such goals.

Present trends in sustainable packaging center on light-weighting, use of recyclates and use of bioplastics and can be summarized as:[116,117]

- Sustainability can be an effective consumer marketing tool, capitalizing on the appeal of doing good. However, being green will not excuse poor design, especially while bioplastics remain more expensive than conventional petrochemically-based plastics.

- World demand for sustainable packaging is projected to rise 6% per year to US$ 212 billion in 2015. Packaging represents about 60% of global bioplastics consumption in 2012 and will remain the largest bioplastics end-use to 2015.
- Greater use of less expensive, lighter and less bulky packaging will favor flexible over rigid packaging, which is projected to grow by 3.5 to 4% annually to 2015.

The use of bioplastics in packaging of fast-moving consumer goods will be described in more detail in Chapter 5 and sustainability in packaging in Section 5.2.

The whole subject of sustainability in packaging has been surveyed and discussed in detail in a June 2012 UK report by PwC entitled *Sustainable Packaging: Myth or Reality*.[118] This report concludes that the term sustainable packaging is no longer relevant, since it has been used as an umbrella term to cover many aspects of sustainability and is too broad to be practically useful and it has now been replaced with a more balanced view of 'efficient packaging', which combines the elements of:

- producing effective packaging using minimum resources;
- protecting the product and minimizing product wastage;
- providing transport and display efficiency; and
- allowing effective disposal and recycling after use.

The evolution of efficient packaging will thus depend on materials costs, new technologies (i.e. nanotechnology) applied to both product and packaging, availability and supply security, consumer demand for convenience, and generation of competitive advantage.

4.11 Recycling of plastic packaging[119,120]

Only mechanical recycling will be considered here, and not chemical recycling of plastics back to monomers or fuels or processing of plastics to provide energy recovery.

Mechanical recycling of plastics is only relevant to thermoplastics that can be re-melted and reprocessed. Mechanical plastics packaging recycling began on a limited scale in the 1970s for HDPE milk bottles and PET carbonated soft drink and water bottles. It has increased dramatically since then because of consumer education programs and as collection systems and recycling programs have become more prevalent, and mandatory in many areas in Europe, and sorting systems and recycling technology have improved. However, the major plastics materials, which can be recycled economically, are still PET and HDPE bottles, although the industries in North America, Europe and Japan have been distorted by the massive amounts of post-consumer bottles exported to China for recycling. In 2008, 63% of post-consumer plastics waste in Europe came from packaging.[121] There is considerable government and NGO ecological pressure to increase recycling and the use of recyclate. For example, EuPR (European Plastics Recyclers) has recently announced the institution of a Europe-wide certification mechanism and audit system, supported by collectors, recyclers and converters, called EuCertPlast, which implements the existing CEN standard 15343 for post-consumer plastics recycling and is expected to encourage the use of recyclates.[122] The aim of using recyclates is to allow partial substitution of virgin polymers in existing applications, so that the recyclate market value is directly linked to virgin polymer prices. However, because of the perception that recyclates are of lower quality than virgin polymers, users are unwilling to pay the same price as for virgin polymers and the potential marketing advantage of using 'green' recyclates is not yet strong enough to eliminate this price differential.

Since there are many different types of thermoplastic with different properties, and these different polymers are generally mutually incompatible and immiscible, mixtures of different polymers often have poor and unacceptable properties, making them unsuitable for many applications. Hence mechanical recycling of plastics waste is generally only feasible for pure plastic streams and mixed wastes require separation into individual types before recycling. The recycling of packaging plastics is most feasible with rigid plastics, although mechanical recycling methods have now been developed for film products such as plastic bags, for mixed plastics and those contaminated with food waste, but this is much more difficult and less economically advantageous, so that these materials are not collected or recycled in many areas.[123] Complex packaging structures with components using different polymers are also generally difficult, impossible or uneconomic to recycle, and problems can arise when an established recycled product stream is changed from a single-polymer to a multi-polymer design, such as the incorporation of barrier or oxygen scavenger layers in PET bottles and the use of multilayer packaging barrier films.

Recent trends for container light-weighting and thinner containers have also been a problem for mechanical recycling, particularly for HDPE and PET bottles and containers and PE films. These thinner and lighter products are more difficult to separate out and to cut and grind, have higher surface-to-mass ratios and hence higher contamination levels, require more washing, have higher moisture levels and hence need more energy for grinding, have low bulk density that make them more difficult to handle after grinding, and are often multi-layers and use multiple polymers that can be difficult to identify and sort.

4.11.1 PET bottle recycling

PET bottles, largely carbonated soft drink and water bottles, are the largest group of plastics recycled. In Europe, the PET container collection in 2011 was 1.75 million tonnes and the collection rate reached a record of 51%, up from 1.60 million tonnes and 48% in 2010.[124] Only three of the 27 EU countries had PET collection rates below the 22.5% target outlined by the Packaging Waste

Directive and a third of countries had collection rates over 70%. About 50% of the recyclate was used to make new packaging containers or sheet and about 39% for fibers. The current PET recycling capacity in Europe is about 2.09 million tonnes, so more material can be processed.

In contrast, figures from National Association for PET Container Resources (NAPCOR) show the US PET bottle recycling rate was 29.1%, up from 22.3% in 2000.[125] Total weight of bottles collected in 2010 was 1.56 billion lbs.

Many different processes have been developed for PET bottle recycling, but they all require thorough decontamination by removing foreign materials such as glass, film, paper, metals, sand and dirt and polymers such as PVC and polyolefins. As with other types of recycling, the success of the process will be highly dependent on the efficiency of decontamination and purification. It is vital to remove contaminants such as PVC, which can cause black specks in the recycled product, or PLA, which can drastically reduce its optical clarity. This decontamination and purification is often accomplished by sequences of mechanical and optical sorting, followed by air sifting to remove paper, grinding to flakes and removal of low density polymers (i.e. PE and PP from bottle caps and closures) by flotation methods, since the less dense polyolefins will float and the denser PET will sink. The separated flake is then subjected to hot washing (generally at ~80°C), alkaline washing and clean water rinsing and drying. It can then be melted and the melt filtered to remove particulate impurities and then pelletized to make the final amorphous recyclate. As with virgin PET, the recyclate will need to be crystallized and dried before thermoplastic processing. In some applications, such as fiber, profile extrusion and cast film production, the conversion may be done directly from the flake rather than going through the pelletizing step.[126–128] Food-grade PET recyclate is also available, from Closed Loop Recycling and from ECO Plastics in the UK, which is working with Coca-Cola in the UK to produce PET recyclate to allow incorporation of 25% rPET in all of Coke's soft drink bottles by the end of 2012. Chapter 11 gives more information on market trends for recycled PET materials.

Major challenges for PET bottle recycling have been barrier bottles, where the incorporation of EVOH barrier resin can lead to chemical changes in the PET resin during the recycling process, and the use of labels. Adhesive labels and adhesive-free shrink-sleeve labels both cause problems in bottle recycling, since they can make the identification of the bottle resin more difficult and the label can be hard to remove and once removed the label material and adhesive can clump. Consequently, the Association of Post-Consumer Plastic Recyclers in the US has developed a guidance document to help PET bottle manufacturers in testing their labels to ensure they do not impede recycling. This document includes quantitative measurements related to clumping of low melting label material residue, since this is a serious problem with the denser shrink sleeve labels that sink in water after hot water washing, and the document provides packaging designers with the clear guidance to mitigate this problem. Another critical problem is caused by inks from the labels carrying over into the recycled PET and causing discoloration. PET bottles with shrink labels (or other opaque labels) can also create problems for material recovery facilities (MRFs), since the automatic bottle sorting equipment often cannot identify the PET resin through the label.[129]

Because of such problems, recycling and environmental organizations are promoting the policy that end-of-life considerations, especially recyclability, disassembly and other appropriate disposal methods, must be vital criteria in product conception and design. With regard to PET bottles, five major European plastics organizations:

- European Association of Plastic Recycling and Recovery Organizations (EPRO),
- European Plastics Recyclers (EUPR),
- PET Containers Recycling Europe (Petcore),
- European Non-Alcoholic Beverages Association (UNESDA), and
- European Federation of Bottled Waters (EFBW),

have cooperated to promulgate the European PET Bottle Platform (EPBP), a voluntary initiative to evaluate new PET bottle designs that may enter the European plastics recycling stream, with the aim of facilitating increased sustainability of the PET recycling industry in Europe.[130]

4.11.2 HDPE bottle recycling

In 2010, 447 000 tonnes (984.1 million lbs) of post-consumer HDPE bottles were recycled in the US, which represents a recycling rate of 29.9%, and the recyclate end uses were non-food bottles 36% (but not food-grade), pipe 32%, railroad ties and lumber 12%, automotive 9%; film and sheet 7%, and others 4%.[131] In addition, 375 000 tonnes (862 million lbs) of non-bottle rigid containers, mostly HDPE and PP, were recycled in 2010 (a 72% increase from 2009), as well as 441 000 tonnes (971 million lbs) of HDPE and LDPE film.[132] Also in 2010, 76% of UK HDPE milk bottles were recycled (93 000 tonnes).[133]

Many of the stages in the recycling process for HDPE bottles (or other containers) are similar to those for PET bottles. It is vital to remove dirt, glass, metals and other non-HDPE plastics, as well as labels and adhesives, and to purify the resin so that it is clean enough to be melt-processed to a finished pellet. Typical stages are chopping and grinding into flake, which is then washed with water and surfactants to remove dirt, labels, adhesives and other foreign matter; hydrocloning and float/sink separation, since HDPE has a specific gravity less than water and will float while many contaminants such as metals, gravel, dirt, PET and PVC will sink; drying; and melt processing by extrusion into pellets. Additives, such as light stabilizers or antioxidants, may also be incorporated at this stage to increase the value of the recyclate. If the recyclate is intended for products made by profile extrusion such as plastic lumber, the pelletization stage may be omitted and the flake directly extruded.[134] Closed Loop Recycling in the UK is the first to manufacture food-grade HDPE recyclate and Sainsbury's is now using 10% recycled HDPE in its own-brand plastic milk bottles and has committed to increasing this to 30% by 2015.[135]

4.12 Sources of further information

CHIELLINI E. (ed.) (2008) *Environmentally Compatible Food Packaging*, Cambridge, UK, Woodhead Publishing Ltd.

GILES G. A. and BAIN D. R. (eds) (2000) *Materials and Development of Plastics Packaging for the Consumer Market*, Oxford, UK, Sheffield Academic Press. ISBN 1-84127-116-0.

HERNANDEZ R. J., SELKE S. E. M. and CULTER J. D. (2000) *Plastics Packaging: Properties, Processing, Applications and Regulations*, Carl Hanser Verlag. ISBN 1-56990-303-4.

LEE D. S., YANG. K. L. and PIERGIOVANNI G. L. (2008) *Food Packaging Science and Technology*, CRC Press.

PARISI S. (2012) *Food Packaging and Food Alterations: The User-oriented Approach*, ChemTec Publishing. ISBN 9871847356062.

PIRINGER O. G. and BANER A. L. (eds) (2000) *Plastic Packaging Materials for Food: Barrier Function, Mass Transport, Quality Assurance and Legislation*, Wiley-VCH. ISBN 3-527-288868-6

ROBERTSON G. L. (ed.) (2009) *Food Packaging and Shelf Life*, CRC Press. ISBN 978-1-4200-7844-2. Available from: www.crcnetbase.com/doi/book/10.1201/9781420078459 [Accessed October 2012].

SELKE S. E. M. (1997) *Understanding Plastics Packaging Technology*, Carl Hanser Verlag. ISBN 3-446-18684-0.

YAM K. L. (ed.) (2009) *The Wiley Encyclopedia of Packaging Technology*, 3rd Edition, Cambridge, UK, Woodhead Publishing Ltd.

YAM K. L. (ed.) (2012) *Emerging Food Packaging Technologies: Principles and Practice*, Cambridge, UK, Woodhead Publishing Ltd.

4.13 References

1. HERNANDEZ R. J., SELKE S. E. M. and CULTER J. D. (2000) 'Major plastics in packaging', in *Plastics Packaging: Properties, Processing, Applications and Regulations*, Carl Hanser Verlag. ISBN 1-56990-303-4, 89.
2. BROOKS D. (2000) 'Types of plastics materials, barrier properties and applications', in *Materials and Development of Plastics Packaging for the Consumer Market*, Giles G. A. and Bain D. R. (eds.), Oxford, UK, Sheffield Academic Press. ISBN 1-84127-116-0. Chapter 2. 16–45.
3. CERESANA RESEARCH (2008) 'Polyethylene – HDPE' (UC-905) September. Available from: www.ceresana.com/en/market-studies/plastics/polyethylene-hdpe [Accessed October 2012].
4. NAMPAK LTD (2012) Available from: http://infinibottle.com/design [Accessed October 2012].
5. CERESANA RESEARCH (2010) 'Polyethylene LLDPE' (UC-1205), March. Available from: www.ceresana.com/en/market-studies/plastics/polyethylene-lldpe [Accessed October 2012].
6. CERESANA RESEARCH (2010) 'Polyethylene LDPE' (UC-1405)', April. Available from: www.ceresana.com/en/market-studies/plastics/polyethylene-ldpe [Accessed October 2012].
7. CERESANA RESEARCH (2008) 'Polypropylene' (UC-805)', June. Available from: www.ceresana.com/en/market-studies/plastics/polypropylene [Accessed October 2012].
8. PACKAGING-GATEWAY.COM (2012) 'ExxonMobil unveils new film for MAP applications', 30 April. Available from: www.packaging-gateway.com/news/newsexxonmobil-unveils-new-film-for-map-applications [Accessed October 2012].

9. MAUL J., FRUSHOUR B. G., KONTOFF J. R., EICHENAUER H., OTT K. H. and SCHADE C. (2007) 'Polystyrene and styrene copolymers', in *Ullmann's Encyclopedia of Industrial Chemistry*, Weinheim, Wiley-VCH. doi:10.1002/14356007.a21_615.pub2.
10. MARK, J. E. (2009) *Polymer Data Handbook*, 2nd Edition, Oxford, Oxford University Press.
11. *HUNTSMAN TECHNICAL BULLETIN* (2001) 1-1.0 'Introduction to Expandable Polystyrene', July.
12. CERESANA RESEARCH (2010) 'Expandable polystyrene' (UC-1305), March.
13. AMERICAN CHEMISTRY COUNCIL (2012) 'Take a Closer Look at Today's Polystyrene Packaging'. Available from: http://plastics.americanchemistry.com/Polystyrene-Packaging-Safe-Affordable-and-Environmentally-Responsible; http://plasticfoodservicefacts.com [Accessed October 2012].
14. THIELE UK (2007) *Polyester Bottle Resins, Production, Processing, Properties and Recycling*, Heidelberg, Germany. ISBN 978-3-9807497-4-9, 85.
15. GUPTA V. B. and BASHIR Z. (2012) Chapter 7 in *Handbook of Thermoplastic Polyesters*, Fakirov S. (ed.). ISBN 978352730113, 320.
16. KÖPNICK H., SCHMIDT M., BRÜGGING W., RÜTER J. and KAMINSKY W. (2005) 'Polyesters', in *Ullmann's Encyclopedia of Industrial Chemistry*, A21, Weinheim, Wiley-VCH. doi:10.1002/14356007.a21_227, 233–8.
17. MORGAN M. (2012) 'Economics of Renewable Polyester', presented at BioPlastek, Arlington VA, 27 March.
18. EASTMAN CHEMICAL COMPANY (2012) Available from: www.eastman.com/brands/eastman_tritan/Pages/Overview.aspx [Accessed October 2012].
19. DELIGIO T. (2010) 'Eastman opens expanded Tritan plant, copolyester's rapid rise continues', *Plastics Today*, 14 May. Available from: www.plasticstoday.com/articles/eastman-opens-expanded-tritan-plant-copolyester-rapid-rise-continues [Accessed October 2012].
20. MIEL R. (2012) 'Tritan tackles entire product line', *Plastics News*, 27 June. Available from: www.plasticsnews.com/headlines2.html?id=25821&q=tritan [Accessed October 2012].
21. 21ST CENTURY PINT (2012) 'Best Practice Guidelines for Drinking Vessels; Plastics'. Available from: www.21stcenturypint.org/plastics [Accessed October 2012].
22. BRANDAU O. (2012) *Bottles, Preforms and Closures, 2nd Edition – A Design Guide for PET Packaging*, May. ChemTec Publishing. ISBN 978-1-4377-3526-0.
23. PETAINER CZECH HOLDINGS S.R.O (2012) Available from: www.petainer.com/Technology.aspx [Accessed October 2012].
24. CALIENDO H. (2012) 'London 2012: toasting victories with recyclable PET bottles', *PlasticsToday*, 9 August. Available from: www.plasticstoday.com/articles/London-2012-Toasting-victories-with-recyclable-PET%20bottles-0809201201 [Accessed October 2012].
25. *PLASTICS AND RUBBER WEEKLY* (2012) 'Petainer – refreshing the beers that other kegs can't reach', 11 May. Available from: www.prw.com/subscriber/newsmail2.html?id=868 [Accessed October 2012].
26. CALIENDO H. (2012) 'FDA bans BPA in baby bottles and sippy cups', *PlasticsToday*, 17 July. Available from: www.plasticstoday.com/articles/FDA-bans-BPA-in-baby-bottles-and-sippy-cups-0717201201 [Accessed October 2012].
27. PIERCE L. M. and CALIENDO H. (2012) 'BPA in Packaging', *Packaging Digest*, 1 July. Available from: www.packagingdigest.com/article/522114-BPA_in_packaging_Defying_the_pressure.php [Accessed October 2012].
28. CALIENDO H. (2012) 'BPA in Packaging', *PlasticsToday*, 2 July. Available from: www.plasticstoday.com/articles/BPA-in-packaging-market-outlook-PC-alternatives-ongoing-lawsuits-0702201202 [Accessed October 2012].
29. MOHAN A. M. (2012) 'FDA rejects petition to ban BPA from food and beverage packaging', *Packaging World Greener Packaging*, 2 April. Available from: www.

greenerpackage.com/material_health/fda_rejects_petition_ban_bpa_food_and_beverage_packaging [Accessed October 2012]; www.greenerpackage.com/node/4765 [Accessed October 2012].
30. GOETZ W. (2003) 'Polyamide for Flexible Packaging Film', *PLACE Conference*, 12–14 May, Rome, Italy. Available from: www.tappi.org/Downloads/unsorted/UNTITLED---epl0348pdf.aspx [Accessed October 2012].
31. GBI RESEARCH (2011) 'Polyvinyl Chloride (PVC) Global Market Dynamics to 2020'; Report Code: GBICH0032MR, January. Available from: http://gbiresearch.com/Report.aspx?ID=Polyvinyl-Chloride-PVC-Global-Market-Dynamics-to-2020-Construction-Industry-Drives-the-Global-Demand&Title=Chemicals&ReportType=Industry_Report [Accessed October 2012].
32. SPINNER J. (2012) 'PVC Package recycling push succeeds in Europe', *Packaging Digest*, 17 August. Available from: www.packagingdigest.com/article/522341-PVC_package_recycling_push_succeeds_in_Europe.php [Accessed October 2012].
33. CALIENDO H. (2012) 'P&G to eliminate PVC packaging', *Packaging Digest*, 24 January. Available from: www.packagingdigest.com/article/520664-P_G_to_eliminate_PVC_packaging.php [Accessed October 2012].
34. EUROPEAN CHEMICALS AGENCY (ECHA) (2012) 'Annex XIV Authorisation List'. Available from: www.echa.europa.eu/web/guest/addressing-chemicals-of-concern/authorisation/recommendation-for-inclusion-in-the-authorisation-list/authorisation-list [Accessed October 2012].
35. PLASTEMART (2012) 'Market trends and formulations improve environmental performance of vinyl compounds', 15 June. Available from: www.plastemart.com/Plastic-Technical-Article.asp?LiteratureID=1808&Paper=market-trends-formulations-improve-environmental-performance-of-vinyl-compounds [Accessed October 2012].
36. VINYL PLUS (2012) Available from: www.vinylplus.eu [Accessed October 2012].
37. LONGWORTH R. and NAGEL H. (1997) 'Packaging', in *Ionomers: Synthesis, Structure, Properties, and Applications*, Tant M. R., Mauritz K. A. and Wilkes G. L. (eds), Springer. ISBN 978-0-7514-0392-3, 365. Available from: http://books.google.com/books?id=w42QO_MOzgUC&pg=PA16 [Accessed October 2012].
38. EISENBERG A. and KIM J. S. (1998) 'Applications', in *Introduction to Ionomers*, Wiley Interscience, 307.
39. FINCH C. A. (ed.) (1992) *Polyvinyl Alcohol Developments*, 2nd Edition, March 1992, John Wiley & Sons. ISBN 978-0-471-99850-1.
40. FINCH C. A. (ed.) (1981) 'Some Properties of PVOH and their possible applications', in *Chemistry and Technology of Water-Soluble Polymers*, Finch, C. A., Plenum Press, 287.
41. FINCH C. A. (2012) *Industrial Water Soluble Polymers*, RSC Spec Publ. no. 186.
42. CHIELLINI E., CORTI A., DEL SARTO G. and DANTONE S. (2006) 'Effect of hydrolysis degree on biodegradation behaviour of poly(vinyl alcohol)', *Polymer Degradation and Stability*, 91(12), 3397–3406.
43. MCCOY M. (2012) 'Selling detergents one load at a time', *Chemical & Engineering News*, 90, 15–19.
44. SMITHERS PIRA (2012) 'The Future of Transparent Barrier Films to 2017', 12 June. Available from: www.smitherspira.com/future-of-transparent-barrier-films-to-2017.aspx [Accessed October 2012].
45. PIRINGER O. G. and BANER A. L. (2000) 'Permeation of gases, water vapor and volatile organic compounds', and '(Appendix I and Tables 1, 2 and 3)', in *Plastic Packaging Materials for Food: Barrier Function*; Piringer O. G. (ed.). *Mass Transport, Quality Assurance and Legislation*, Wiley-VCH. ISBN 3-527-288868-6. 240 and 470.
46. ROBERTSON G. L. (ed.) (2009) *Food Packaging and Shelf Life*, CRC Press. ISBN; 978-1-4200-7844-2. Available from: www.crcnetbase.com/doi/book/10.1201/9781420078459 [Accessed October 2012].
47. SOARUS LLC (2012) 'Soarnol EVOH'. Available from: www.soarus.com/enterprise/index_1.html [Accessed October 2012].

48. IHS CHEMICAL (2011) 'Ethylene-vinyl alcohol resins', *Chemical Economics Handbook*, October. Available from: www.ihs.com/products/chemical/planning/ceh/ethylene-vinyl-alcohol.aspx [Accessed October 2012].
49. KURARAY AMERICA INC. (2012) EVAL Business Unit, 'About EVAL'. Available from: www.eval.eu/us/about-eval.aspx [Accessed October 2012].
50. MITSUBUSHI GAS CHEMICAL INC. (2012) 'Gas Barrier Technologies: Nylon-MXD6'. Available from: www.gasbarriertechnologies.com/mxd6.html [Accessed October 2012]; www.mgc.co.jp/eng/products/nop/nmxd6/index.html [Accessed October 2012].
51. MITSUBISHI GAS CHEMICAL INC. and NANOCOR (2012) 'Multilayer Containers Featuring Nano-Nylon MXD6 Barrier Layers with Superior Performance and Clarity'. Available from: www.nanocor.com/tech_papers/NOVAPACK03.pdf [Accessed October 2012].
52. NANOCOR (2012) 'IMPERM Grade 105 Superior Gas Barrier Resin'. Available from: www.nanocor.com/tech_sheets/I105.pdf [Accessed October 2012] and references therein.
53. HONEYWELL INTERNATIONAL INC. (2012) 'Honeywell Aegis® Barrier Nylon Resins'. Available from: www51.honeywell.com/sm/aegis [Accessed October 2012].
54. ASSOCIATION OF POSTCONSUMER PLASTICS RECYCLERS (2011) *PET Bottle Critical Guidance Document*, Edition 5, 2 May. Available from: www.plasticsrecycling.org/images/stories/doc/pet_critical_guidance_doc_approved_5-2-11_ed_5.pdf [Accessed October 2012].
55. KUREHA CORPORATION (2011) 'Kuredux Polyglycolic Acid (PGA) Resin for Packaging and Industrial Applications', April. Available from: www.kureha.com/pdfs/Kuredux_brochure.pdf [Accessed October 2012].
56. KUREHA CORPORATION (2011) 'Kuredux Polyglycolic Acid (PGA), Technical Guidebook', April. Available from: www.kureha.com/pdfs/Kuredux_technical.pdf [Accessed October 2012].
57. PLANTIC TECHNOLOGIES (2012) 'Ecoplastic'. Available from: www.plantic.com.au/technologies/ecoplastic.html [Accessed October 2012].
58. BLACK T. (2012) 'Plantic EcoPlastic – the World's Most Advanced Bioplastic', *Bioplastics: The Re-Invention of Plastics Conference*, San Francisco, 13 June. InnoPlast Solutions, Available from: www.bioplastix.com [Accessed October 2012].
59. SPECIALCHEM (2012) 'Braskem in Association with Plantic uses Green Plastic to Develop Ultra-high Barrier Packaging', *SpecialChem*, 21 May. Available from: www.omnexus.com/news/news.aspx?id=30461&lr=dom12818la6&li=61011797 [Accessed October 2012].
60. WHEYLAYER CONSORTIUM (2012) Available from: www.wheylayer.eu; www.wheylayer.eu/download/final-press-release-for-web.pdf
61. SIDEL, 'Actis™ (2012) Plasma Coating Machine'. Available from: www.sidel.com/en/Our-products/Coating/ACTIS?fromId=195 [Accessed October 2012]. And Sidel, 'ACTIS™ Barrier Process'. Available from: www.sidel.com/en/content/download/168/762/version/3/file/Actis_EN.pdf [Accessed October 2012].
62. MITSUBISHI SHOJI PLASTICS CORPORATION (2012) 'Method of Manufacturing Gas Barrier Film Coated Plastic Container', WIPO Patent Application WO/2005/007925.
63. KHS PLASMAX GMBH (2012) 'Optimal Barrier Solutions to Sensitive Products'. Available from: www.khsplasmax.com/images/publikationen/InnoPET-Plasmax_GB.pdf [Accessed October 2012].
64. TOYO SEIKAN KAISHA LTD (2012) 'SIBARD Silicon Oxide Coated PET Bottle'. Available from: www.toyo-seikan.co.jp/e/technique/petbottle/barrierbottle [Accessed October 2012].
65. AMCOR LTD (2012) 'CeramisA®'. Available from: www.amcor.com/businesses/flexibles_industrial/Amcor_Ceramis.html [Accessed October 2012]. 'CeramisA® Product Range'. Available from: www.amcor.com/businesses/flexibles_industrial/Ceramis.html [Accessed October 2012]. 'Ceramis A® High Barrier Films'. Available

from: www.amcor.com/products_services/Amcor_Ceramis_High_Barrier_Films.html [Accessed October 2012].

66. MICHAELI W., BAHROUN K. V., FRAGSTEIN F. and BEHM H. (2010) 'Plasma assisted barrier coating', *Bioplastics Magazine*, 5(4), 24.
67. SMART PLANET TECHNOLOGIES (2010) 'EarthCoating'. Available from: www.smartplanettech.com/earthcoating [Accessed October 2012].
68. SHERMAN L. M. (2012) 'Sub-micron additives make strides', *Plastics Technology*, July, 26.
69. HERNANDEZ R. J., SELKE S. E. M. and CULTER J. D. (2000) 'Additives and compounding', *Plastics Packaging: Properties, Processing, Applications and Regulations*, Carl Hanser Verlag. ISBN 1-56990-303-4; 135.
70. ROONEY M. L. (2000) 'Plastics in active packaging', in *Materials and Development of Plastics Packaging for the Consumer Market*, Giles G. A. and Bain D. R. (eds.), Oxford, UK, Sheffield Academic Press. ISBN 1-84127-116-0. Chapter 7. 105–29.
71. COOKSEY K. (2010) 'Oxygen scavenging packaging systems', Wiley Online Library, *Encyclopedia of Poly Science and Technology*, 26 April. Available from: http://onlinelibrary.wiley.com/doi/10.1002/0471440264.pst570/pdf [Accessed October 2012] and references therein.
72. BRODY A. (2006) 'State of the Art in Active/Intelligent Food Packaging', *Institute of Food Technologists Food Packaging Summit*, 7–9 May. Available from: http://www.ift.org/knowledge-center/read-ift-publications/science-reports/research-summits/~/media/Knowledge%20Center/Science%20Reports/Research%20Summits/Packaging/Packaging_ActiveIntelligentPackaging_Brody.pdf [Accessed October 2012].
73. YE Y. and KING R. E. (2006) 'Additives for Polyolefin Film Products: An Overview of Chemistry and Effects', Tappi. Available from: www.tappi.org/content/enewsletters/eplace/2006/06PLA04.pdf [Accessed October 2012] and references therein.
74. KAUFMAN J., LACOSTE A., SCHULOK J., SHEHADY E. and YAM K. L. (2000) 'An Overview of Oxygen Scavenging Packaging and Applications', *Packaging Net.com* 27 October. Available from: www.packagingnetwork.com/doc.mvc/An-overview-of-oxygen-scavenging-packaging-an-0003#oxygen [Accessed October 2012].
75. CONSTAR INTERNATIONAL LLC (2011) 'Oxbar®'. Available from: www.constar.net/tech-barrier-oxbar.php [Accessed October 2012].
76. Plastemart (2011) 'Active packaging for fruit and vegetables', *Plastemart*, 16 March. Available from: www.plastemart.com/upload/literature/active-packaging-technologies-for-fruits-vegetables-important-to-prevent-wastage-multi%20layer-plastic-films.asp [Accessed October 2012].
77. MILLIKEN AND COMPANY (2012) 'Hyperform for Nucleation'. Available from: http://millikenchemical.com/pages/products/hyperform-for-nucleation/default.aspx [Accessed October 2012].
78. TULLO A. (2010) 'Making clearer polypropylene', *C&EN*, 88(36), 34–7.
79. JACOBY P. (2012) 'Using β-nucleation to improve the thermoforming characteristics of polypropylene', *Plastics Engineering*, July/August, 32–3. Available from: www.nxtbook.com/nxtbooks/wiley/pe_20120708/#/36 [Accessed October 2012].
80. RIKA INTERNATIONAL LIMITED (2012) 'NJSTAR® NU-100'. Available from: www.ppadditives.com/docs/English/brochure/NJSTAR%20NU-100.pdf [Accessed October 2012].
81. SCHEIRS J. (2012) 'Additives for the Modification of Poly(Ethylene Terephthalate) to Produce Engineering-Grade Polymers'. Available from: www.excelplas.com/admin/media/pages/43/PET_chapter.pdf [Accessed October 2012] and references therein give detailed information on nucleants and other additives for PET.
82. DRAHL C. (2012) 'How nanosilver zaps germs', *Chemical & Engineering News*, 90(30) 19 July.
83. KRIBOSHE S. (2012) 'New Anti-Bacterial Packaging to Prolong Pepsi's Shelf-Life', *NoCamels*, 26 August. Available from: http://nocamels.com/2012/05/will-israeli-company-prolong-pepsis-shelf-life [Accessed October 2012].

84. WILSON C. L. (ed.) (2007) *Intelligent and Active Packaging for Fruits and Vegetables*, CRC Press. ISBN 13:978-0-8493-9166-8.
85. YAM K. L. (ed.) (2012) *Emerging Food Packaging Technologies: Principles and Practice*, Cambridge, UK, Woodhead Publishing Ltd.
86. KERRY J. (2008) *Smart Packaging Technologies for Fast Moving Consumer Goods*, Wiley.
87. WALDMAN K. (2012) 'Waiting for the Wrap-ture', *Slate Magazine*, 5 June. Available from: www.slate.com/articles/health_and_science/future_tense/2012/06/smart_food_packaging_to_keep_fruit_fresh_and_detect_spoilage_.html
88. PHOTONIC BIOSYSTEMS INC. (2012) Available from: www.photonicsystems.com/index.html [Accessed October 2012].
89. SCENTSATIONAL TECHNOLOGIES (2012) 'ScentSational Packaging'. Available from: www.scentsationaltechnologies.com/technology.cfm [Accessed October 2012].
90. CALIENDO H. (2012) 'A smarter package?', *PlasticsToday*, 18 July. Available from: www.plasticstoday.com/articles/Bemis-and-Thinfilm-to-develop-intelligent-packaging-platform-0718201201 [Accessed October 2012].
91. DESCHAMPS M. J. (2012) 'Packaging firms look to solve green challenges', *European Plastics News*, 27 February. Available from: www.europeanplasticsnews.com/subscriber/headlines2.html?cat=1&id=1164&q=Deschamps [Accessed October 2012].
92. HERNANDEZ R. J., SELKE S. E. M. and CULTER J. D. (2000) 'Thermoforming', in *Plastics Packaging: Properties, Processing, Applications and Regulations*, Carl Hanser Verlag. ISBN 1-56990-303-4; 239.
93. CALIENDO H. (2012) www.reportsnreports.com [Accessed October 2012]; *PlasticsToday*, 16 May. Available from: www.plasticstoday.com/articles/Thermoformed-plastics-market-projected-to-grow-at-four-percent-CAGR-through-2017-0516201201 [Accessed October 2012].
94. *PLASTICS NEWS* (2012) 'Thermoformers Market Review and Outlook 2012'. Available from: www.plasticsnews.com/rankings/product.html?pid=116 [Accessed October 2012]. and DeRosa A, 'Thermoforming consolidation to continue', *Plastics News*, 4 June 2012, Available from: www.plasticsnews.com/headlines2.html?id=25607&q=derosa [Accessed October 2012].
95. MCTIGUE L. (2012) 'PP food containers preserve oxygen barrier after retort', *Packaging Digest*, 1 March. Available from: www.packagingdigest.com/article/520616-PP_food_containers_preserve_oxygen_barrier_after_retort.php [Accessed October 2012].
96. AMI CONSULTING (2012) 'Thin Wall Packaging'. Available from: www.webpackaging.com/packaging-suppliers/featured-content/news/param/news/2528075/New-AMI-Consulting-report-details-trends-in-thin-wall-packaging-markets-in-Europe [Accessed October 2012].
97. PLASTEMART.COM (2012) 'Thin-walled plastics packaging reduces weight, opens up design opportunities', 15 June. Available from: www.plastemart.com/Plastic-Technical-Article.asp?LiteratureID=1810&Paper=thinwalled-plastics-packaging-reduces-weight-opens-up-design-opportunities [Accessed October 2012].
98. PLASTEMART.COM (2012) 'Material Developments in Thin-Wall Packaging,' 15 June. Available from: www.plastemart.com/Plastic-Technical-Article.asp?LiteratureID=1809&Paper=material-developments-in-thin-wall-packaging [Accessed October 2012].
99. ZIMAKIEWICZ M. (2012) 'Competition hots up in thin wall packaging', *Film and Sheet Extrusion*, May, 9–10.
100. FLEXIBLE PACKAGING ASSOCIATION INDUSTRY REPORT (2011) Available from: www.flexpack.org/FPA_update/Articles/201109/FPA_Update_0911_FINAL.pdf [Accessed October 2012].
101. HERNANDEZ R. J., SELKE S. E. M. and CULTER J. D. (2000) 'Flexible packaging', *Plastics Packaging: Properties, Processing, Applications and Regulations*, Carl Hanser Verlag. ISBN 1-56990-303-4; 229.

102. DIXON J. (2000) 'Development of flexible plastics packaging', in *Materials and Development of Plastics Packaging for the Consumer Market*, Giles G. A. and Bain D. R. (eds.), Oxford, UK, Sheffield Academic Press. ISBN 1-84127-116-0. Chapter 6. 79–104.
103. TAYLOR J. D. (2012) 'Bioplastics in Flexible Packaging', *Proceedings of the BioPlastek Conference*, Arlington VA, 27 March.
104. FLEXIBLE PACKAGING EUROPE (2012) 'The Perfect Fit'. Available from: www.flexpack-europe.org/tl_files/FPE/downloads/ThePerfectFit.pdf [Accessed October 2012].
105. WAGNER J. R. (2009) *Multilayer Flexible Packaging*, ChemTec Publishing. ISBN 978-0-8155-2021-4, provides detailed information on materials, additives, coatings, processing and applications of PE and PP based multilayer flexible packaging.
106. YORK E. B. (2012), 'More foods going to pouch packaging', *Chicago Tribune*, 6 March. Available from: http://articles.chicagotribune.com/2012-03-06/business/ct-biz-0306-pouches-20120306_1_pouch-soup-giant-packaging-digest [Accessed October 2012].
107. ALLIED DEVELOPMENT (2012) 'Stand-up Pouches 2012 to 2016'. Available from: www.allied-dev.com/Studies/Stand-up_Pouches_Details.aspx [Accessed October 2012].
108. SPINNER J. (2012) 'Stand-up pouch sales to reach $4 billion by 2016', *Packaging Digest*, 26 June. Available from: www.packagingdigest.com/article/print/522096-Stand_up_pouch_sales_to_reach_4_billion_by_2016.php [Accessed October 2012].
109. CONNOLLY K. B. (2012) 'Pouches Flex their Marketing Muscles', *Foodprocessing.com*, 7 May. Available from: www.foodprocessing.com/articles/2012/pouches-flex-marketing-muscles.html?page=print [Accessed October 2012].
110. CALIENDO H. (2012) 'Wine producers serving greener packaging', *Packaging Digest*, 23 March. Available from: www.packagingdigest.com/article/521299-Wine_producers_serving_greener_packaging.php [Accessed October 2012].
111. HERNANDEZ R. J., SELKE S. E. M. and CULTER J. D. (2000) 'Extrusion, film and sheet', in *Plastics Packaging: Properties, Processing, Applications and Regulations*, Carl Hanser Verlag. ISBN 1-56990-303-4; 185.
112. YAM K. L. (2009) *Encyclopedia of Packaging Technology*, John Wiley & Sons. ISBN 978-0-470-08704-6.
113. SOROKA W. (2002) *Fundamentals of Packaging Technology*, IoPP. ISBN 1-930268-25-4.
114. VERGHESE K., LEWIS H. and FITZPATRICK L. (eds.) (2012) *Packaging for Sustainability*, Springer Verlag.
115. UPS (2012) 'Sustainable supply chain data book', *Environmental Leader*, 5. Available from: www.environmentalleader.com/sustainable-supply-chain-data-book [Accessed October 2012].
116. LAIRD K. (2012) 'Green matter, packaging the message', *PlasticsToday*, 12 January. Available from: www.plasticstoday.com/blogs/green-matter-packaging-message-0112201201 [Accessed October 2012].
117. FREEDONIA GROUP (2012) 'World Bioplastics to 2015', 3 January. Available from: www.freedoniagroup.com/DocumentDetails.aspx?DocumentId=563324 [Accessed October 2012].
118. PRICEWATERHOUSECOOPERS LLP (PWC) (2012) 'Sustainable packaging: myth or reality', June. Available from: www.pwc.com/en_GX/gx/sustainability/research-insights/assets/pwc-sustainable-packaging-report.pdf [Accessed October 2012].
119. HERNANDEZ R. J., SELKE S. E. M. and CULTER J. D. (2000) 'Environmental considerations', *Plastics Packaging: Properties, Processing, Applications and Regulations*, Carl Hanser Verlag. ISBN 1-56990-303-4; 375.
120. DENT I. S. (2000) 'Recycling and reuse of plastics packaging for the consumer market', in *Materials and Development of Plastics Packaging for the Consumer Market*, Giles G. A. and Bain D. R. (eds.), Oxford, UK, Sheffield Academic Press. ISBN 1-84127-116-0. Chapter 10. 177–202.

121. EUPR (European Plastics Recyclers (2009), 'The Compelling Facts About Plastics 2009', Available from: www.plasticsrecyclers.eu/uploads/media/eupr/Compelling/Brochure_UK_FactsFigures_2009_22sept_6_Final-20090930-001-EN-v1.pdf [Accessed October 2012].
122. EUPR (European Plastics Recyclers) (2012) 'EuCertPlast Audit Scheme, version 3.3'. Available from: www.eucertplast.eu/uploads/downloads/eucertplast-audit-scheme.pdf [Accessed October 2012]; www.eucertplast.eu/en/ [Accessed October 2012].
123. WRAP (Waste and Resources Action Programme) (2008) 'Domestic Mixed Plastics Packaging Waste Management Options', Foster S., June. ISBN: 1-84405-396-2, Available from: www.wrap.org.uk/sites/files/wrap/Mixed_Plastic_Final_Report.5479.pdf [Accessed October 2012].
124. EUROPEAN PLASTICS RECYCLERS (EUPR) and PETCORE (PET Containers Recycling Europe) (2011) 'PET Collection Report, 2011'. Available from: www.plasticsrecyclers.eu/docs/press%20release/071312%20%20Petcore%20and%20EuPR%20Press%20Release%20-%202011%20European%20PET%20Collection%20Statistics.pdf [Accessed October 2012].
125. NATIONAL ASSOCIATION FOR PET CONTAINER RESOURCES (NAPCOR) (2010) '2010 Report On Post-Consumer PET Container Recycling Activity'. Available from: http://www.napcor.com/pdf/2010_Report.pdf [Accessed October 2012].
126. THIELE UK (2007) '*Polyester Bottle Resins Production, Processing, Properties and Recycling*', Heidleberg, Germany, PETplanet Publisher GmbH. ISBN 978-3-9807497-4-9; 259.
127. PET-RECYCLING FORUM (2006) 'Current technological trends in polyester recycling', *9th International Polyester Recycling Forum Washington*, Sao Paulo. ISBN 3-00-019765-6.
128. FAKIROV S. (2002) *Handbook of Thermoplastic Polyesters*, Weinham, Wiley-VCH. ISBN 3-527-30113-5; 1223.
129. ASSOCIATION OF POSTCONSUMER PLASTIC RECYCLERS (APR) (2012) 'Shrink Label Learning Protocol', 15 March. Available from: www.plasticsrecycling.org/images/stories/doc/apr_shrink_sleeve_protocol_report_3_15_12.pdf [Accessed October 2012].
130. 'European PET Bottle Platform' (2012) Available from: www.petbottleplatform.eu/technical.php [Accessed October 2012].
131. AMERICAN CHEMISTRY COUNCIL AND ASSOCIATION OF POSTCONSUMER PLASTICS RECYCLERS (2010) 'End uses for US post-consumer recycled HDPE bottles, 2010', *Plastics News*. Available from: www.plasticsnews.com/fyi-charts/packaging.html?id=24861 [Accessed October 2012].
132. CARTEAUX W. R., RUSSEL S. and HOCHU C. (2012) 'One-resin focus saps recycling', *Plastics News*, 3 September, 6 and 20.
133. CLARK A. (2011) 'Recyclers milk bottle market at record level', *Plastics and Rubber Weekly*, 8 November. Available from: www.prw.com/subscriber/headlines2.html?cat=1&id=7&q=HDPE+recycling [Accessed October 2012].
134. CLEAN WASHINGTON CENTER (2012) 'Best Practice Guide for HDPE Recycling'. Available from: www.cwc.org/hdpe/hdpe1.htm [Accessed October 2012].
135. CLARK A. (2010) 'Closed Loop Recycling skims 10% off Sainsbury's milk', *Plastics and Rubber Weekly*, 30 April. Available from: www.prw.com/subscriber/headlines2.html?cat=1&id=1272621301&q=closed+loop+sainsbury [Accessed October 2012].

5

Developments in bioplastic materials for packaging food, beverages and other fast-moving consumer goods

T. A. Cooper, ARGO Group International, USA

DOI: 10.1533/9780857098979.108

Abstract: This chapter discusses the current situation, developments and future trends in the use of biodegradable and biobased plastics in packaging for food, beverages and other fast-moving consumer items. Areas covered include the production, properties and processing of the various types of biodegradable polymer used for packaging, biobased versions of conventional polymers used in packaging and developments in the synthesis of monomers, chemicals and polymers by biobased routes. Particular emphasis is given to biobased polyethylene, polypropylene and polyethylene terepthalate. Definitions and standards of bioplastics, biodegradability, compostability and biobased content are also discussed.

Key words: bioplastic packaging, biodegradable, compostable and biobased polymers and packaging, renewably sourced polymers and packaging, end-of-life, new carbon.

5.1 Introduction

Bioplastics have become an increasingly visible group of materials over the past 15 years and general awareness of them is rapidly rising as they more frequently appear in consumer products. Bioplastics development has been driven by drawbacks of conventional plastics, particularly concerning use of fossil resources, recycling, littering and pollution. This chapter will discuss the current situation, developments and future packaging trends with biobased and biodegradable plastics.[1,2]

The origins of the bioplastics industry go back to the mid-nineteenth century, when the first plastics ever made were biobased; for example, celluloid, which was originally developed to replace ivory in 1863, and casein-based plastics. However, this was soon overtaken by the development of the petrochemical industry, which quickly became the main source for plastic materials and technology. Consequently, the modern interest in bioplastics derives from about 15 years ago, starting predominantly with the development of biodegradable and compostable thermoplastics, such as polylactic acid (PLA) and polyhydroxyalkanoates (PHA). These biodegradable materials were mostly 'new' polymers, compared to conventional petrochemical polymers such as polyethylene, as well as blends with natural products such as cellulose and starches. However, these new polymers are not necessarily biobased and many, particularly some aliphatic polyesters, are petrochemical-based. The interest at that time was solely on end-of-life considerations, to reduce waste going to landfill and control litter and marine pollution, although the emphasis is now changing back to renewable sourcing, as with the first bioplastics from the nineteenth century.[3,4]

These 'new' bioplastics were designed for packaging and other applications to be thermoplastically processable using conventional methods such as extrusion, film blowing and casting, lamination, sheet casting and thermoforming, injection-molding, blow-molding and foaming, both for monolithic and multilayer systems.[5] They can also be combined with paperboard by lamination or extrusion coating. As with conventional polymers, specific bioplastics are selected for particular applications, depending on the requirements for: mechanical, thermal, optical and tactile properties; chemical, oil, fat and grease resistance; barrier properties towards oxygen, carbon dioxide, water vapor and organoleptics; adhesion and printability; sterilizability; and cost. Also, as with conventional polymers, there is considerable ability to tailor-make packaging to fit specific property combinations and optimize protection of the packaged contents, while reducing costs by minimizing materials use and maximizing processing speeds.[6–10]

5.2 Definition and rationale for bioplastics

The word 'bioplastic' is presently somewhat confusing, since it is used as a portmanteau term to refer to several different types of material.[1,5,8] These include:

- Biodegradable or compostable plastics and polymers, which can be biobased or petroleum-based.
- Biobased plastics and polymers (i.e. made from renewable rather than fossil sources) that are not necessarily biodegradable (e.g. conventional polymers such as biopolyethylene and polyamide-11, which are made from biobased monomers).
- Plastics and polymers containing both petroleum-based and biobased components (these can be biodegradable or non-biodegradable).
- Plastics and polymers used for biomedical applications (i.e. prosthetics, bioabsorption or controlled release).

Only the first three categories will be considered in this chapter.

It is important to emphasize that biobased or renewably-sourced polymers may not be biodegradable; for example polyamide-11, which is used in automotive applications because of its excellent resistance to chemical degradation. Conversely, biodegradable polymers can be biobased or petrochemical-based. Thus many aliphatic polyesters, which are readily biodegradable, are currently synthesized from petrochemical sources, although there is considerable present effort to develop biobased routes. Many partially biobased polymers, where one monomer may be biobased and another petrochemical-based, are also becoming available, such as polytrimethylene terephthalate (PTT), polyethylene terephthalate (PET), and castor oil-based polyamides and block copolyether-amides.

Renewable sourcing means that the polymers are made from resources that can be renewed within one to two years, for example plant products.[11–13] This is completely different from petrochemical products, where renewal is millions of years. However, materials that are fossil-based and hence not renewably sourced can still be considered sustainable if they can be reused or recycled indefinitely at the end of their designed use.

The packaging industry is being massively affected by the trend towards renewable materials and 'sustainability', by perceived consumer demands on manufacturers and retailers to adopt more sustainable systems and provide solutions to environmental problems, and by the use of 'greenness' as a marketing tool. According to the Consumer Goods Forum Global Packaging Project (2011):[14] 'Sustainability encompasses strategies and practices that aim to meet the needs of stakeholders today while seeking to protect, support and enhance the human and natural resources that will be needed in the future.' This has led to the present situation where sustainability is no longer just 'nice to have' or exclusively part of corporate social responsibility but is seen as a business necessity to attract consumers and protect market share. A 2012 survey by Accenture of 250 senior executives, in eight leading mature and emerging economies, showed that 44% think sustainability is critical to their business and 78% think that it is vital to their future growth. Sixty-two percent say their sustainable investments are motivated by customer expectations for sustainable products and services and 60% by the opportunity to drive growth. Only 41% were motivated by regulatory compliance and only 29% by the need for energy and material cost reductions.[15]

This trend is also being driven by government demands arising from their environmental and carbon emissions policies and by large consumer goods corporations such as Wal-Mart, with its 'Sustainability Scorecard', and P&G, which is causing pressure along their supply chains to conform. This is equally true in the food industry, where packaging is perceived as a prime sustainability factor to allow minimization of carbon footprint and energy consumption, while simultaneously avoiding a significant adverse effect on packaging functionality and cost. Companies such as Coca-Cola, PepsiCo, Danone, Unilever, P&G, Sainsbury's, M&S and Co-op perceive such packaging as needing the use of biobased and/or recycled materials in addition to light-weighting. Further information on sustainability in packaging is presented in Chapter 4, Section 4.10.

Thus the rationale for the development of bioplastics, particularly in packaging, has not generally resulted from any deficiencies in the properties of petrochemically-based plastics. Instead it derives from factors such as:

- Ongoing attacks on conventional plastics for environmental, societal and political reasons and increasing legislative, regulatory and consumer concerns over carbon footprint, climate change, environmental pollution and litter. This has resulted in increasing legislation in many areas, restricting uses and disposal of plastics and often promoting bioplastics. Legislative initiatives for bioplastics have been occurring worldwide, including mandatory biodegradable shopping carrier bags in Italy;[16] exemption of biodegradable plastics from the recycling directives in Germany; reduced taxes on bioplastics for packaging in The Netherlands; a mandate that 20% of all plastics will be renewably sourced by 2020 in Japan; the US Federal Farm Bill, which directs all Federal Agencies to maximize biobased plastics procurement based on biobased content, price and performance – the 'BioPreferred program' – and sets up the 'USDA Certified Biobased Product' labeling program; and a plethora of enacted and proposed plastic bag and packaging bans, restrictions and taxes around the world in, for example, San Francisco, Santa Clara, Long Beach, Los Angeles and many other California locations, Toronto, Ireland, Wales, Macedonia, the EU, India, the UAE, and many more.
- Lack of recycling infrastructure in many parts of the world, difficulties in waste plastics collection and sorting, opposition to incineration for energy recovery, and increasing landfill restrictions for conventional plastics.
- Pressures, particularly in the US, to reduce oil dependence, particularly imported oil, and theories of peak oil that predict dwindling petroleum reserves. Of course, most petroleum is used for fuel, and petroleum-based plastics only use about 2.5 billion barrels of oil annually (<5% of all oil used), so that this reaction is overkill as far as plastics is concerned.

5.3 Classification of bioplastics

Bioplastics can be classified into two major types:

1. *Biodegradable plastics*: where the emphasis is on disposal or 'end-of-life' processes and is independent of the carbon sources used to make the plastic. Biodegradability is shown by test methods such as ASTM D5511 and ASTM D5526 (anaerobic) and ASTM D5338 and ASTM D5209 (aerobic). Biodegradable plastics, which degrade under specific composting conditions and timescales can be certified compostable in conformity to standards such as ASTM D-6400 in the US or EN 13432 in Europe. This will be discussed in more detail later.
2. *Biobased plastics*: where the emphasis is on the origin of the carbon atom constituents, irrespective of the disposal method, as measured by carbon-14 content using the ASTM D6866 test method.

It should be noted that these two classifications are not mutually exclusive and specific plastics, for example PLA and PHA, may meet both categories.

When current thermoplastic bioplastics were first being developed and introduced around the end of the 1990s, emphasis was totally on biodegradability and 'end-of-life'. However, over the past ten years, it has been recognized that while biodegradability may be attractive for specific applications such as food service items and compostable bags, particularly where composting or anaerobic digestion facilities are available, consumers are unlikely to pay more for biodegradable plastics than for conventional plastics. Many market studies have since been carried out on biodegradable and biobased plastics. Most forecast similar trends but may have different market size projections. However, these reports all agree that the composition of bioplastics production will transition over the next few years from a predominance of biodegradable plastics to a predominance of biobased plastics and the future market driver will predominantly be renewable sourcing rather than biodegradability. Consequently, the industrial and market emphasis has switched to developing biobased materials and the use of renewable feedstocks to supplement or replace petrochemical feedstocks.

Major reasons for this transition from biodegradable to biobased materials have been:

- High prices and lack of performance properties of biodegradable plastics, particularly heat distortion temperature, barrier properties and moisture sensitivity, although these disadvantages are now being rapidly overcome, particularly for PLA. In addition, some suppliers focused inordinately on biodegradability as a key differentiating property, but this was often an insufficient advantage for users to specify biodegradable plastics.
- Processing requirements, conditions and equipment for biodegradable plastics are often different from conventional plastics, so that the biodegradable plastics are not just a drop-in.
- The continuing lack of curbside collection and disposal infrastructure for biodegradable and compostable plastics, while 'conventional' polymers (which would include biobased PE and PET) can increasingly be recycled.
- An increasing demand for durable biobased products for electronic, household, consumer and automotive applications, which was estimated to be about 40% of global bioplastics use by 2011 and growing.[17]
- The rapid development of biobased routes to existing monomers and chemicals, allowing the production of 'conventional' petroleum-based polymers, plasticizers and other chemicals from renewable sources.

Although they are achieving significant commercial success, bioplastics are still in an early development stage and fill only a small industry niche. Further growth will be driven by improvements in properties and availability and decreased prices. Biodegradable resins were the major part of the bioplastics market in 2011, but non-biodegradable biopolymers will grow faster and increase from 13% in 2011 to more than 40% by 2021. This will be mainly because of high-volume production of biobased polyethylene (PE), followed by the commercialization of biobased PET, polypropylene (PP) and polyvinyl chloride (PVC). So long as their economics are comparable, these biobased resins will gain rapid market acceptance, since they are chemically identical to their petrochemical-based counterparts. Biobased polyethylene was introduced by Braskem in 2009 and is anticipated to show the highest growth through 2016 as production capacity expands and price is reduced to compete with petrochemical-based PE.

Despite this, PLA is still expected to be the most-used bioplastic during this period, although there are issues with the inability of PLA (and some other biodegradable products) to decompose in home compost systems, by anaerobic digestion or in soil or marine environments, and with their potential to contaminate recycling streams for other polymers, notably PET. However, the use of biodegradable polymers, such as PLA, will be facilitated by an expanding industrial composting infrastructure, greater acceptance and familiarity by plastics processors, and development of synthetic routes from non-food feedstocks to eliminate or alleviate the 'food-versus-fuel' debate and problems of the energy- and pesticide-intensiveness of corn (maize) production (Chapter 11).[11–14,18–20]

5.4 Biodegradability, compostability and anaerobic digestibility

Biodegradation is the enzymatic, highly-specific, stepwise fragmentation and degradation of an organic material (substrate) by microorganisms. Complete biodegradation leads to water, carbon dioxide under aerobic conditions, carbon dioxide and methane under anaerobic conditions, and residual biomass.

The term 'biodegradable' is an extremely loose description.[5] There are no defined criteria, specified environments, products or time limits, so that describing or labeling a plastic or polymer as biodegradable gives insufficient information on suitable disposal methods. Also, a material may be biodegradable in one set of environmental conditions but not in others; for example, a material that is biodegradable over long time frames such as years will not meet the requirements for industrial compostability where a much shorter time limit is specified (see ASTM D6400 and EN13432, as described below). A material that is biodegradable in aerobic environments may not be degradable in anaerobic environments (as in anaerobic digestion) in a useful time frame. Also a material that is compostable or biodegrades in aerobic conditions may not be soil or marine biodegradable. Consequently, making a packaging material 'biodegradable' does not confer a license to litter, although labeling a packaging as biodegradable may induce littering behavior by users.

Standards are crucial to biodegradable plastics. The key factor is the timescale, since almost all plastics can biodegrade over a long time period, perhaps centuries. However, to be practically useful, a material must be disposed of and biodegrade rapidly over a timescale comparable to its production and disposal rate, to prevent accumulation of solid waste in the environment.[21,22]

To meet the essential requirements specified under the European Directive on Packaging and Packaging Waste (94/62/EC), 'organic recycling' is defined as 'the aerobic (composting) or anaerobic (biomethanization) treatment, under controlled conditions and using microorganisms, of the biodegradable parts of packaging waste, which produces stabilized organic residues and methane. Landfill shall not be considered a form of organic recycling.' EN 13432 is the European standard meeting these requirements, 'Requirements for packaging recoverable through composting and biodegradation – Test scheme and evaluation criteria for the final acceptance of packaging'.[1,2,23]

For a material to be considered compostable, it must meet the conditions of ASTM D6400 in the US, EN 13432 in Europe or equivalent specifications in other geographic areas.[24] ASTM D6400 and EN 13432 are similar. EN 13432 defines both the test program and the assessment criteria that compostable packaging must meet, and its requirements can be outlined as follows:

1. *Test conditions*: oxygen level >10%, temperature 60–75°C, pH 5.0 or greater, biweekly or more frequent turning of the compost.
2. *Fragmentation and disintegration*: within 12 weeks, 10% or less of the plastic may have a particle size >2 mm, and must be invisible in the final compost. When the test materials are composted with solid waste, the mass of any residues still larger than 2 mm in size after 3 months must be <10% of the initial mass.
3. *Biodegradation*: namely microbial conversion into carbon dioxide and water (for all organic constituents present at >1% level): there must be 90% mineralization in less than 6 months (the 90% level is to allow for the formation of biomass).
4. *Compost quality*:

   a. *Ecotoxicity by plant growth (summer barley and cress)*: must show no negative effect from the compost.
   b. *Heavy metal and fluorine content*: levels must be under defined maximum limits and show no negative effects on the composting process or compost quality, as demonstrated by the plant growth test above.

All of these conditions must be met for certification under EN 13432. EN 14995 is technically equivalent to EN 13432, but is for non-packaging materials such as food service cutlery and compostable waste collection bags.

ASTM D6400 is similar to EN 13432 except that the limit of biodegradation, otherwise 90% in 180 days, is reduced to 60% for homopolymers and random copolymers and also that the test duration is extended to 365 days if the test is conducted with radiocarbon labeling to measure the evolution of radio-labeled carbon dioxide.

It is important to understand that what is certified by these methods is not just a material but a fabricated product such as a film, sheet or molded component. The certification generally includes determining the maximum thickness of material that can be degraded within one composting cycle, so that it is much easier for thin films to pass the specification than thick sections. Consequently, there can often be difficulties with thick-section items such as caps and closure passing compostability certifications, even when the rest of the packaging does so.[25] Certifications are made on the basis of independent testing by the Biodegradable Products Institute (BPI) in the US, AFOR in the UK, DIN and Vincotte in Europe, the Japanese Bioplastics Association (GreenPla) in Japan, and equivalent bodies in other countries.[24] These standard test methods, specifications and certifications provide a clear legal framework for the definition of biodegradability and compostability. EN 13432 has become significantly more important in Italy with the ban on the sale of non-biodegradable carrier bags on 1 January 2011 and is now effectively used to differentiate between biodegradable and non-biodegradable packaging.[16]

Certification under these industrial compostability standards does not necessarily mean a product will compost in a home backyard composter.[26] Home composting takes place under different conditions from EN 13432 or ASTM D6400, generally at much lower temperature and with no controlled turning or oxygen or moisture content. Separate compostability standards and logos are now being developed and introduced for home compostability by organizations such as Vincotte and ASTM International, whose Subcommittee D20.96 on Environmentally Degradable Plastics and Biobased Products is currently developing a new standard WK 35342, 'Specification for Home Composting of Biodegradable Plastics', with the goal of reducing the amounts of materials going to landfill.[27] Home compostable packaging will need to be designated by the brand owners and packaging producers so that it can be differentiated from non-biodegradable materials.

A major reason for the recent swing from emphasis on biodegradability or compostability to biobased sourcing is that for such biodegradable or compostable materials to demonstrate any advantage, a composting (or anaerobic digestion[28]) infrastructure must be available, including not only the composting operation itself but also collection, identification and sorting. While biodegradability and compostability are still promoted for 'green' marketing, most locations do not have such commercial composting facilities for general use. Thus biodegradable or compostable plastics make most sense for applications such as food packaging and service items contaminated by food residues, and waste bags for food scraps and organic matter (yard waste) where the bag can be composted along with the contents and there is sufficient volume in a given geographic area (a critical mass) to make collection economic, for example in densely populated areas, large restaurant facilities and functions such as the 2012 Olympics and other sports venues and events.[29] Recent litigation in Vermont in the US has attacked the labeling of packaging or products as compostable or industrially compostable where there are no local facilities for it to be composted in as being misleading or even fraudulent.[30]

In addition, it can be difficult for both the consumer and the industrial composter to identify compostable packaging. Thus there has been continual controversy over the use of PLA in bottles, which was originally seen as a major application, because of potential contamination of the PET waste stream. Only a small amount (1%) of PLA in PET will markedly degrade PET optical properties, and near infrared scanning equipment to separate them is expensive.[31] Thus many composters will not take plastics at all but simply screen them out to landfill, since they have no way to select out the compostable plastics in an economic fashion and cannot risk having undegraded plastic in their compost product.

Finally, in many areas, current emphasis is now being placed on anaerobic digestion rather than composting, since composting simply returns carbon dioxide to the atmosphere and generates a fertilizer compost while anaerobic digestion also produces methane, which can be used as a fuel to power the operation or to generate electricity.[21,28] Bioplastic compositions designed for compostability do not necessarily degrade well under anaerobic digestion (e.g. PLA), although many anaerobic digester facilities also have a final composting stage, which may obviate this problem.

5.5 Major types of biodegradable and compostable plastics used in packaging[4,7,9,30–35]

There has been a long learning period with the use of 'new' biodegradable bioplastics, such as PLA, PHAs, and starch- and cellulose-based materials, and it has required extended and still ongoing development programs to design usable formulations that minimize or obviate some of the property shortcomings and differences in processability of the bioplastics. Currently, biodegradable and compostable polymers are mostly used for products with short shelf life and storage times, particularly rigid and flexible food packaging. Biodegradable and biobased packaging is also appearing in the cosmetics and personal care industries, although in this case high aroma barrier and resistance to oily, greasy and moist products is also required to provide the extended shelf life necessary.

Major current commercial biodegradable plastic types are:

- *Polylactic acid (PLA)*: this is made from a biosourced monomer, lactide dimer (derived from lactic acid), by chemical polymerization.
- *Polyhydroxyalkanoates (PHA)*: this is a family of aliphatic polyesters, the most important being polyhydroxybutyrate (PHB), polyhydroxybutyrate-co-valerate (PHBV) and polyhydroxybutyrate-co-hexanoate (PHBH). Properties vary with the specific structure. These materials are generally made by direct fermentation of starch or plant-based fatty acids. There has also been considerable research on making them directly from transgenetically modified plants such as switchgrass and tobacco.
- *Biodegradable aliphatic and aliphatic/aromatic copolyesters*: the current major types are polybutylene succinate (PBS), polybutylene succinate-co-adipate (PBSA), polycaprolactone (PCL) and polybutylene adipate-co-terephthalate (PBAT). These polymers have been petrochemical-based up to now, but biobased sourcing is becoming available for PBS, PBSA and (for the butylene glycol and adipate monomers) PBAT.
- Starch blends and derivatized starch blends.
- Cellulose and cellulose derivatives and blends.
- Polyvinyl alcohol (PVOH) and blends.

PVOH is currently petrochemical-based, but a factory to make biobased PVOH from biobased vinyl acetate is starting up in China. In most cases, these materials are used as blends rather than single materials, for example PLA/copolyester, PLA/starch, starch/copolyester and starch/PVOH blends.

Many other biobased materials are under development, including protein-based compositions (e.g. keratin-, zein-, casein-, whey- and soy-based plastics) and bacterial-, algal- and fungal-based materials.

Suppliers can be classified into primary polymer producers and compounders who provide formulated compounds for thermoplastic processing. Some companies (e.g. Kingfa) do both. Most compounders make a range of materials derived from different base polymers.[4,35]

Some primary polymer producers are:

- *NatureWorks, Purac, Synbra, Hisun*: PLA.
- *Teijin, Purac, Sekisui Plastics*: PLA Stereoblends.
- *Tianjin Bio Green/DSM; Metabolix*: PHAs.
- *Tianan Biologic*: PHB, PHBV.
- *Biomer; Ecomann:* PHB.
- *Meredian, Kaneka*: PHBH.
- *BASF, Kingfa, China New Materials*: Aliphatic/Aromatic Polyesters.
- *Kingfa, Xinfu, Sinoven, Mitsubishi, Showa Denko, Hexing Aliphatic*: Polyesters.
- *Perstorp, Daicel*: Polycaprolactone.

Some compounders and converters are:

- *Cereplast, Biome/Biotec GmbH, Novamont, Cardia*: PLA blends; Starch blends.
- *Plantic, Rodenburg, Ingredion, Resirene Teknor Apex:* Starch blends.
- *FKuR*: PLA and cellulosic blends.
- *Cortec*: PLA and PHA blends.
- *BASF, Teknor Apex, RTP*: PLA blends.
- *Innovia Films*: Cellulosic films.

Neither of these lists is comprehensive.

5.5.1 Polylactic acid (PLA)[1,32–39]

Polylactic acid (PLA) polymerization

Polylactic acid (PLA) was the original 'synthetic' compostable bioplastic and is still the largest volume compostable synthetic plastic made today. Lactic acid is produced by microbiological processes, at present generally by cornstarch fermentation. Original attempts to homopolymerize it by conventional condensation polymerization gave only low molecular weight products because of back-hydrolysis caused by the liberated water. Consequently, the lactic acid is prepolymerized to form oligomers, which are then depolymerized to form lactic acid dimers. The dimers are then catalytically polymerized by ring-opening polymerization to form the high molecular weight PLA homopolymer. Other processes are under development to consolidate this to fewer stages or into a single fermentation step. Since lactic acid forms both D- and L-enantiomers, the lactide dimers can be D-D-, D-L- or L-L-forms and the polymers can have both L- and D-stereoisomeric structures.[33,37–41]

Polylactic acid (PLA) production

As previously noted in Chapter 1, the major polylactic acid (PLA) producer is NatureWorks, which has a 140 000 tonnes/yr plant in Blair, Nebraska, making PLA from corn (maize), which is sold under the trade name 'Ingeo'. Following a $150 million investment in the company (which was previously wholly-owned by Cargill) by PTT Chemical, the largest chemical company in Thailand, a second plant, which will more than double capacity, is coming onstream in Thailand in 2015, and will use sugar cane and cassava as feedstocks. Other producers are Purac, which is building a 75 000 tonne/yr lactide plant for PLA manufacture in Thailand, scheduled to begin production in 2012, and Hisun and some other smaller producers in China and Japan, so that 2015 worldwide capacity is expected to be about 400 000 tonnes. A survey by the Nova Institut gives projections as high as 800 000 tonnes/yr by 2020, with at least seven production sites with a capacity greater than 50 000 tonnes/yr.[42] Pricing is now comparable with many other packaging materials at less than $1/lb and about $0.85/lb under contract. Purac and Teijin are also producing D-, L-stereocomplex materials for high temperature applications.[43–45]

Polylactic acid (PLA) processing and properties[34,46]

Polylactic acid (PLA) crystallinity and other properties are highly dependent on the stereoisomer composition, but present commercial grades are mostly L-, D-copolymers containing 2–10% of the D-stereoisomer, which acts as a nucleating agent for the L-stereoisomer and increases the crystallization rate. Commercial PLA is a highly transparent, glossy rigid thermoplastic with many attractive properties for film, sheet, thermoformed, injection- and blow-molded products and fibers, lying between crystal polystyrene and PET. It has high tear strength like polystyrene and is easily printable and sealable. However, a major disadvantage is its low glass transition temperature (Tg) (~57°C) and significantly lower temperature resistance than PET, since the heat distortion temperature of the poly-L-lactide is about 60°C, making it unsuitable for hot-cup applications. This can be increased up to about 190°C by the incorporation of fillers or nucleating agents, blending with other polymers as noted below, and forming a more highly crystalline stereocomplex by controlling the stereochemistry and blending the poly-D-lactide with the poly-L-lactide (although at higher cost). A 1:1 L-, D-stereocomplex blend provides the maximum temperature stability. High heat distortion temperature versions have now become commercially available.[44,45]

PLA is compostable under industrial composting conditions (EN 13432, ASTM D-6400),[24] but not by home composting, and it is not readily anaerobically digestible, soil biodegradable or marine biodegradable. Biodegradation of the D-polylactide isomer is slower than for the L-isomer, because of the higher crystallinity.

PLA shows a poor barrier towards oxygen, carbon dioxide and moisture vapor, although it provides a high flavor and aroma barrier and high resistance to grease and oil.[47] Although the gas barrier properties can be improved by plasma SiO_x deposition, they are still not as good as PET (Chapter 4, Section 4.3.8). Co-extrusions incorporating PLA as one component have been used to make barrier materials, but are not generally biodegradable, although they can have 60 to 80% renewable content because of the PLA.[48] However, EVAP in the UK has recently developed compostable transparent multilayer films with high barrier to oxygen and carbon dioxide and high renewable content, which are constructed of PLA-based outer layers and an undisclosed biodegradable inner barrier layer. These films have structures such as PLA/Tie Layer/Barrier Polymer/Tie Layer/FKuR BioFlex A 4100 CL (a PLA-based blend), which gives a compostable structure certified under EN 13432.[49,50] These structures are breathable to water vapor and provide significantly increased shelf life for products such as bread and vegetables. Since about 90% of the carbon footprint of packaged foods derives from the food itself, not the packaging, use of this packaging can ensure reduced food waste and carbon footprint.

Commercial PLA has suffered from poor processability and low melt strength at the extrusion temperature of 180–210°C, which is considerably lower than for conventional polymers. It has low thermal conductivity (like PET) but a higher density than PP and PE.[51,52] Apart from the problems of low heat distortion temperature and poor barrier properties noted above, other initial problems with PLA have included:

- Brittleness, poor impact strength and low temperature flexibility.
- Physical aging (ductile to brittle transition in 24 hours) due to low Tg.
- Slow crystallization (Tm 140–185°C, depending on stereochemistry).
- High sensitivity to hydrolysis by atmospheric moisture.[53]

PLA melt strength decreases with molecular weight, so that efficient drying is required before processing. PLA must be dried at 65–85°C to 200 ppm for short residence time processes and to 50 ppm for long residence time processes to avoid molecular weight degradation, sheet embrittlement and sagging, and strict storage conditions are required for pellets and processed parts to prevent moisture pickup.

Most of these problems have been alleviated or eliminated by the development over the past ten years of PLA formulation and compounding technology which provides enhanced impact strength, lower brittleness and increased flexibility to make the materials more like PET, and there is continual development of PLA compounds with improved properties. Many of these compounds have incorporated other biodegradable polymers, such as biodegradable polyesters and thermoplastic starch, polymeric additives such as impact modifiers and chain extension agents, plasticizers, nucleating agents and functional fillers and reinforcements.[54,55]

A wide range of enhanced-property formulated PLA compounds and of additives for PLA systems is now commercially available and only a few examples can be given here. One of the earliest modifications was to incorporate low

Tg aliphatic and aliphatic/aromatic polyesters such as PBS and PBAT, which can eliminate physical aging and low temperature brittleness problems and reduce moisture sensitivity,[56,57] while barrier properties towards oxygen and carbon dioxide can be increased to levels better than PP but not as good as APET without a barrier layer.[58] These improved formulations were translucent or opaque and had lost the excellent clarity of the base PLA. However, several companies, including FKuR and Biome, have now developed such materials that retain at least contact clarity. Commercial suppliers of PLA compounds include FKuR, Biome (Biotec GmbH), BASF, Novamont, Cortec, Heritage Plastics, Cardia, Kingfa and Cereplast.

Other PLA additives employed for property improvement in packaging applications include Rhein Chemie BioAdimide 100, to enhance hydrolytic stability and increase service life, and BioAdimide 500 XT, which is a chain extender used to increase melt viscosity and facilitate melt processing.[59] Other chain extenders are available from BASF (Joncryl)[60] and Goulston.[61] Arkema Biostrength[61] additives are incorporated to increase melt strength and melt elasticity to facilitate film blowing and foaming and improve deep draw thermoforming, which processes require high draw ratios or controlled, rapid expansion of the melt. Arkema core-shell impact modifiers increase impact resistance, including low-temperature impact resistance, and also reduce chipping and cracking during thermoforming. Other additives can be used to decrease the tensile and flex modulus for blown film production. Incorporation of 9% of DuPont Biomax® Strong 120, which is an ethylene copolymer, produces nine-fold improvement in flex fatigue of amorphous PLA, increased elongation, and a significant brittleness improvement in crystallized and oriented PLA sheet, as well as reducing processing energy while still maintaining contact clarity.[63] PLA melt strength can also be improved by 50 to 100% by the incorporation of 2–4% of linear high molecular weight acrylic or methacrylic copolymers, which are highly miscible with PLA and give optically transparent blends.[64]

Other additives available are heat distortion temperature improvers and crystallization rate accelerators. For example, Teknor Apex Terraloy biodegradable PLA compounds are about 90% biobased and meet FDA food contact requirements. They are designed for extrusion and thermoforming and avoid the usual tradeoff with PLA compounds between heat distortion temperature and impact strength. Previously developed compounds either had high impact strength or higher HDT, not both at once. Typical data are HDT 100°C and Izod impact strength 69 J/m compared to HDT 55°C and Izod impact strength 16 J/m for standard extrusion-grade PLA.[65]

NatureWorks has also introduced a high-heat, high-impact grade called Ingeo™ 3801X, which incorporates an impact modifier, crystallization accelerant, and reinforcing and nucleation agents and has made the formulation details openly available to Ingeo™ customers to facilitate its use and modification for injection-molded products. Ingeo™ 3801X is 80% biobased (71% PLA resin, 10% impact modifier, 9% crystallization accelerant, 9% reinforcing agent, 1% nucleant) and gives injection-molding cycle times similar to styrenic resins. Mechanical properties show a modulus of 3.10 GPa and notched Izod impact strength greater than 107 J/m, while the heat distortion temperature is up to 120°C.[66]

PLA can be blended with many other polymers, such as PBS, polycarbonate (PC) and polymethyl methacrylate (PMMA). NatureWorks has licensed US Patent No. 5,883,199,[57] which covers blends of PLA with aliphatic polyesters, including PBS from Metabolix (which previously licensed it from the University of Massachusetts). In February 2012, NatureWorks and BioAmber, a US-based bio-succinic acid producer, formed a joint venture called AmberWorks, to develop biodegradable PLA/PBS blends and NatureWorks is presently sampling developmental grades for injection-molding and thermoforming. Although PLA blends with PC and PMMA are not biodegradable, they can be used for making durable goods with increased biobased content. Thus, in April, 2012, NatureWorks and the Arkema subsidiary Altuglas International, a PMMA producer, introduced a new Plexiglas Rnew PLA/PMMA alloy. This material is extremely attractive and crystal clear, exhibiting a single Tg, and could be suitable for high-quality rigid packaging. Many other bioplastic developments based on PLA are underway around the world.

Polylactic acid (PLA) packaging applications[6,7,9,32,67]

Polylactic acid (PLA) packaging films have had major problems of brittleness, thermal stability and high processing costs and can be unacceptably noisy due to the PLA stiffness and crystallinity. Frito-Lay Sun Chips packaged in compostable multilayer PLA bags were so noisy they were removed from the market.[68] These bags were 90% PLA and comprised a 20 micron thick, 6-layer film construction of PLA/print layer/adhesive/metal/PLA/PLA sealant and were certified compostable according to ASTM D6400.[69] To rectify the noise problem, the multilayer structure has been reformulated using a more rubbery adhesive layer and Toray is also developing a quieter PLA film using a comonomer to make the polymer more amorphous.

Biaxially-oriented PLA films, such as Tafhleef Industries' NATIVIA™, include both transparent and metalized films with thicknesses of 20 to 40 μ and are heat sealable. BOPLA films are currently used for flexible packaging of fresh produce and bakery goods, frozen foods, snacks and confectionary packaging, cookies and cereal and nutrition bars, and for lidding films. Non-food packaging includes toys, office supplies and window films for envelopes. Future improvements are expected in packaging noise reduction, barrier films, shrink films and thinner films down to 15 μ. Processing conditions required to make BOPLA are different from BOPP because of the hygroscopic nature and moisture sensitivity of PLA, the different melt strength, stretch ratios and thermal properties, and differences in the wind-up due to the high PLA film modulus.[70,71]

PLA use is expanding in both rigid and flexible fresh food packages, such as clamshells and bags for pre-washed salads. Clear Lam has introduced flexible modified PLA/polyolefin laminates for cheese wrapping. The development of higher heat-resistant and high-impact PLA allows it to be used as a renewable replacement for PS and PP in high-heat thermoformable applications, such as microwaveable frozen food trays, hot food take-out containers and hot beverage cups lids. PLA is also being used in applications previously employing high-impact polystyrene (HIPS). Examples are use in yogurt cups by Stonyfield Farms in the US and Danone Activia yoghurt in Germany beginning in May 2011.[72] Since these products are generally stored and sold cold, the low PLA heat distortion temperature is not a problem and food-contact approved additives are used to reduce brittleness. The low PLA film shrinkage can be counteracted by applying a silicone layer to it to improve demolding, stacking and destacking. Danone projects that use of PLA will decrease the product packaging carbon footprint by 25% and use 43% less fossil-fuel resources than polystyrene packaging, according to a life-cycle assessment made by the Institute for Energy and Environmental Research in Heidelberg, Germany. Danone now plans to expand its use of PLA to include other Activia products, such as drinks, yogurt fruit puree and larger size packaging. PLA is also used for thermoformed cold drink cups, such as the Fabri-Kal Greenware and the Pactiv EarthChoice cups. It has similar shrink to PET but processing adjustments were necessary because of the much lower melting temperature and greater stickiness than PET, the need for precise process temperature control, and the extreme sensitivity to contamination.

PLA, like PET, can also be injection stretch blow-molded.[52,73] In contrast to PP and HDPE, which are blow-molded closer to their melt temperature, the preform reheat temperature is 15–25°C above the Tg. PLA strain hardens during orientation and its temperature sensitivity results in a narrow processing window so that a reheat additive (to help absorb energy from the heat lamps in the reheat process) may be incorporated in the injection-molded preform. PLA has a natural stretch ratio rather lower than PET but can often be run successfully on existing PET equipment. It also flows well into detailed mold designs producing attractive containers. Modified PLA blends with increased melt strength can also be extrusion blow-molded. Despite early hopes by the industry, PLA is not a good polymer for bottles compared with PET, because of its poor barrier properties, low heat distortion temperature and the controversy over contamination of the PET recycle stream. Consequently, NatureWorks is no longer promoting PLA to replace PET in carbonated soft drink and water bottles. However, PLA is starting to displace PET in some thermoforming applications.

There has also been considerable work on PLA foam packaging, and technology has now been developed to avoid the problems of the poor rheological properties and low melt strength of PLA to produce foams with moderate mechanical and thermal properties using branching and chain extension agents and specialized extrusion equipment.[74,75] Foam types include extruded foams, using direct gas injection or chemical blowing agents, which can then be thermoformed,[76] and batch foaming processes to make expanded PLA (EPLA), which produce pre-expanded beads that can then be molded into the final article using heat or steam (similar to expanded polystyrene (EPS)). The mechanical and thermal conductivity properties of EPLA are similar to those of EPS. Crystallinity, molecular weight and melt elasticity are all important factors in the foaming process, with amorphous and higher molecular weight PLA types giving lower foam densities. Synbra

Technology has developed low-density EPLA foams called BioFoam®, which use carbon dioxide as the blowing agent.[77] Sekisui Plastics has introduced Bioceller™ heat-resistant PLA bead foam based on D-, L-stereocomplexes[78] while Sealed Air and Dyne-a-Pak have introduced PLA foam trays[79] and Cereplast has developed a PLA formulation for compostable foam trays.[80]

5.5.2 Polyhydroxyalkanoates (PHAs)[4,32–34,81–83]

Polyhydroxyalkanoates (PHAs) are a family of aliphatic polyesters. Those currently commercially available include PHB, PHBV and PHBH. By manipulation of the molecular structure, properties ranging from hard, brittle plastics to soft elastomers can be obtained with Tg in the range −20 to +5°C. PHB is a hard, brittle material like polystyrene. PHBV has physical and optical properties similar to LDPE, HDPE or PP, depending on the chemical structure. PHBH is more like PET.

Polyhydroxyalkanoate (PHA) polymerization[33,34,82,84]

Polyhydroxyalkanoates (PHAs) are currently manufactured by bacterial fermentation directly from starch from genetically-modified crops. Polymerization occurs during the fermentation process and no additional synthesis steps are required. The bacteria deposit the PHA polyesters in their cell structures as energy storage reserves, and the resulting polymers only need extraction and isolation. This is different from the situation with PLA, where lactic acid is made by fermentation and then chemically polymerized as the lactide dimer. The bacterial fermentation is currently performed as a batch process, since this generates higher intracellular PHA concentrations than continuous processes. The cells are then separated by centrifugation and filtration and polymer extracted either by solvents or by solvent-free systems. The separation and extraction methods have a significant effect on the final product purity and suitability for food contact uses. Plasticizers and additives, such as nucleants, can also be incorporated in the final products to aid processability Because of the batch manufacturing and the complexity of the synthesis and separation processes, PHAs are more expensive than many other biopolymers, currently at about $2.50/lb. Methods for direct production of PHA in transgenic plants such as switchgrass and tobacco have also been investigated but are far from commercial as yet. Processes are also under development for the production of PHAs from non-food sources such as from agricultural, organic and industrial waste[85] and waste water.[86]

Polyhydroxyalkanoate (PHA) structure, properties and processing

The monomer building blocks of polyhydroxyalkanoates (PHAs) can be branched or unbranched 3-, 4- or 5-hydroxyalkanoic acids or homologs with substituted side chains. In addition, the PHAs can be homopolymers or copolymers of these various monomers and can have a broad range of molecular weights, and polymer blends can be made with controlled microstructures. Consequently the PHA family includes a wide variety of biopolymers with different property combinations.

PHB has been the most investigated and has a linear isotactic structure and high crystallinity of 60–70%, so that it is too brittle for many applications. It also provides only a small processing window between its melting and decomposition temperatures, and can be degraded by high moisture levels or excessive temperatures or residence time in the processing equipment. Secondary crystallization and plasticizer loss over time can also lead to degradation in mechanical properties. Consequently, PHBV copolymers are more often used than PHB, since the longer side chain imparts greater ductility and lower Tg and crystallinity High levels of valerate comonomer over about 40% make the polymer soft and sticky and unsuitable for processes such as injection-molding. Hexanoate copolymers containing about 10% hexanoate are flexible and ductile and above 15% become soft and elastic, so that this intermediate range is suitable for processes such as blow-molding, film blowing and casting, thermoforming and extrusion coatings.

PHAs are semi-crystalline (0–60%), crystallize more rapidly than PLA and have a high crystalline melting point (80–175°C). They exhibit higher impact resistance and lower temperature flexibility than PLA and also have considerably higher heat distortion temperatures up to 110°C. Furthermore, PHAs are much less moisture sensitive than PLA on storage and processing and are resistant to oils and hot water. Their barrier properties towards moisture, oxygen and carbon dioxide are better than PLA, particularly for PHBH.

Also, unlike PLA, PHAs are industrially compostable, home compostable, soil and marine biodegradable, anaerobically digestible and biodegrade in wetlands, septic systems and municipal solid waste systems. Testing at OWS has shown that Metabolix Mirel PHA is anaerobically biodegradable under ASTM D5511 high-solids test conditions at 52°C. After only 15 days, 100% biodegradation relative to the cellulose control was reached and generated six times more biogas per unit weight than typical biomass substrates, including food and municipal organic waste. Even when run at 37°C, this test showed 70–100% biodegradation in 42 days.[87,88] Mirel also meets the ASTM D7081 standard for marine biodegradation and is Vincotte certified 'OK Compost Home and OK Compostable'.[89]

PHAs, especially PHBV and PHBH, have good properties for packaging because of their high temperature stability, ability to gain FDA food contact approval, and their ability to thermoform with high detail reproduction accuracy and sharpness. However, they require high crystallinity and must be demolded above 60°C; insufficient crystallinity is attained if the crystallization stage is passed through too quickly in the thermoformer and the final thermoformed article stability is only reached after several days.[90] PHAs have broad packaging-use potential wherever degradability is useful, including agricultural, horticultural and single-use food-service applications. PHAs can also be blended with other biopolymers, for example with PLA to increase its temperature resistance, with starch and chitosan.[91–94]

PHAs generally injection-mold well and can also be reheat stretch blow-molded on PET equipment to make single-serve and other containers. The material properties, such as crystallization rate and mechanical properties, can be tailored to the application, since PHAs are more rubbery than PLA or PET. Reheat stretch blow-molding is different from injection stretch, since in injection stretch both the preform and bottle are produced on a single machine, whereas in reheat stretch they are made separately. PHAs are also potential replacements for PE and PP in extrusion blow-molding applications and PHBV shampoo bottles were introduced in the US and Germany in the mid-1990s but subsequently removed from the market, probably because of price.[73,95–97]

Current PHA producers include Metabolix, Ecomann, Tianan, Biomer, Tianjin Green/DSM, and Meredian. Further information on PHAs is given in Chapter 11.

5.5.3 Aliphatic and aliphatic/aromatic copolyesters[4,32–34,98–101]

Unlike aromatic polyesters such as PET, aliphatic polyesters and aliphatic-aromatic copolyesters are readily biodegradable and compostable. Many types are commercially available aside from PLA, and the PHAs are described above and polyglycolic acid (PGA) was previously discussed in Chapter 4, Section 4.3.4. These aliphatic and aliphatic-aromatic polyesters include:

- *Polybutylenesuccinate (PBS)*: made by Showa Denko (Bionolle) SK Chemical, Kingfa, Mitsubishi, BioAmber, Sinoven (DNP Green), Hexing and Xinfu.
- *Polybutylene succinate adipate (PBSA)*: made by Showa Denko, Kingfa, Hexing and Mitsubishi.
- *Polycaprolactone (PCL)*: made by Perstorp and Daicel.
- *Polybutylene adipate-co-terephthalate (PBAT)*: made by BASF (Ecoflex), Kingfa and China New Materials.

Small amounts of chain extenders are used in the condensation polymerization of the diols with the diacids, to make PBS, PBSA and PBAT with high molecular weight. In contrast, PCL is made by catalytic ring-opening polymerization of ε-caprolactone.

PCL is semi-crystalline and melts at around 60°C. It is semi-rigid, has a modulus similar to LDPE, low tensile strength and high elongation at break (i.e. bubble-gum). It is soluble in a wide range of solvents and compatible with many polymers. PBS, in contrast, melts at about 110°C, has a Tg of about −30°C and crystallinity of 35–45%. Mechanical properties are similar to HDPE or PP. The introduction of the adipic acid comonomer in PBSA decreases its melting point to about 90°C, its Tg to about 45°C and its crystallinity to 20–35%. Flex modulus is about 330 MPa compared to about 650 MPa for PBS and the elongation at break is higher while the tensile strength is lower. PBAT shows good mechanical and thermal properties when the proportion of terephthalic acid comonomer is greater than 35%, but its biodegradability is markedly decreased at levels of 55% or higher.

These polyesters have properties generally similar to polyolefins. However, they are relatively expensive (>$2/lb) and are mostly used in combination with other polymers such as PLA and starch as components of biodegradable plastic formulations, particularly for film uses. They are still mostly petrochemical-based, although biobased routes are becoming available including partially biobased PBS from Showa Denko and Sinoven, while PTT and Mitsubishi are building a 20 000 tonnes/yr plant for bio-PBS in Thailand. BASF also makes partially biobased PBAT (Ecoflex FS), but is not selling it commercially except in formulated compounds.

5.5.4 Thermoplastic starch and starch blends

Natural starches are crystalline mixtures of linear amylose and branched amylopectin polymers, and the amylose/amylopectin ratio depends on the origin of the starch, the type of plant it is derived from (corn, potato, tapioca, pea, etc.) and whether the plant has been genetically modified. Starches can be chemically modified and derivatized by many different methods and many engineered thermoplastic and water-soluble starches are now commercially available.[102–105]

Starches are environmentally friendly and among the cheapest of biodegradable polymers. However, they suffer from high sensitivity to water and poor mechanical properties, and because of their high crystallinity and hydrogen bonding levels, they generally degrade at a lower temperature than the melting point, making it difficult to process them as thermoplastics without modification. The starch structure can be disrupted by heating with water (cooking) and modification with polyhydric alcohols such as glycerol and sorbitol, carboxylic acids and anhydrides, polyglycols, epoxides, ureas and many other plasticizers and reagents to produce thermoplastic and plasticized starches, which can be processed by standard thermoplastic methods such as extrusion.[104,106]

Such thermoplastic and plasticized starches, as well as chemically-modified and derivatized starches such as acetylated, oxidized or hydrolyzed starches and high-amylose starch, can be advantageously blended with other biodegradable polymers such as aliphatic and aliphatic/aromatic polyesters, PLA, PHAs and PVOH to make bioplastic compounds that have useful mechanical properties, are not brittle, and can be extrusion blown or cast into film or sheet, extrusion coated, thermoformed and injection-molded for food packaging, food service and other applications. A wide range of property combinations can be tailor-made for specific applications. Depending on the other compounding ingredients, these formulations are generally biodegradable and industrially compostable. However, starch blends of this type can suffer from problems of moisture sensitivity and limitations on the amount of starch that can be incorporated in the formulation without causing undesirable deterioration of physical or thermal properties or increase in melt viscosity.[4,32,54,58,93,107,108]

These types of starch blend were first commercialized by Novamont and Biotec GmbH. Other current producers are Biome, Cereplast, Cardia, Plantic, Teknor Apex and Rodenburg. They can be used for a wide variety of film products, including compostable waste bags. Some recent innovations are inflatable pillow packaging (Cardia/Sealed Air) and compostable bubble wrap (Cardia).[109] Plantic have developed a range of thermoformable starch blend products based on hydroxypropylated high-amylose starch for dry food packaging.[110] In general, the higher the amylose level in the starch, the higher the strength and elongation of the blends obtained. As described in Chapter 4 (Section 4.3.5), starch also provides a high barrier to oxygen and carbon dioxide and Plantic have commercialized barrier packaging based on hydroxypropylated high-amylose starch.

A second type of starch blend system has been developed by Cereplast, Cardia, Teknor Apex and Resirene. In these cases, the starch, generally at a 30–50% level, is blended with polyethylene, PP, polystyrene or ethylene-vinyl acetate copolymers (EVA) to form a two-phase system which retains most of the properties of the matrix polymer. The aim here is to use the biobased starch to increase the renewable content of the blend and reduce the content of petrochemically-based polyolefin or EVA. Because the starch has a higher density than the polyolefin, even at the 50% level by weight, it is completely encapsulated by the polyolefin. Cereplast uses an inexpensive raw starch but incorporates a compatibilizing agent such as maleated PP in the formulation.[111] Teknor Apex uses a thermoplastic starch with a controlled small particle size, which is blended with the polyolefin by a specially designed process licensed from Cerestech.[112–114]

Starches are also used for foam structures commonly used in protective packaging. An example of a starch composite foam is the Novamont Mater-Bi Wave resilient closed-cell foam sheet. This can be used as a replacement for polystyrene and polyethylene foams and has good antistatic properties. The starch is expanded with water, extruded into sheets and then cut into the desired shape. Densities from 35 to 400 kg/m^3 are available.[115]

5.5.5 Cellulosics and cellulosic blends[4,32–34,98,116]

Cellulose is also a plant polysaccharide containing long, linear chains of cellobiose units. Because of its crystallinity, insolubility and infusibility, it must be chemically modified to make it useful as a packaging plastic, often by esterification. Two types of cellulosic material are used in packaging for fast-moving consumer goods: cellulosic films and compounded cellulosic bioplastics.

Renewable cellulosic (cellophane) films are made from wood pulp by reaction with alkali and carbon disulfide and then neutralization in acid and film casting. These films are rapidly compostable (generally home compostable) and have excellent deadfold properties. They also have good barrier properties towards oxygen, carbon dioxide and aromas, but show poor barrier towards moisture vapor unless moisture barrier coatings are applied. Consequently they have often been used in applications where moisture breathability is advantageous. Despite the high permeability of uncoated cellulose films to water vapor, they still provide an excellent barrier to aroma and flavor components and microorganisms. Use of cellulose in barrier packaging systems is also described in Chapter 4 (Section 4.3.6). Cellulose films also show advantages such as inherent dead fold and anti-static properties, high gloss and transparency, resistance to grease and oils and a wide heat-seal range. See Chapters 1 and 11 for more on market developments in these sectors.

A major supplier of cellulosic films is Innovia Films (www.innoviafilms.com) under the brand name NatureFlex. Innovia has developed unique proprietary coating technologies for cellulosic films. For example, its new NatureFlex NK cellulosic film, which incorporates a small amount of PVdC to increase the moisture barrier, is transparent, flexible, compostable, high-barrier, printable and heat-sealable. Its moisture vapor barrier properties are the best of any currently available biopolymer films and are close to those of oriented PP. Applications include twist-wrap, flow wrap, overwraps, laminations and vertical form fill and seal uses for moisture sensitive products. Other types of NaturFlex film are also used in laminates with PLA formulations for food packaging, including stand-up pouches. NatureFlex™ NVR, a two-side coated, heat-sealable film with an intermediate moisture barrier and high mineral oil barrier, is highly suitable for box overwrap and individual flow wrap applications such as tea packaging.[117]

Cellulose esters can be made with varying degrees of substitution, but both their mechanical properties and susceptibility to biodegradation deteriorate as the degree of substitution (DS) is increased while their solubility in polar solvents increases. Major types are the acetates, followed by propionates and butyrates, often in the form of mixed esters. Commercial cellulose acetate is made by reaction of cellulose (wood pulp) with acetic anhydride and has a DS from 1.7 to 3.0, most commonly about 2.0 to 2.5, and a DS value of about 2.5 allows solubility in acetone. Cellulose acetate has a very high Tg (~190°C), which results in a narrow processing window, so that it must be cast as a film from solution or compounded with plasticizers such as phthalates, or triphenyl or tricresyl phosphates for flame retardancy, or other low molecular weight components for it to be processed as a thermoplastic without undergoing degradation. The propionates and butyrates both have wider processing windows than the acetates and are compatible with a wide range of plasticizers, as well as forming miscible blends with a variety of aliphatic polyesters.[118]

Thermoplastic cellulosic blends are supplied by FKuR under the name Biograde. They are biodegradable and have high renewable content. Optical, mechanical and modulus properties are comparable to polystyrene and the blends are food contact approved. They exhibit extremely high heat resistance of up to 115°C and are primarily designed for injection-molding applications (including thin-wall molding and long flow paths) and for sheet extrusion and deep-draw thermoforming on standard equipment.[119]

5.5.6 Protein-based bioplastics

There has been considerable development work over the past five years on protein-based bioplastics including their use in packaging, since they show high barrier properties towards oxygen and moisture. Plant- and animal-based proteins have been employed and the use of whey proteins as barrier materials is described in Chapter 4 (Section 4.3.7).[120–124]

5.5.7 Polyvinyl alcohol (PVOH)

Polyvinyl alcohol (PVOH) is discussed in Chapter 4 (Section 4.2.10). It is biodegradable, particularly in solution, but so far petrochemically-based. However,

Inner Mongolia Mengwei (a subsidiary of Wanwei Updated High-Tech Material Industry Co.) in China, is reported to be building a 50 000 tonne/yr plant to make biobased PVOH. The most direct route for renewable sourcing is conversion of biobased ethylene (from biobased ethanol) to biobased vinyl acetate, polymerization to biobased polyvinyl acetate and then hydrolysis to biobased PVOH.[125]

5.6 Biodegradable plastics processing[52,126,127]

The introduction of the new biodegradable plastics described above inevitably caused challenges for processors, who did not want to modify their existing equipment or drastically alter processing conditions. Some biodegradable plastics are relatively easy to process, such as PLA blends with copolyesters, where little modification may be necessary except for using lower processing temperatures than for conventional polymers. Others, such as uncompounded PLA and formulations containing PVOH and starch, can be more difficult because of problems with moisture sensitivity and degradation by heat and shear. In these cases, low-shear processing, avoiding high-shear screw mixing sections and high-speed screws, good temperature control and good drying may be essential. With many biodegradable materials, low melt strength may also cause problems for making blown film and cast film and sheet and extra cooling may be needed. Bag making also generally requires more careful temperature control than for conventional polymers such as polyolefins and the processing window is narrower.

For injection-molding, the biodegradable plastics generally process on standard equipment using medium to high injection speeds, but good temperature control is vital for the mold and to avoid either degradation (if it is run too hot or the material is left in the machine for too long) or freeze up (if it is run too cold or the flow front is not kept moving). In addition, because of the low melt strength, most biodegradable polymers must be injection stretch blow-molded and conventional PET machines can be used, although at lower temperatures of 155–180°C, with the mold being warmed to avoid condensation. For all of these processes, efficient purging is critical, since the biodegradable plastics should not be left hot in the equipment when it is not running and are not compatible with the polyolefins which may normally be used in the equipment. Higher modulus materials, such as PLA blends, can be thermoformed on standard equipment, but at lower temperatures than for PP or PS. Shrinkage is lower than for PP, more comparable to PET. Since the biodegradable plastics are generally polar, printing does not usually require corona treatment or other surface modification.

5.7 Major packaging uses for biodegradable and compostable plastics

The most suitable packaging applications for biodegradable and compostable plastics are those where the biodegradability or compostability provides significant benefits, collection is easy (e.g. from large restaurants, stores and institutions, the 2012 London Olympic Park, etc.) or there are tax incentives or legal constraints (as in Italy).[16] These applications include:

- Single-use compostable refuse bags for organic waste (e.g. kitchen and yard waste).
- Single-use compostable thermoformed and injection-molded food service products, food trays and food and beverage cartons or marine-biodegradable versions for marine use.
- Grocery carrier bags.
- Agricultural and horticultural uses, such as fertilizer bags, potato bags, plant pots, seed trays and mulch films.
- Loose-fill packaging.

Large amounts of starch and cellulose-based packaging were used in the 2012 London Olympics, all compliant with EN13432. This allowed the food packaging, service-ware and food waste to be sent together to an in-vessel composter (or anaerobic digester) without the need for separation of the food scraps.[29,128]

5.8 Biobased plastics

As discussed in Sections 5.4 and 5.5, 'biodegradability' of plastics refers specifically to end-of-life processes and biodegradable plastics are not necessarily 'biobased', which refers to the origin of the carbon atoms.

The US Department of Agriculture defines biobased plastics as organic plastics composed wholly or significantly of recently fixed (new) carbon from biological sources, such as renewable plant, forestry, animal, algal, bacterial or marine materials, based on C^{14} content measurement as defined by ASTM D6866.

Such biobased plastics are not necessarily biodegradable (e.g. nylon-11 and 'green' polyethylene), but can be biodegradable (e.g. starch and cellulose), compostable (e.g. PLA and PHAs), durable or recyclable.

5.8.1 Biobased measurements and certifications

Biobased measurement and certification is based on C^{14} content as determined by the ASTM D6866 test method. Carbon has three isotopes: C^{12} (stable) constitutes about 99% of all carbon, while C^{13} (also stable) is about 1%. C^{14}, which is radioactive and has a half-life of 5568 years, is present to only about one part per trillion in atmospheric carbon but is continually renewed by cosmic radiation. There is constant exchange between atmospheric carbon and animal and plant life, which keeps its C^{14} content in balance. However, once an organism dies, the exchange stops and the C^{14} level diminishes at a precise rate due to radioactive decay and can be used for radiocarbon dating. Fossil carbon sources, such as petroleum deposits that are many millions of years old contain no C^{14} since it has all long since decayed. Consequently, the proportion of new carbon in a material can be determined by obtaining the C^{14} content by mass spectrometry or scintillation counting.

This analysis can be important for meeting the requirements for the BioPreferred, DINCertco, Vincotte and other certifications (see below) and also for legal reasons. For example, since 2005, the German packaging ordinance has exempted certified compostable biopolymers containing at least 75% renewable content from take-back obligations by suppliers, and it is highly likely that such special regulations covering bioplastics will increase in scope so that the renewable percentage will become more important. However, there are issues with the ASTM D6866 method in that it relates only to biogenic carbon and does not consider other elements such as hydrogen, oxygen or nitrogen. Consequently, natural minerals used as inorganic fillers (e.g. calcium carbonate) are classified as not biobased since they contain no C^{14}, while conversely glass-filled biocomposites can qualify as 100% biobased since only the carbon content is considered. Blends of bioplastics can also give ambiguous results, since the renewable carbon levels of the components cannot be individually measured and corrections must be applied to allow for the carbon contents of the individual components to calculate an overall carbon content for the blend.[22,129,130]

The incidence of voluntary certification systems for products made from renewable resources is also growing. The USDA 'BioPreferred Program', which was launched in February 2011, sets a minimum level of 25% new carbon for most products for them to be considered biobased. The first 'Certified Biobased Product' labels were awarded to 11 companies in March 2011, and by March 2012 there were 212 certified biobased bioplastic products (23% of all certified products) from 57 companies.[131]

The US Department of Energy also has a sustainability goal of obtaining 25% of industrial chemicals from biological processes by 2025. Today, over 90% of raw materials for chemical-based products come from petroleum or natural gas. Meeting this goal will necessitate efficient and economic production of renewable chemicals, monomers and polymers, which are identical to existing petrochemical counterparts. These biobased chemicals, monomers and polymers can then be dropped into existing syntheses, processes and recycling infrastructure using current know-how and manufacturing facilities to provide products identical to those already available from petrochemical sources.

In Europe, the DIN Certco (Germany) and Vincotte (Belgium) programs both use ASTM D6866 for evaluation. DIN Certco provides quality marks for different biobased content ranges (20–50%, 50–85% and >85%). There must also be a minimum organic content as determined by loss on ignition. The Vincotte certification mark represents the renewable carbon level using different numbers of stars on the left of the logo (1 star, 20–40%, two stars, 40–60%, 3 stars, 60–80% and 4 stars >80%). Extension of these methods using new tests to measure isotopic ratios of hydrogen, nitrogen and oxygen is also under consideration.[130]

While not part of the D6866 test specification, methods are also available, such as the Picarro cavity ring-down laser absorption spectroscopy, which can measure C^{13}/C^{14} ratios and use these to determine the origin of components in materials such as biodegradable and biobased chemicals, monomers and plastics. This technique can distinguish between different types of plant origin, for example between wheat, corn (maize) and potato starch and also between biobased components and those derived from fossil sources such as petroleum and natural gas.[132]

5.8.2 Evolving feedstocks for biobased plastics

Much of the original work on biobased plastics involved polymers which are also biodegradable and their production was based on starches, either for chemical or biological synthesis (e.g. PLA and PHAs) or in the formulation of starch blend compounds. Over the past five years, with increasing oil prices and rising consumer 'green' concerns, there has been expanded effort in the development of economic processes for the synthesis of biobased monomers (i.e. from biobased feedstocks) and biobased conventional polymers, rather than specialized biodegradable polymers such as PLAs and PHAs. However, the commercialization of biobased polymers still has to overcome hurdles such as product quality and development of economies of scale to compete with the petrochemical products, which have been under development for many years. Present emphasis has now turned to non-food crop feedstocks, and particularly cellulosic biomass, to avoid competition with food sources, although this is much more of a consideration in the production of biofuels, particularly ethanol, since worldwide fuel consumption is an order of magnitude higher than plastics.[133]

Current biobased feedstocks can be summarized as follows:

- *Polysaccharides*: starches, sugars, pectin, chitosan, gums, etc.
- *Cellulosics and lignocellulosics*: wood, straw, stover, plant biomass, and other non-food sources.
- *Fats and oils*: animal fats (fatty acid glycerides), vegetable oils such as soy, castor, palm, waxes and triglycerides, and algal oils.
- *Proteins*: materials under development include animal proteins such as keratin, casein, whey, gelatin, collagen, spider silk, meat and bone meal; vegetable proteins such as soy, zein, gluten; and algal and fungal proteins.
- *Carbon dioxide*: this is perhaps the ultimate in non-food biobased feedstocks. Routes to aliphatic PCs by catalytic reaction with epoxides are under development, as well as algal photolytic conversion of carbon dioxide to carbohydrates and hydrocarbons. Carbon dioxide is also a potential precursor of ethylene and propylene.[134] Carbon monoxide is also being used as a precursor.[135]

5.8.3 Biobased monomers

The list of monomers and chemicals for which synthesis routes are available or under development from the feedstocks listed above is now becoming extremely long, as is the number of companies involved. Some of the more relevant ones for future plastics packaging biopolymer development and production are listed below. However, because of the rapid worldwide developments in this field, these lists have changed and grown longer almost weekly over the past two years and no claim is made to completeness.

a. Those already available commercially or in pilot plant quantities:

Olefins:

- *Ethylene*: Braskem, Dow/Mitsui, India Glycols, Solvay.
- *Isobutylene*: Gevo, Butamax.

Acids and derivatives:

- *Lactic acid*: Cargill, Purac, Myriant, Galactic – Total, and others.
- *Succinic acid*: DSM, BioAmber, Roquette, Mitsubishi, Mitsui, Myriant, METEX, PTT, BASF.
- *2–5-Furan dicarboxylic acid and furanic esters*: Avantium.
- *Castor oil/12 hydroxy stearic acid*: Arkema, India Glycols, Suzhou Hipro.
- *Amino undecanoic acid*: Arkema.

Glycols and polyhydric alcohols:

- *Ethylene glycol from ethylene*: Braskem, India Glycols.
- *Propane 1,3 diol*: Braskem, DuPont/Tate & Lyle, ADM, BASF/Oleon.
- *Butanediol*: Genomatica, LanzaTech (from carbon monoxide), Myriant, BioAmber, Mitsui, DuPont, Tate & Lyle.
- *Sorbitol/Isosorbide*: Cargill, ADM, Roquette.
- *Xylitol*: Reluceo, Cargill, ADM, Roquette.
- *Glycerol*: Multiple producers.
- *Soy-based polyols*: Dow, Cargill, Bayer.

b. Those under development:

Olefins:

- *Propylene*: Braskem, Arkema, DuPont/Tate & Lyle, DSM, Dow, Mazda.
- *Butylene*: Genomatica (from n-butanol).
- *Vinyl chloride*: Solvay (via ethylene).
- *Isoprene*: Goodyear, Genencor.
- *Butadiene*: Versalis/Genomatica/Novamont, LanzaTech, Synthos.

Acids and acid derivatives:

- *Terephthalic acid*: Virent (from plant sugars and cellulose via p-xylene), Gevo (from bio-isobutanol via p-xylene), Annelotech (biomass via p-xylene), and Amyris (formerly Draths) (from glucose via muconate esters).
- *Adipic acid*: Verdezyne (from alkanes and fatty acids using yeast strain), Rennovia (from glucose via glucaric acid).
- *Sebacic acid*: Verdezyne, Arkema.
- *Malic acid*: Novozymes.
- *Aspartic, glutamic, glucaric, citric, ascorbic and itaconic acid*: multiple producers.

- *Levulinic acid*: Sergetis.
- *Dodecanedioic acid*: Verdezyne.
- *3-Hydroxy propionic acid*: Cargill, Codexis.
- *4-Hydroxybutyrolactone and Ketals*: Segetis, DSM.
- *Acrylic acid*: Ceres, Dow, Arkema, OPX, BASF-Cargill-Novozymes.
- *Methyl methacrylate*: Gevo (via isobutylene), Dow (via ethanol), Arkema.
- *Caprolactam*: Amyris (formerly Draths) (via lysine or muconic acid).
- *Caprolactone*: Verdezyne, Amyris (formerly Draths) (via muconic acid).

Diamines:

- *Hexamethylene diamine*: Verdezyne, Rennovia, Amyris (formerly Draths) (via muconic acid).

Many of these biobased chemicals and monomers are themselves intermediates in the production of further biobased derivatives. An important example is succinic acid, which can be derived from plant sugars and can then be transformed into a wide variety of other chemicals and polymers. Those of interest for packaging polymers include:

- 1,4-butanediol for copolymerization with succinic and other acids to produce aliphatic polyesters such as PBS, PBSA and PBAT.
- PTMEG and polyester polyols.
- Adipic acid and hexamethylenediamine for conversion to PBSA, PBAT and PA-6,6.

BioAmber operates the world's first biosuccinic acid production plant in Pomacle, France, which has a 350 000 liter fermenter. In addition, together with Mitsui, it is building a 17 000 tonne/yr plant in Sarnia, Ontario, to start up in late 2013 and plans to expand this plant to make 34 000 tonnes/yr of succinic acid together with a 23 000 tonnes/yr butanediol in late 2014, the latter based on Evonik catalyst technology.[136]

Other examples of platform intermediates are bio-isobutylene, which can be converted into p-xylene (via iso-octene) for terephthalic acid production, to methyl methacrylate and to polyisobutylene and butyl rubber; and muconic acid from plant sugars, which can be converted into terephthalic acid, caprolactone and polycaprolactone, caprolactam and polyamide-6, adipic acid and hexamethylene diamine to make polyamide-6,6, cyclohexane dimethanol for making glycol-modified PET (PETG), and chemicals.[137]

5.9 Biobased polymers for packaging

Given this plethora of biobased starting materials and monomers already available or under development, a major increase in commercial biobased plastics that can be used in packaging is anticipated over the next decade, along with many others for non-packaging uses such as engineering plastics. This, of course, will depend on such factors as development of efficient catalysts and economic biobased feedstock conversion and isolation systems, changes in petrochemical prices, and ability to produce biobased monomers at competitive prices to petrochemical counterparts. Some biobased materials of primary interest for packaging are discussed below.

5.9.1 Polyolefins: biobased polyethylene, polypropylene and polyvinyl chloride

Braskem went onstream in September 2010 with a 200 000 tonnes/yr bio-polyethylene plant in Triunfo, Rio Grande do Sul, Brazil (a $290 million capital investment). This uses ethanol from sugar cane to manufacture its 'I'm Green' plant-based polyethylene, which is now shipped worldwide for HDPE and LLDPE applications. According to Braskem, its feedstock production has no impact either on the Amazon rainforest or on global food production. The current level of 200 000 tonnes/yr requires only 1.7% of Brazilian ethanol production and 0.02% of Brazilian arable land, and the sugar cane growing areas are 2000 km from the rainforest and do not impinge on it. One hectare of land grows 82.5 tonnes/yr of sugar cane, which produces 7200 liters of ethanol, which is converted to 3 tonnes of 'green' ethylene and hence 3 tonnes of 'green' PE. Braskem will be expanding its 'green' PE production both in Brazil and elsewhere, but is taking care to avoid disrupting food supplies or impacting the environment and has instituted its Vision 2020 sustainability program to be the world leader in sustainable chemistry. Its biobased PE is chemically identical to petrochemical-based PE and is a 'drop-in' for standard PE processing, conversion and recycling processes.

Braskem has set up agreements with FKuR, Plantic and Tecnaro GmbH for new packaging and other applications in Europe and is also working with Coca-Cola (the HDPE version of the PlantBottle), P&G, Johnson & Johnson, Danone, Nestlé and others to supply 'green' PE. Braskem currently sells this biobased PE at a 50–60% price premium, because the production costs are presently somewhat higher than for petrochemical PE, but this will drop as more biobased PE capacity comes on-line and the technology is further developed.[138] A 50:50 joint venture of Dow and Mitsui is following with a 400 000 tonnes/yr sugar cane to bioPE plant, also in Brazil, which should come onstream in 2015. Since the whole production chain from sugar cane growing to PE manufacture will be controlled by the joint venture, this should be cost competitive with petrochemical PE.[139,140]

Other olefinic materials are also under development, including PP from ethanol and glycerol. Braskem is planning an ethanol-based PP plant and a second ethanol-based PE plant, although the construction timeline is at present uncertain. The capacity of the bioPP plant is projected to be about 30 000 tonnes/yr and will produce propylene from ethanol using existing technology via ethylene dimerization followed by metathesis. Future plants are expected to use improved technology resulting from the development of fermentation processes to make biopropanol, which is then dehydrated to propylene.[141] Both PE and PP made from sugar-cane ethanol have an extremely favorable carbon footprint and Braskem estimate that from cradle to polymer plant gate, the production of 200 million tonnes/yr of bioPE saves 920 000 tonnes of carbon dioxide per year compared to petrochemical PE, while the production of the same amount of bioPP would save 850 000 tonnes of carbon dioxide per year compared to petrochemical PP.[138] Mazda in Japan is also developing production of biobased PP from cellulosic biomass.

Other olefinic materials under development include polyvinyl chloride and polyvinyl acetate from bioethylene and PMMA from bioethylene and isobutylene. Solvay previously announced a 60 000 tonnes/yr bio-based ethylene plant for PVC production, but then delayed this with the 2008 economic crisis and has now restarted the project.[140] The Inner Mongolian Mengwei production of biobased PVOH (Section 5.5.7) also requires the production of biobased vinyl acetate as the starting point to produce biobased polyvinyl acetate which is then converted to PVOH.

5.9.2 Polyesters

Biobased PBS using bioderived succinic acid is already commercially available from Sinoven (DNP Green) and Showa Denko. Biobased polybutylene succinate adipate and polybutylene adipate are also under development, while BASF has developed partially biobased PBAT for internal use but is not selling it commercially except in formulated materials. Novamont is also developing its 'biorefinery' to make biodegradable polyesters and other materials from plant oils.[142]

Partially biobased PET, where the ethylene glycol component is biobased but not the terephthalic acid, has been developed by Coca-Cola for water and carbonated soft drink bottles – its PlantBottle® – and will be discussed in detail below. Polyethylene furanoate (PEF) has also been developed to the pilot plant scale by Avantium as an improvement on or replacement for PET, and will also be discussed in more detail later.

SK Chemicals Ecozen R is a new high-temperature copolyester for bottles, which contains an undisclosed biobased monomer and is currently 9–30% biobased. It has improved high temperature resistance (heat distortion temperature 85–110°C depending on the grade) to compete with heat-set PET, making it suitable for hot-fill and pasteurized applications and dishwasher-resistant containers, and could be a viable alternative to PC. It is also compatible with traditional PET and PETG.[143]

Partially biobased PTT is manufactured by DuPont from biobased propylene glycol and fossil-based terephthalic acid. Present applications are mostly in fibers and engineering resins, but packaging applications are also possible. Partially and wholly biobased polyamides have also undergone extensive development, but mostly for engineering resins rather than packaging uses. However, biobased PA-6 and PA-66 are certainly feasible to be made by biobased routes, such as PA-6 from biobased caprolactam via biobased lysine, and PA-6,6, via biobased adipic acid and biobased hexamethylene diamine.[144]

Present and potential manufacturers of biobased but non-biodegradable 'conventional' polymers are a roll call of major polymer companies worldwide. Those related to packaging materials include:

- *Braskem, Dow/Mitsui, DSM, Solvay, Arkema*: Polyethylene, PP, PVC.
- *Coca-Cola, Braskem*: PET & HDPE PlantBottle.
- *BASF, DSM, Arkema, DuPont, SK Chemicals*: Polyesters, polyamides.
- *Dow, Arkema*: Methacrylics.
- *Arkema, BASF-Cargill-Novozymes*: Acrylates.
- *DSM*: PCs.

These developments all show increasing synergy and common elements with biofuels development, although the product volumes will be smaller. Apart from PEF and PTT, these polymers can be considered 'conventional' rather than 'new polymers' and will be chemically identical to the petrochemical-based versions, except that their monomers are derived from renewable sources and plant products rather than from petroleum or natural gas.

5.9.3 Biobased routes to terephthalic acid

There have been intensive efforts to replace PET in bottles. Approaches tried have included biobased and biodegradable replacements such as PLA, which has been unsuccessful due to poor properties, particularly heat distortion temperature, end-of-life concerns such as lack of composting facilities, and contamination of the PET recycling stream. Attempts to drastically increase recycled content in PET bottles have also been uncertain because of unreliable availability of post-consumer PET bottles for recycling, particularly because of exports to China. Consequently, both Coca-Cola and PepsiCo are promoting development of 100% biobased PET to replace petroleum-based PET. This will have broad consequences not only for PET bottles but also for other PET products, such as fibers and films and for other plastics using terephthalic acid as a comonomer, such as polybutylene terephthalate (PBT), PTT and PBAT

PET contains two major monomers – ethylene glycol and terephthalic acid (Chapter 4, Section 4.2.4). Biobased ethylene glycol is already commercially available made from biobased ethylene by standard petrochemical processes. However, terephthalic acid is at present made from fossil-derived p-xylene. Multiple organizations are now working on biobased routes to terephthalic acid, particularly eventually from non-food biomass. Many of the routes under investigation actually involve the production of biobased p-xylene, which can then be converted to terephthalic acid and PET by the normal processes and facilities already used for the petrochemical-based materials.

Major companies involved include:

- *Virent*: Virent has developed its BioForming® Platform, which converts plant products such as corn, sugar cane and cellulosic biomass to mixtures of soluble plant sugars and cellulose derivatives. These are processed catalytically to reduce the oxygen content and to yield mono-oxygenated species such as alcohols, ketones and aldehydes. These compounds are then converted into non-oxygenated hydrocarbons, including bio-p-xylene as well as other aromatics such as benzene, toluene and biofuels, using conventional petrochemical continuous catalytic condensation and hydrotreating processes. The product mix also provides aromatic intermediates, which could be used to make biobased polystyrene, PCs and polyurethanes. A 10 000 gallon/yr pilot plant is in operation and a commercial plant is under development to be onstream in 2015.[145]

- *Gevo*: Gevo has developed a process for making biobased isobutanol. This is converted to isobutylene, which is then subjected to cyclodimerization and hydrogen abstraction to give p-xylene via iso-octene. Gevo has also developed routes for conversion of isobutylene to methyl methacrylate (for PMMA) and to polyisobutylene and butyl rubber.
- *Anellotech*: the Anellotech Catalytic Fast Pyrolysis technology, using inexpensive zeolite catalysts in a fluidized bed reactor, converts biomass (wood waste, corn stover, sugar cane bagasse, etc.) to a mixture of butene, toluene and xylenes (BTX), which is then isomerized to form p-xylene. This differs from the conventional petrochemical process only in the first stage, where the CFP process is used to generate BTP from biomass rather from naphtha.
- *Sabic*: Sabic's process converts biomass containing terpenes or terpenoids, such as limonene (from citrus fruit), to terephthalic acid.
- *Amyris (Draths)*: the Draths route avoids p-xylene and converts glucose to muconate esters and then to terephthalic acid. The goal for commercialization is 2015. This platform also provides routes to other important monomers such as caprolactam (for PA-6), hexamethylene diamine and adipic acid (for PA-6,6), caprolactone (for PCL), cyclohexane dimethanol (for PETG), and others.[137]

5.9.4 Polyethylene furanoate (PEF)

In addition to the above development work on biobased routes to terephthalic acid, Avantium, originally a Shell spin-off, has developed a process called 'YXY technology' to produce biobased polyethylene furanoate (PEF). This technology catalytically converts carbohydrates from plants, grains, energy crops, lignocellulosic matter, waste streams, waste paper or agricultural residues into a wide variety of bio-based polymers. Thus sugars from biomass are catalytically dehydrated to furane derivatives, which are then oxidized to furane dicarboxylic acid (FDCA). The furane dicarboxylic acid can then be polymerized with ethylene glycol to make PEF. Avantium considers PEF to be the next-generation polyester. It exhibits a higher Tg and better heat distortion temperature than PET, higher modulus than PET, six-fold better oxygen barrier, three-fold better carbon dioxide barrier and twice the barrier versus water vapor. In addition, Avantium claims that PEF is recyclable in regular PET streams, with 5% PEF not affecting the recyclability of PET, and that it has a smaller carbon footprint than petrochemical PET and bioethylene.[146] Production cost for the furane dicarboxylic acid is similar to terephthalic acid, and the PEF can be manufactured on existing PET production equipment and infrastructure.

Avantium currently produces PEF polyesters at its 40 tonnes/yr pilot plant in Geleen, The Netherlands, which opened in December 2011. PEF can be used to produce bottles by injection stretch and reheat stretch blow-molding using standard PET molds, and fibers and films using standard production equipment, and food contact testing is being carried out. The focus is on making a commercial PEF bottle for Coca-Cola and Avantium plans to expand to a 40 000 tonnes/yr commercial plant by 2015, to be producing on a world scale and be cost competitive to oil-based PET by 2018. It also partnered with Danone in March 2012 for the production of PEF bottles, with Teijin for the production of FDCA-based polyaramide fibers, with Rhodia for the production of FDCA-based polyamides for fibers, and with Solvay for the production of FDCA-based engineering polyamides.[147,148]

In mid-December 2011, Coca-Cola formed partnerships with Gevo, Virent and Avantium to develop a 100% plant-based resin by 2016 for Coca-Cola's PlantBottle® packaging. PepsiCo is also working on 100% plant-based PET.

5.10 Examples of adoption of biobased materials in packaging markets

Many large consumer and multinational companies have started to use biobased materials as part of their sustainability programs. Some major examples are:

- Johnson & Johnson, which began using biobased polyethylene containers for sunscreen products in Brazil in 2009.
- Procter & Gamble, which unveiled its long-term vision on 27 September 2010.[149] This includes topics involving packaging such as:
  - Use 100% renewable materials or recyclate for all products and packaging and eliminate use of all petroleum-based virgin plastic and PVC-based plastic.
  - Design products that maximize consumer satisfaction while maximizing resource conservation.
  - Use 100% renewable energy for manufacturing.
  - Have zero manufacturing or consumer waste going to landfill.

P&G's perceived challenges with bioplastics for packaging are that the performance/price ratio of bio-sourced resins is currently less favorable than for petroleum-based resins, while mainstream consumers will not pay more for bioplastics. A *Shoppers Trend Report* survey showed that while 46% of respondents are more likely to buy an eco-friendly product, the majority (59%) will not pay more for it.[150] Consequently, P&G's approach is to embed bioplastics in the early concept stage of new package development and to give preference to biobased 'conventional' resins (e.g. HDPE). This started with the use of bio-based HDPE packaging for 'Pantene' shampoo and 'Cover Girl' cosmetics in the USA in 2011. P&G also set up a co-operative effort with the US World Wildlife Fund (WWF-US) on the use of renewable materials, packaging, forestry, energy and water. Dr Jason Clay, Senior Vice-President of Market Transformation at WWF-US, stated that 'P&G's commitment to use renewable bio-derived plastic in its global beauty and grooming product packaging is an important step forward in its efforts to improve the environmental profile of its products.'

P&G packaging-related goals to be reached by 2020 include:

- Replace 25% of all petroleum-based materials with 'sustainably sourced' renewables.
- Reduce packaging. Current programs have reduced packaging materials usage by 300 000 tons since 2007.
- Treat environmental sustainability as part of the base business to earn gratitude, admiration and trust that lead to opportunity, partnerships and growth.

In May 2010, P&G also launched a sustainability scorecard to measure and improve the environmental performance of its key suppliers using similar measures to the Wal-Mart packaging scorecard introduced in 2007. This will assess suppliers' environmental footprints and encourage improvement by measuring year-to-year energy and water use, waste disposal and greenhouse gas emissions. Chapter 1 provides more information on developments by P&G and others in this sector and on the sustainability programs of leading fast-moving consumer goods producers.

5.10.1 Coca-Cola PlantBottle®

There are currently two types of PlantBottle®, one made from bio-based HDPE and the other from partially biobased PET. These are drop-in applications, since the biobased HDPE and PET are chemically identical to the petrochemical plastics.

The PET PlantBottle was first introduced in late 2009 in Denmark, Western Canada and the US Northwest, followed by Japan, and has now spread to 20 countries and been used for 10 billion bottles in two years. Coke's goal is for all its plastic packaging to be PlantBottle by 2020. Heinz has also licensed the PET PlantBottle for its 20 oz (600 ml) ketchup size in the USA and Canada, and all 20 oz Heinz Ketchup bottles in restaurants and sold by retailers now use PlantBottle packaging. This is the first change to the iconic Heinz bottle since it first moved to plastic in 1983. Volvic in Europe is also moving to partially biobased PET for its water bottles. In addition, Coca-Cola introduced 100% biobased HDPE PlantBottles in 2011 for Odwalla juice products, and bioHDPE PlantBottle packaging is now also used by Danone in Europe for Actimel yoghurt drinks and P&G in the US for shampoo and cosmetics. More information on the Coca-Cola PlantBottle® is given in Chapters 1 and 11.

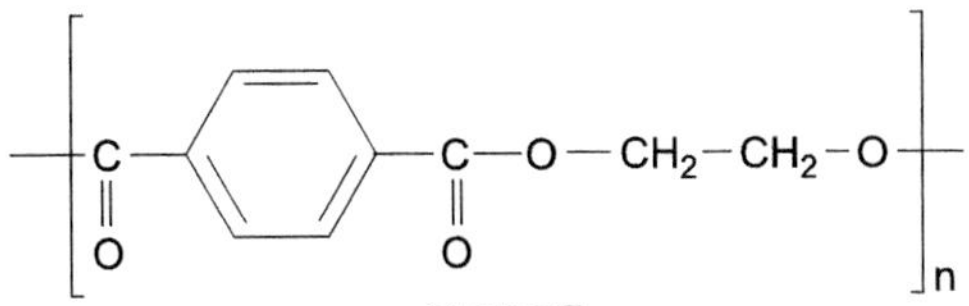

Fig. 5.1 Structure of biobased polyethylene terephthalate.

The partially biobased PET used in the present PlantBottle® contains up to 30% by weight (the ethylene glycol component) from renewable (sugar cane) sources, the exact proportion depending on whether the oxygen atoms in the chain backbone are considered to have come from the ethylene glycol or the terephthalic acid component. However, more strict accounting conforming to ASTM D6866, which requires summing the renewable carbon atoms coming from the ethylene glycol and expressing this as a proportion of the total number of carbon atoms, indicates that the current material actually contains 20% of renewable carbon and thus is 20% biobased.

Coca-Cola's goal is to eventually make fully recyclable PET and HDPE bottles from 100% renewable raw materials and plant-based waste. For PET, this requires development of a commercially viable biobased process for terephthalic acid, as has been described above (Section 5.9.3), and Coke is now providing funding to Gevo and Virent to accelerate this as well as to Avantium for the development of PEF bottles (Section 5.9.4) (Fig. 5.1).

Coca-Cola, Ford Motor, H.J. Heinz, Nike, and Procter & Gamble have formed the Plant PET Technology Collaborative (PTC) to build on Coca-Cola's current PlantBottle technology and accelerate the development of sustainably-sourced 100% plant-based PET and drive its use in a range of industries. The PTC also aims to promote the development of common methodologies and standards for plant-based plastics worldwide, including lifecycle analysis and universal terminology.[151]

5.10.2 PepsiCo

PepsiCo is also working on the development of 100% biobased PET with an unnamed biotechnology company and has made laboratory quantities from plant sources such as switchgrass, pine bark and corn husks. PepsiCo intends to have a pilot unit in operation in 2012. The goal is that future production will also use agricultural by-products from its food business, such as oat hulls, potato peels and orange peels, which it already has available.[152]

5.10.3 Some other developmental biobased packaging materials

Polycarbonates

Aliphatic polycarbonates (PCs) are under development based on carbon dioxide feedstock. Companies active in carbon dioxide-based polymers include Evonik,

BASF and Bayer. SK Innovation in Korea, Novomer and Empower Materials in North America, and Nomer in Norway are developing routes to aliphatic PCs, mostly the generation of propylene polycarbonate (PPC) by metal-catalyzed reactions with propylene oxide, while Bayer has made polyurethanes from carbon dioxide-based polyols and DSM has investigated use of carbon dioxide in the synthesis of PCs. PPC is biodegradable, transparent, has elasticity and shows a memory effect, and contains 43% carbon dioxide by weight. PPC polymers have also been demonstrated to be useful, biodegradable and biobased modifiers for PLA and PHB to decrease brittleness.[134,153,154]

Mushroom-based packaging
Ecovative Technologies has developed a mushroom root filament (mycelium)-based biomaterial called EcoCradle Mushroom Packaging and is working with Sealed Air to commercialize it. This uses a wide range of agricultural wastes as a feedstock, which is cleaned and inoculated with mycelium, which then grows through the feedstock and bonds it. While it is still growing, the material is transferred into molds and grows to the mold shape and size, after which the growth is stopped by dehydration and heat treatment. This material is viewed as a biodegradable alternative to expanded plastic packaging foams.[155]

Edible packaging
Edible packaging may represent the ultimate in sustainable packaging. Many common types are starch-, polysaccharide- or protein-based.[7,156,157] However, a recent development is WikiCells, which are transportable foods and drinks with edible containers, developed by David Edwards at Harvard. These currently include yogurts, soups and fruits designed to sit on store shelves without plastic packaging but with a thin edible packaging analogous to grape skin. The skins are gelacious natural food membranes, which can be made in a wide variety of shelf-stable forms composed of natural polymers such as chitosan and particles from chocolate, nuts and dried fruit, which are held together by ionic bonding. Some WikiCells also have an outer biodegradable shell, made for example from chocolate or sugar cane bagasse. The flavored skin is edible along with the contents and becomes part of the food, avoiding disposal and recycling problems. This type of packaging can contain liquids, emulsions, foams and solids.[158]

5.11 Major concerns with biobased plastics

Despite the impressive advances that are being made in biobased chemicals, monomers and polymers, there are still important issues that need to be resolved in any significant transfer to biobased from petrochemical-based plastics. These include:

- *Economics*: Can manufacture from biological feedstocks allow equivalent or better economics than petrochemical processes? The price of biobased monomers, polymers and other chemical products must be at least equivalent to those from petrochemicals in the long run to be viable. However, petroleum prices are likely to continue increasing and biobased processes have had a much shorter development time than the century available so far for petrochemical-based materials, so that further advances can be anticipated.

- *Use of foodstuffs as feedstocks*: To avoid any controversy over this, there is major work on routes from non-food biomass and crop waste to eliminate competition with food. Parallel concerns are the eutrophication of water supplies and soil from increased use of fertilizers and pesticides, and whether bioplastics really do reduce the carbon footprint relative to petroleum-based materials, given agricultural and fuel inputs to grow the feedstocks.[85,86,133,159,160]
- *Use of genetic modification*: of plants and other biological sources to increase disease resistance, improve yields and reduce costs. This is of particular concern in Europe but less so elsewhere.
- *For the most advantageous use of bioplastics*: as for petrochemical-based plastics, adequate recycling and other end of use infrastructures (such as composting and anaerobic digestion for biodegradable materials and conversion to fuel and/or energy for non-recyclable materials) will be necessary.

All of these concerns, except the last, equally apply to biofuels. The adoption of bioplastics addresses political, societal and environmental issues rather than technical issues, which are already adequately met by petroleum-based polymers.

5.12 Sources of further information

BELGACEM M. N. and GANDINI A. (2008) *Monomers, Polymers and Composites from Renewable Resources*, Amsterdam, Elsevier.

CHIELLINI E. (ed.) (2008) *Environmentally Compatible Food Packaging*, Cambridge, UK, Woodhead Publishing Ltd.

ENDRES H. J. and SIEBERT-RATHS A. (2011) *Engineering Biopolymers: Markets, Manufacturing, Properties and Applications*, Munich, Hanser.

FAKIROV S. and BHATTACHARYYA D. (eds.) (2007) *Handbook of Engineering Biopolymers, Homopolymers, Blends and Composites*, Munich, Hanser.

LENDLEIN A. and SISSON A. (eds) (2011) *Handbook of Biodegradable Polymers: Isolation, Synthesis, Characterization and Applications*, New York, Wiley.

MATHERS R. and MEIERS M. (eds) (2011) *Green Polymerization Methods: Renewable Starting Materials, Catalysis and Waste Reduction*, New York, Wiley.

PILLA S. (ed.) (2011) *Handbook of Bioplastics and Biocomposites Engineering Applications*, Salem MA, Scrivener Publishing – Wiley.

PLACKETT D. (ed.) (2011) *Biopolymers: New Materials for Sustainable Films and Coatings*, New York, Wiley.

RAVENSTIJN J. T. J. (2010) *The State of the Art on Bioplastics: Products, Markets, Trends and Technologies*, Lüdenscheid, Germany, Polymedia Publisher GmbH

THIELEN M. (2012) *Bioplastics – Basics, Applications, Markets*, 1st edition, Polymedia Publishers GmbH (an introductory book), ISBN 978-3-9814981-1-0.

VERGHESE K., LEWIS H. and FITZPATRICK L. (eds) (2012) *Packaging for Sustainability*, London, Springer.

YU L. (2008) *Biodegradable Polymer Blends and Composites from Renewable Resources*, New York, Wiley.

5.13 References

1. THIELEN M. (2012) *Bioplastics – Basics, Applications, Markets*, 1st edition, Polymedia Publishers GmbH, ISBN 978-3-9814981-1-0.
2. RAVENSTIJN J. T. J. (2010) *The State of the Art on Bioplastics: Products, Markets, Trends and Technologies.*
3. ENDRES H. J. and SIEBERT-RATHS A. (2011) Chapter 1 'Introduction' and Chapter 2 'State of Knowledge', in *Engineering Biopolymers: Markets, Manufacturing, Properties and Applications*, Munich, Hanser. Available from: www.hanserpublications.com/Products/290-engineering-biopolymers.aspx [Accessed October 2012].
4. SHEN L., HAUFE J. and PATEL M. K. (2009) 'Product overview and market projection of emerging bio-based plastics', PRO-BIP, Final Report November. Available from: http://en.european-bioplastics.org/wp-content/uploads/2011/03/publications/PROBIP2009_Final_June_2009.pdf [Accessed October 2012].
5. EUROPEAN BIOPLASTICS (2012) 'Fact Sheet: What are bioplastics?' August 2012. Available from: http://en.european-bioplastics.org/wp-content/uploads/2011/04/fs/Bioplastics_eng.pdf [Accessed October 2012].
6. KAEB H. (2011) 'Bioplastics in packaging', *Bioplastics Magazine*, 6(2), 46.
7. CHIELLINI E. (ed.) (2008) *Environmentally Compatible Food Packaging*, Cambridge, UK, Woodhead Publishing Ltd.
8. PILLA S. (ed.) (2011) *Handbook of Bioplastics and Biocomposites Engineering Applications*, Part 2, Chapters 5–8, Salem MA, Scrivener Publishing – Wiley.
9. GONTARD N. et al. (2011) 'Food packaging applications of biopolymer-based films', in *Biopolymers: New Materials for Sustainable Films and Coatings*, Plackett D. (ed.), Chapter 10, New York, Wiley, 211–32.
10. AULIN C. and LINDSTRÖM T. (2011) 'Biopolymer coatings for paper and paperboard', in *Biopolymers: New Materials for Sustainable Films and Coatings*, Plackett D. (ed.), Chapter 12, New York, Wiley, 255–76.
11. EUROPEAN BIOPLASTICS (2011) 'Fact Sheet: Renewable Resources for the Production of Bioplastics', September 2011. Available from: http://en.european-bioplastics.org/wp-content/uploads/2011/04/fs/Renewable_resources_eng.pdf [Accessed October 2012].
12. BELGACEM M. N. and GANDINI A. (2008) *Monomers, Polymers and Composites from Renewable Resources*, Amsterdam, Elsevier.
13. YU L. (2008) *Biodegradable Polymer Blends and Composites from Renewable Resources*, New York, Wiley.
14. CONSUMER GOODS FORUM GLOBAL PACKAGING PROJECT (2011) *A Global Language for Packaging and Sustainability*, September 2011, revised edition. Available from: http://globalpackaging.mycgforum.com/allfiles/FinalReport_2011.pdf [Accessed October 2012].
15. ACCENTURE (2012) 'Long-term Growth, Short-term Differentiation and Profits from Sustainable Products and Services'. Available from: www.accenture.com/SiteCollectionDocuments/PDF/Accenture-Long-Term-Growth-Short-Term-Differentiation-and-Profits-from-Sustainable-Products-and-Services.pdf [Accessed October 2012].
16. KAEB H. (2011) 'Bagislation in Europe', *Bioplastics Magazine*, 6(6), 28–31.
17. EUROPEAN BIOPLASTICS (2012) 'Production Capacity for Bioplastics'. Available from: http://en.european-bioplastics.org/market [Accessed October 2012].
18. FREEDONIA GROUP (2011) 'World Bioplastics – 2015 and 2020', study 2823, November 2011. Available from: www.freedoniagroup.com/brochure/28xx/2823smwe.pdf [Accessed October 2012].
19. LAIRD K. (2011) 'Sustainability as economic driver; bioplastics in full bloom', *PlasticsToday*, 14 December. Available from: www.plasticstoday.com/articles/sustainability-economic-driver-bioplastics-full-bloom [Accessed October 2012].

20. *plastemart.com* (2012) 'US Demand for Bioplastics to Reach 550 Million lbs in 2016, Global Production to Touch the Million Tonne Mark Soon', 14 July. Available from: www.plastemart.com/Plastic-Technical-Article.asp?LiteratureID=1825&Paper=us-bioplastics-demand-reach-550-mln-lbs-2016-global-production-million-ton-mark [Accessed October 2012].
21. ENDRES H. J. and SIEBERT-RATHS A. (2011) 'End-of-life options for biopolymers', in *Engineering Biopolymers: Markets, Manufacturing, Properties and Applications*, Chapter 5, Munich, Hanser. Available from: http://www.hanserpublications.com/Products/290-engineering-biopolymers.aspx [Accessed October 2012].
22. ENDRES H. J. and SIEBERT-RATHS A. (2011) 'The regulatory framework for biopolymers', in *Engineering Biopolymers: Markets, manufacturing, properties and applications*, Chapter 3, Munich, Hanser. Available from: http://www.hanserpublications.com/Products/290-engineering-biopolymers.aspx [Accessed October 2012].
23. INNOCENTI F. D. (2011) 'The role of standards for biodegradable plastics', *Bioplastics Magazine*, 6(4), 36–8.
24. EUROPEAN BIOPLASTICS E.V. (2009) 'Fact Sheet: Industrial Composting', November 2009. Available from: http://en.european-bioplastics.org/wp-content/uploads/2011/04/fs/FactSheet_Industrial_Composting.pdf [Accessed October 2012].
25. KISICO VERPACKUNGSTECHNIK GMBH (2012) 'Caps and closures from bio resources', *Bioplastics Magazine* 7(4) 32–3.
26. EUROPEAN BIOPLASTICS E.V. (2010) 'Fact Sheet: Home Composting', March 2010. Available from: http://en.european-bioplastics.org/wp-content/uploads/2011/04/fs/FactSheet_Home_Composting.pdf [Accessed October 2012].
27. ASTM INTERNATIONAL (2012) 'Work Item: ASTM WK35342 – New Specification for Home Composting of Biodegradable Plastics'. Available from: www.astm.org/WorkItems/WK35342.htm [Accessed October 2012].
28. EUROPEAN BIOPLASTICS E.V. (2010) 'Fact Sheet: Anaerobic Digestion', March 2010. Available from: http://en.european-bioplastics.org/wp-content/uploads/2011/04/fs/FactSheet_Anaerobic_Digestion.pdf [Accessed October 2012].
29. THE NATIONAL NON-FOOD CROPS CENTRE (NNFCC) (2012) 'London 2012 Olympic and Paralympic Games to feature food packaging made from plant starch, cellulose; all compostable packaging to be collected in separate, orange bins', Industry Intelligence Inc. 28 June. Available from: www.industryintel.com/news/read/3366938496/London-2012-Olympic-and-Paralympic-Games-to-feature.html?goback=%2Egde_160429_member_129039084
30. STATE OF VERMONT, OFFICE OF THE ATTORNEY GENERAL, IN RE FABRI-KAL CORPORATION (2012) 'Assurance of Discontinuance', 2 June. Available from: www.atg.state.vt.us/assets/files/Fabri-Kal%20AOD.pdf [Accessed October 2012].
31. VERESPEJ M. (2009) 'Recycling tests show sorting equipment can separate PLA from PET bottles', *Plastics News*, 2 July; 'NAPCOR concerned about the impact of PLA bottles on PET recycling', *Plastics News*, 24 July; and Davies S., 'PLA will be key part of sustainability', *Plastics News*, 4 September.
32. EUROPEAN BIOPLASTICS (2011) 'Fact Sheet: Better Packaging with Bioplastics', July/August. Available from: http://en.european-bioplastics.org/wp-content/uploads/2011/04/fs/Packaging_eng.pdf [Accessed October 2012].
33. ENDRES H. J. AND SIEBERT-RATHS A. (2011) 'Manufacturing and chemical structure of biopolymers', in *Engineering Biopolymers: Markets, Manufacturing, Properties and Applications*, Chapter 4, Munich, Hanser. Available from: http://www.hanserpublications.com/Products/290-engineering-biopolymers.aspx [Accessed October 2012].
34. ENDRES H. J. and SIEBERT-RATHS A. (2011) 'Engineering property profiles of biopolymers', in *Engineering Biopolymers: Markets, Manufacturing, Properties and Application*, Chapter 5, Munich, Hanser. Available from: http://www.hanserpublications.com/Products/290-engineering-biopolymers.aspx [Accessed October 2012].

35. ENDRES H. J. and SIEBERT-RATHS A. (2011) 'Market characterization for biopolymers' and Appendix, In *Engineering Biopolymers: Markets, Manufacturing, Properties and Applications*, Chapter 8, Munich, Hanser. Available from: http://www.hanserpublications.com/Products/290-engineering-biopolymers.aspx [Accessed October 2012].
36. AURAS R., LIM L., SELKE S. E. M. and HIDETO T. (eds) (2010) *Poly(lactic acid): Structures, Production, Synthesis and Applications*, New York, Wiley.
37. THIELEN M. (2012) 'Basics of PLA', *Bioplastics Magazine*, 7(1), 54–6.
38. HENTON D. E., GRUBER P., LUNT J. and RANDALL J. (2012) 'Polylactic Acid Technology'. Available from: www.jimluntllc.com/pdfs/polylactic_acid_technology.pdf [Accessed October 2012].
39. SÖDERGÅRD A. and INKINEN S. (2011) 'Production, chemistry and properties of polylactides', in *Biopolymers: New Materials for Sustainable Films and Coatings*, Plackett D. (ed.), Chapter 3, New York, Wiley, 43–63.
40. GROOT W., VAN KRIEKEN J., SLIEKERSL O. and DE VOS S. (2010) 'Production and purification of lactic acid and lactide', in *Poly(lactic acid): Structures, Production, Synthesis and Applications*, Auras R., Lim L. T., Selke S. E. M. and Tsuji H. (eds.), Chapter 1, New York, Wiley, 1–18.
41. SÖDERGÅRD A. and STOLT M. (2010) 'Industrial production of high molecular weight poly(lactic acid)', in *Poly(lactic acid): Structures, Production, Synthesis and Applications*, Auras R., Lim L. T., Selke S. E. M. and Tsuji H. (eds.), Chapter 3, New York, Wiley, 27–41.
42. NOVA-INSTITUT, BIO-BASED NEWS (2012) 'Growth in PLA bioplastics: a production capacity of over 800 000 tonnes expected by 2020'. Available from: www.nachwachs-enderohstoffe.info/nachricht.php?id=20120807-02 [Accessed October 2012].
43. HIRATA M. and KIMURA Y. (2010) 'Structure and properties of stereocomplex-type poly(lactic acid)', in *Poly(lactic acid): Structures, Production, Synthesis and Applications*, Auras R., Lim L. T., Selke S. E. M. and Tsuji H. (eds.), Chapter 5, New York, Wiley, 59–65.
44. FAMBRI L. and MIGLARESI C. (2010) 'Crystallization and thermal properties', in *Poly(lactic acid): Structures, Production, Synthesis and Applications*, Auras R., Lim L. T., Selke S. E. M. and Tsuji H. (eds.), Chapter 9, New York, Wiley, 113–24.
45. TSUJI H. and IKADA Y. (2008) 'Stereocomplex between enantiomeric poly(lactide)s', in *Biodegradable Polymer Blends and Composites from Renewable Resources*, Yu L. (ed.), Chapter 7, New York, Wiley.
46. PEREGO G. and CELLA G. D. (2010) 'Mechanical properties', in *Poly(lactic acid): Structures, Production, Synthesis and Applications*, Auras R., Lim L. T., Selke S. E. M. and Tsuji H. (eds.), Chapter 11, New York, Wiley, 141–53.
47. ALMENAR E. and AURAS R. (2010) 'Permeation, sorption and diffusion in poly(lactic acid)', in *Poly(lactic acid): Structures, Production, Synthesis and Applications*, Auras R., Lim L. T., Selke S. E. M. and Tsuji H. (eds.), Chapter 12, New York, Wiley, 155–79.
48. ZIRKEL L. (2012) 'PLA for thermoforming', *Bioplastics Magazine*, 7(2) 18–20.
49. EVAP (2012) 'EVAP Film'. Available from: www.evap.co.uk/site/index-2.html [Accessed October 2012].
50. Mctigue L. (2012) 'Tesco trials packaging that doubles shelf life of produce', *Packaging Digest*, 16 May. Available from: www.packagingdigest.com/article/print/521806-Tesco_trials_packaging_that_doubles_shelf_life_of_produce.php [Accessed October 2012].
51. DORGAN J. R. (2010) 'Rheology of poly(lactic acid)', in *Poly(lactic acid): Structures, Production, Synthesis and Applications*, Auras R., Lim L. T., Selke S. E. M. and Tsuji H. (eds.), Chapter 10, New York, Wiley, 125–39.
52. LIM L. T., CINK K. and VANYO T. (2010) 'Processing of poly(lactic acid)', in *Poly(lactic acid): Structures, Production, Synthesis and Applications*, Auras R., Lim L. T., Selke S. E. M. and Tsuji H. (eds.), Chapter 14, New York, Wiley, 189–215.

53. TSUJI H. (2010) 'Hydrolytic degradation', in *Poly(lactic acid): Structures, Production, Synthesis and Applications*, Auras R., Lim L. T., Selke S. E. M. and Tsuji H. (eds.), Chapter 21, New York, Wiley, 343–81.
54. YU L., PETINAKIS E., DEAN K. and LIU H. (2010) 'Poly(lactic acid)/starch blends', in *Poly(lactic acid): Structures, Production, Synthesis and Applications*, Auras R., Lim L. T., Selke S. E. M. and Tsuji H. (eds.), Chapter 15, New York, Wiley, 217–26.
55. DETYOTHIN S., KATHURIA A., JARUWATTANAYON W., SELKE S. E. M. and AURAS R. (2010) 'Poly(lactic acid) blends', in *Poly(lactic acid): Structures, Production, Synthesis and Applications*, Auras R., Lim L. T., Selke S. E. M. and Tsuji H. (eds.), Chapter 16, New York, Wiley, 227–71.
56. MCCARTHY S. P. (2006) 'Processing and characterization of bioplastics', presented at the *Plastics Technology Bioplastics Processing Conference*, 5–7 December, Charlotte NC.
57. MCCARTHY S. P., GROSS R. A. and MA W., University of Massachusetts, *US Patent No. 5,883,199*.
58. WHITCHURCH G. J. and COOPER T. A. (2006) 'Some recent advances in thermoplastically-processable biodegradable polyvinyl alcohol-, starch- and polylactide-based polymer blends', presented at the *International Degradable Plastics Symposium, Society of the Plastics Industry*, Chicago, IL, 16 June 2006.
59. RHEIN-CHEMIE (2012) 'Bioadimide™ in Bioplastics, Expanding the Performance of Biopolyesters'. Available from: www.bioadimide.com [Accessed October 2012].
60. BASF (2012) 'Chain Extender'. Available from: www.performancechemicals.basf.com/ev/internet/functionaladditives/en_GB/content/microsites/functionaladditives/chain_extender/index [Accessed October 2012].
61. KUTSENKO M., FOX S. and MONIZ D. (2011) 'Reclamation of polyester waste: cost effective improvement in IV and melt viscosity', presented at the *Society of Plastics Engineers Global Plastics Environmental Conference*, Atlanta, 17–19 October.
62. ARKEMA (2012) 'Biostrength'. Available from: www.arkema-inc.com/index.cfm?pag=1295&gclid=COnLnPfKlbICFQjf4Aod0EUAHw [Accessed October 2012].
63. CASARINO C. M. (2011) 'Clearing the way for wider PLA use', *Bioplastics Magazine*, 6(2), 16–17. Available from: http://biomax.dupont.com [Accessed October 2012].
64. LO C. and DONNELLY Z. (2012) 'Make PLA better: improving processing and properties', *Bioplastics Magazine*, 7(2), 36–7.
65. TEKNOR APEX (2010) 'High-heat, impact-resistant PLA', *Bioplastics Magazine*, 6(5) 25; 'New additives for PLA', *Bioplastics Magazine*, 5(4), 14.
66. NATUREWORKS LLC (2010) Press release: 'High Heat Injection Molding Breakthrough for NatureWorks Ingeo™ Bioresin', 14 January.
67. OBUCHI S. and OGAWA S. (2010) 'Packaging and other commercial applications', in *Poly(lactic acid): Structures, Production, Synthesis and Applications*, Auras R., Lim L. T., Selke S. E. M. and Tsuji H. (eds.), Chapter 28, New York, Wiley, 457–67.
68. BREGAR B. (2011) 'Frito-Lay exec: Sun Chips bag not a flop', *Plastics News*, 23 May. Available from: www.plasticsnews.com/headlines2.html?id=11052300101&q=bregar+pla+sun+chip+bag [Accessed October 2012].
69. DELIGIO T. (2010) 'Frito-Lay perseveres to develop biodegradable PLA bag', *PlasticsToday*, 22 April. Available from: www.plasticstoday.com/articles/frito-lay-overcame-numerous-challenges-development-biodegradable-pla-bag [Accessed October 2012].
70. TAGHLEEF INDUSTRIES (2010) 'New BOPLA film', *Bioplastics Magazine*, 5(3), 41; 'New developments in PLA packaging films', *Bioplastics Magazine*, 5(6), 24.
71. BRUECKNER GROUP (2010) 'How to produce BOPLA films', *Bioplastics Magazine*, 5(6), 25.
72. DANONE (2011) 'Danone and WWF introduce Activia in PLA cups', *Bioplastics Magazine*, 6(3) 32.
73. YODER L. (2011) 'Blow molding of bioplastics', *Bioplastics Magazine*, 6(4), 48–50.

74. PARKER K., GARANCHER J. P., SHAH S., WEAL S. and FERNYHOUGH A. (2011) 'Polylactic acid (PLA) foams for packaging applications', in *Handbook of Bioplastics and Biocomposites Engineering Applications*, Pilla S. (ed.), Chapter 6, Salem MA, Scrivener Publishing – Wiley.
75. MATUANA L. M. (2010) 'Foaming', in *Poly(lactic acid): Structures, Production, Synthesis and Applications*, Auras R., Lim L. T., Selke S. E. M. and Tsuji H. (eds.), Chapter 17, New York, Wiley, 273–91.
76. WEGNER J. E. and GROESELING M. (2010) 'Foaming agents and chain extenders for PLA foam', *Bioplastics Magazine*, 5(1), 22–3.
77. NOORDEGRAAF J. (2010) 'Cradle to cradle certified PLA foam', *Bioplastics Magazine*, 5(1), 24–5.
78. SEKISUI PLASTICS (2010) 'Heat-resistant PLA bead foam', *Bioplastics Magazine*, 5(1), 29.
79. KUNNEMANN D. (2010) 'PLA foam trays', *Bioplastics Magazine*, 5(1), 30–1.
80. KELLY W. and LARRIVEE L. (2010) 'A true compostable foam', *Bioplastics Magazine*, 5(1), 32.
81. CHODAK, I. (2008) 'Polyhydroxyalkanoates: origin, properties and applications', in *Monomers, Polymers and Composites from Renewable Resources*, Belgacem M. N. and Gandini A. (eds.), Chapter 22, Amsterdam, Elsevier.
82. POLLET E. and AVEROUS L. (2011) 'Production, chemistry and properties of polyhydroxyalkanoates', in *Biopolymers: New Materials for Sustainable Films and Coatings*, Plackett D. (ed.), Chapter 4, New York, Wiley, 65–86.
83. SUDESH K. and ABE H. (2010) *Practical Guide to Microbial Polyhydroxyalkanoates*, Shrewsbury, UK, Smithers-Rapra.
84. IENCZAK J. L. and FALCÃO DE ARAGÃO G. M. (2011) 'Biotechnologically produced biodegradable polyesters', in *Handbook of Biodegradable Polymers: Isolation, Synthesis, Characterization and Applications*, Lendlein A. and Sisson A. (eds.), Chapter 2, New York, Wiley, 23–44.
85. MARANG L., JIANG Y., TAMIS J., MORALEJO-GÁRATE H. and VAN LOOSDRECHT M. C. M. (2012) 'Microbial community engineering: producing bioplastic from waste', *Bioplastics Magazine*, 7(4), 22–4.
86. DE VEGT O., WERKER A., FETTER B., HOPMAN R., KRINS B. and WINTERS R. (2012) 'PHA from waste water', *Bioplastics Magazine*, 7(4), 26–8.
87. DARBY D. (2011) 'More responsible end-of-life options', *Bioplastics Magazine*, 6(4) 40.
88. KEAN R. (2012) 'Good for AD', *Bioplastics Magazine*, 7(2), 43–5.
89. WHITEHOUSE R. (2011) 'Advancements in marine biodegradable bioplastics', presented at the *Society of Plastics Engineers Global Plastics Environmental Conference*, Atlanta GA, 17–19 October.
90. ILLIG GMBH (2012) 'Thermoforming of bioplastics', *Bioplastics Magazine*, 7(2), 56–7.
91. JAVADI A., PILLA S., GONG S. and TURNG L. S. (2011) 'Biobased and Biodegradable PHBV-based polymer blends and biocomposites: properties and applications', in *Handbook of Bioplastics and Biocomposites Engineering Applications*, Pilla S. (ed.), Chapter 14, Salem MA, Scrivener Publishing – Wiley, 372–96.
92. CHEN G. Q. and HUA T. (2008) 'Polyhydroxyalkanoate blends and composites', in *Biodegradable Polymer Blends and Composites from Renewable Resources*, Yu L. (ed.), Chapter 8, New York, Wiley.
93. SHOGREN R. (2008) 'Starch-poly(hydroxyalkanoate) composites and blends', in *Biodegradable Polymer Blends and Composites from Renewable Resources*, Yu L. (ed.), Chapter 9, New York, Wiley.
94. CHEN C. and DONG L. (2008) 'Biodegradable blends based on microbial poly (3-hydroxybutyrate) and natural chitosan', *Biodegradable Polymer Blends and Composites from Renewable Resources*, Yu L. (ed.), Chapter 10, New York, Wiley.
95. DARBY D. (2010) 'Injection molded PHA: pens and pencils', *Bioplastics Magazine*, 5(3), 16.

96. TAYLOR K. (2010) 'PHA bioplastics for caps and closures', *Bioplastics Magazine*, 5(4), 18.
97. RADUSCH H. J. (2007) 'Unconventional processing methods for poly(hydroxybutyrate)', in *Handbook of Engineering Biopolymers, Homopolymers, Blends and Composites*, Fakirov S. and Bhattacharyya D. (eds.), Chapter 24, Munich, Hanser, 717–44.
98. VROMAN I. and TIGHZERT L. (2009) 'Biodegradable polymers', *Materials*, 2, 307–44.
99. ALBERTSSON A. C. and VARMA I. (2002) 'Aliphatic polyesters: synthesis, properties and applications of degradable aliphatic polyesters', in *Degradable Aliphatic Polyesters, Advances in Polymer Science*, 157, 2.
100. DENCHEV Z. Z. (2007) 'Biodegradation studies of polymer blends and composites comprising biopolymers', in *Handbook of Engineering Biopolymers, Homopolymers, Blends and Composites*, Fakirov S. and Bhattacharyya D. (eds.), Chapter 27, Munich, Hanser, 799–836.
101. DOI Y. and STEINBÜCHEL A. (eds) (2004) 'Polyesters III, applications and commercial products', in *Biopolymers*, vol. 4, Wiley-VCH.
102. BERTOLINI A. C. (ed.) (2010) *Starches: Characterization, Properties and Applications*, Boca Raton, FL, CRC Press; Hoboken, NJ, Wiley.
103. WURTZBURG O. B. (ed.) (1986) *Modified Starch: Properties and Uses*, Boca Raton, FL, CRC Press; Hoboken, NJ, Wiley.
104. JANSSEN L. P. B. M. and MOSCICKI L. (eds) (2009) *Thermoplastic Starch*, Boca Raton, FL, CRC Press; Hoboken, NJ, Wiley.
105. GOTLIEB K. F. and CAPELLE A. (2005) *Starch Derivatization*, Science Publishers.
106. ABBOTT A. P., BALLANTYNE A. D. and CONDE J. P. (2012) 'Salt modified starch: sustainable, recyclable plastics', *Green Chemistry*, 14, 1302–7, doi: 10.1039/c2gc16568f.
107. CZIGÁNY T, ROMHÁNY G. and KOVÁCS J. G. (2007) 'Starch for injection molding purposes', in *Handbook of Engineering Biopolymers, Homopolymers, Blends and Composites*, Fakirov S. and Bhattacharyya D. (eds.), Chapter 3, Munich, Hanser, 81; Mohd Ishak Z. A., Taib R. and Ishiaku U. S., 'Plastics filled with tropical starches: mechanical properties and degradation behavior', in *Handbook of Engineering Biopolymers, Homopolymers, Blends and Composites*, Fakirov S. and Bhattacharyya D. (eds.), Chapter 4, Munich, Hanser, 109.
108. GUPTA K. (2011) 'Starch-based composites for packaging applications', in *Handbook of Bioplastics and Biocomposites Engineering Applications*, Pilla S. (ed.), Part 2, Chapter 8, Salem MA, Scrivener Publishing – Wiley, 189–266.
109. GYCKYE L. (2012) 'Cardia launches compostable bubble wrap', *Packaging News*, 18 May. Available from: www.packagingnews.co.uk/news/cardia-launches-new-wrap/?utm_medium=email&utm_source=Packaging+News&utm_campaign=1364470_PN+DAILY+180512&dm_i=JLP,T8TY,2XY9KS,2E0BN,1 [Accessed October 2012].
110. PLANTIC TECHNOLOGIES LTD (2011) 'Starch Technology'. Available from: www.plantic.com.au/technology/innovation/starch-technologies.html [Accessed October 2012].
111. BAGRODIA S. (2008) 'Advanced materials from novel biobased resins', presented at the *Society of Plastics Engineers Global Plastics Environmental Conference*, Orlando, FL, 11–12 March.
112. INMAN H. (2010) 'Starch-based thermoplastics', *Plastics Engineering*, April, 42.
113. ANDERSON G. and TAM E. (2011) 'Recycled polyethylene and thermoplastic starch blend for extrusion blow molded bottle', presented at the *Society of Plastics Engineers Global Plastics Environmental Conference*, Atlanta, 17–19 October.
114. TEKNOR APEX (2010) 'Starch blends with enhanced performance', *Bioplastics Magazine*, 5(2), 22.
115. BIRON M. (2012) 'What's the latest in foams?' *SpecialChem*, 17 July. Available from: www.specialchem4polymers.com/resources/articles/article.aspx?id=7769 [Accessed October 2012].

116. BELGACEM M. N. and GANDINI A. (2011) 'Production, chemistry and properties of cellulose-based materials', in *Biopolymers: New Materials for Sustainable Films and Coatings*, Plackett D. (ed.), Chapter 8, New York, Wiley, 151–78.
117. LEBENSBAUM ULRICH WALTER GMBH AND INNOVIA FILMS (2012) 'Compostable film for organic tea', *Bioplastics Magazine*, 7(4), 36.
118. MOORE G. F. and SAUNDERS S. M. (1997) 'Advances in biodegradable polymers', *Rapra Review Reports No. 98*, 9(2), 25–8.
119. BONTEN C. (2008) 'Generation zero', *Bioplastics Magazine*, 3(5), 28–9.
120. GÄLLSTEDT M., HEDENQVIST S. and TURE H. (2011) 'Production, chemistry and properties of proteins', in *Biopolymers: New Materials for Sustainable Films and Coatings, Biopolymers: New Materials for Sustainable Films and Coatings*, Plackett D. (ed.), Chapter 6, New York, Wiley, 107–32.
121. REDDY M. M., MOHANTY A. K. and MISRA M. (2012) 'Proteineous meals for bioplastics', *Bioplastics Magazine*, 7(4), 37–9.
122. GREWELL D. (2012) 'Bioplastics from proteins', *Bioplastics Magazine*, 7(4), 40–1.
123. VERBEEK J. (2012) 'Bioplastics from the slaughterhouse: animal-based protein for thermoplastic products', *Bioplastics Magazine*, 7(4), 42–3.
124. SAMARASINGHE S., EASTEAL A. J. and LIN A. P. Y. (2007) 'Corn gluten meal-based blends and composites', in *Handbook of Engineering Biopolymers, Homopolymers, Blends and Composites*, Fakirov S. and Bhattacharyya D. (eds.), Chapter 18, Munich, Hanser, 537–73.
125. CCM INTERNATIONAL (2011) 'Wan-Wei updated high-tech expands PVA capacity', *Biomaterials China News*, 7. Available from: www.cnchemicals.com [Accessed October 2012].
126. WHITCHURCH G. (2007) 'The processing of bio-plastics', presented at *Biopolymers and Biodegradable Plastics: An Analysis of their Future Role*, British Plastics Federation, 3 October 2007.
127. KOVATS A. (2011) 'The handling of various forms of dry ingredients in bioplastics manufacturing and processing applications', in *Handbook of Bioplastics and Biocomposites Engineering Applications*, Pilla S. (ed.), Part 2, Chapter 2, Salem MA Scrivener Publishing – Wiley, 18–42.
128. NATIONAL NON-FOOD CROP CENTRE (NNFCC) (2011) Newsletter, Issue 22, 29 September. Available from: www.nnfcc.co.uk/publications/nnfcc-newsletter-issue-22.-the-london-2012-olympic-games/
129. CARUS M. and SCHOLZ L. (2010) 'How to measure the biobased content', *Bioplastics Magazine*, 5(3), 54.
130. KITZLER A. S., ENDRES H. J. and SCHETTLER A. (2012) 'Basics of the ^{14}C method', *Bioplastics Magazine*, 7(2), 50–3.
131. LEWIS K. (2012) 'BioPreferred®: creating new markets for bioplastics', presented at the *Society of the Plastics Industry Business of Plastics Conference*, Orlando, 2–5 April.
132. GREENE I. (2012) 'Protect your brand: fast, easy and inexpensive screening of bioplastics', presented at *Bioplastics: the Reinvention of Plastics*, San Francisco, June. Available from: www.picarro.com [Accessed October 2012].
133. HIGSON A. and AYLOTT M. (2012) National Non-Food Crops Centre (NNFCC) Briefing Document, 'The Changing Face of the Planet: the Role of Bioenergy, Biofuels and Biobased Products in Global Land Use Change', May. Available from: www.slideshare.net/AHigson/the-role-of-bio-based-plastics-on-global-land-use-change
134. HAAS G. R. (2009) 'Green and sustainable polymers: high-value aliphatic polycarbonates from CO_2 and epoxides', Doctoral Dissertation, University of Ulm, Germany.
135. BOMGARDNER M. M. (2012) 'Biobased chemicals without biomass', *Chemical & Engineering News*, 90, 25.
136. BIOAMBER (2011) 'BioAmber to open biochemical/bioplastics plant in Ontario', *PlasticsToday*, 30 August. Available from: www.plasticstoday.com/articles/bioamber-

open-biochemicalbioplastics-plant-ontario0830201102 [Accessed October 2012]. 'BioAmber offers further details on Mitsui succinic acid and BDO ventures', *PlasticsToday*, 8 November. Available from: www.plasticstoday.com/articles/bioamber-offers-further-detail-mitsui-succinic-acid-and-bdo-ventures-110820111 [Accessed October 2012]. 'Production campaign at BioAmber converts bio-succinic acid into biobased 1,4-BDO', *SpecialChem*, 28 March 2012. Available from: www.specialchem4adhesives.com/resources/latest/displaynews.aspx?id=5753 [Accessed October 2012].

137. FAULKNER B. (2011) 'Renewable monomers enable game-changing 100% biobased polymers', presented at the *Bioplastics: Re-Shaping an Industry Conference*, Las Vegas NV, 1–3 February.
138. KAHN J. A. (2012) 'Green polyolefins: commercial update on PE and green PP developments', presented at *BioPlastics: the Re-Invention of Plastics Conference*, San Francisco, 13–15 June.
139. PLASTICSTODAY (2011) 'Dow and Mitsui announce Brazilian biobased polyolefin venture', *PlasticsToday*, 21 July. Available from: www.plasticstoday.com/articles/dow-and-mitsui-announce-brazilian-bio-based-polyolefin-venture [Accessed October 2012].
140. JAGGER A., APLA (2011) 'Brazil grows biobased materials effort', *ICIS Chemical Business*, 7 November. Available from: www.icis.com/Articles/2011/11/07/9505416/apla-brazil-grows-bio-based-materials-effort.html [Accessed October 2012].
141. CASCONE R. (2012) 'Many paths to bio-polypropylene: which are real', presented at *BioPlastek 2012*, Arlington VA, 29 March.
142. SCOTT A. (2012) 'Italy's biotech surge', *Chemical & Engineering News*, 90, 22–4.
143. HYUN K. J. (2011) 'New bottle material', *Bioplastics Magazine*, 6(4), 18. Available from: www.polymertech.co.kr/images/freebowl_download/Ecozen%20Introduction%20ver2.3.pdf [Accessed October 2012].
144. THIELEN M. (2010) 'Basics of bio-polyamides', *Bioplastics Magazine*, 5(3), 51–3, and references therein.
145. KORMULA D. (2011) 'Completing the puzzle: 100% plant-derived PET', *Bioplastics Magazine*, 6(4), 14.
146. EERHART A. J. J. E., FAAIJ A. P. C. and PATEL M. K. (2012) 'Replacing fossil based PET with biobased PEF; process analysis, energy and GHG balance', *Energy Environmental Science*, 5, 6407–22.
147. MIRASOL F. (2012) 'Big potential for bioplastics packaging', *ICIS Chemical Business UK*, 30 April. Available from: www.icis.com/Articles/2012/04/30/9554300/big-potential-for-bioplastics-packaging.html [Accessed October 2012].
148. MAGNUS P. (2012) 'The world's next-generation polyester: 100% biobased polyethylene furanoate (PEF)', *Bioplastics Magazine*, 7(4), 12–15.
149. HOCKENSMITH D. (2010) 'P&G outlines broad sustainability plans', 4 October. Available from: www.plasticsnews.com/headlines2.html?id=10100400104 [Accessed October 2012]. See also: Boswell E. *et al.* (2012) (The Procter & Gamble Company, USA). PCT Int. Appl. WO 2012102778 A1 20120802, 21 August.
150. SPINNER J. (2012) 'Earth day study: most shoppers won't pay more for green packaging', *Packaging Digest*, 20 April. Available from: www.packagingdigest.com/article/521561-Earth_Day_study_Most_shoppers_won_t_pay_more_for_green_packaging.php [Accessed October 2012].
151. CALIENDO H. (2012) 'Five major US brands collaborating on plant-based PET', *PlasticsToday*, 5 June. Available from: www.plasticstoday.com/articles/Five-major-US-brands-collaborating-on-plant-based-PET-0605201202 [Accessed October 2012].
152. VERESPEJ M. (2011) 'Consumer preferences driving PepsiCo sustainability efforts', *Plastics News*, 11 August. Available from: www.plasticsnews.com/headlines2.html?id=22824&q=consumer+preferences+pepsico [Accessed October 2012].

153. SPECIALCHEM (2012) 'CO_2 polymers', 11 July. Available from: www.omnexus.com/news/news.aspx?id=30815&lr=dom12851la2&li=61011797#utm_source–L&utm_medium=EML&utm_campaign=dom12851la2 [Accessed October 2012].
154. LAIRD K. (2012) 'Green matter: carbon capture redefined; CO_2 as feedstock', *PlasticsToday*, 12 July. Available from: www.plasticstoday.com/blogs/green-matter-carbon-capture-redefined-co2-feedstock-071620121 [Accessed October 2012].
155. ANTHONY S. (2012) 'Evaluating claims for mushroom-based packaging', *Packaging World Greener Packaging*, 13 April. Available from: www.greenerpackage.com/compost_biodegrade/evaluating_claims_mushroom-based_packaging [Accessed October 2012].
156. FERNÁNDEZ-PAN I., IGNACIO J. and CABALLERO M. (2011) 'Biopolymers for edible films and coatings in food applications', in *Biopolymers: New Materials for Sustainable Films and Coatings*, Plackett D. (ed.), Chapter 11, New York, Wiley, 233–54.
157. EMBUSCADO M. E. and HUBER K. C. (eds) (2009) *Edible Films and Coatings for Food Applications*, London, Springer.
158. CAI S. L. (2012) 'Edible packaging', *Chemical and Engineering News*, 90, 56. Available from: www.seas.harvard.edu/news-events/calendars/biotechnology/wikicells-bottles-that-we-eat [Accessed October 2012].
159. PRASAD M. S. S. and KUMAR S. (2012) 'Bioplastics from agro waste', *Bioplastics Magazine*, 7(4), 16–17; Gonzaláz R. and Sibila M. A., 'Bread 4 PLA: biodegradable food packaging from bakery industry waste', *Bioplastics Magazine*, 7(4), 18–19; and Fernyhough A. and Markotsis M., 'Bioplastic products from kiwi waste', *Bioplastics Magazine*, 7(4), 20–21.
160. ENDRES H. J. and SIEBERT-RATHS A. (2011) 'Life-cycle assessment for biopolymers', in *Engineering Biopolymers: Markets, manufacturing, properties and applications*, Chapter 7, Munich, Hanser.

6

Innovations and trends in metal packaging for food, beverages and other fast-moving consumer goods

D. A. Abramowicz and L. Jenkins, Crown Packaging Technology, USA, K. Ambrose, I. Bucklow and T. Benge, Crown Packaging Technology, UK, B. Fields, Crown Packaging Technology, USA, G. Hughes, A. Noke, J. Bilko, A. Ioannides and C. Ramsey, Crown Packaging Technology, UK

DOI: 10.1533/9780857098979.153

Abstract: Metal packaging has enjoyed a rich history, providing easy access to nutritious, shelf-stable food and beverages for many decades. It has remained relevant through innovations that provide consumers convenience and other value-added features and brands' unique ways to display and market their products at the retail level. Innovations and key trends for metal packaging manufacturing processes, can body design and closure/end design are discussed. Current and emerging technologies for new metal container products, printing, coatings and materials are included, as well as a discussion of new metal packaging concepts and future trends. Strong environmental credentials and a steady stream of innovations will position metal packaging as a key format for protecting foods and beverages for the long term.

Key words: metal packaging, metal forming and shaping, printing and decoration, coatings and compounds, shelf differentiation/shelf presence, branding, easy opening, sustainability.

6.1 Introduction

Metal packaging is one of the most trusted of containers, with a rich history dating over 200 years. It is still a vibrant packaging format today, as new technologies provide consumers with convenience and other value-added features and brands with distinctive shelf-differentiation. This chapter includes a comprehensive look at the metal packaging industry, including manufacturing technologies used to produce the containers and lids (ends), materials used to coat, protect and decorate the containers, new products and new manufacturing processes. A discussion of new packaging concepts addresses the growing importance of cost-effective designs that speak to a range of consumer needs. A description of the strong environmental case for metal packaging is also included, as these containers and ends function as permanent resources on Earth. These solid environmental credentials and a steady stream of innovations position metal packaging as a key format for protecting foods and beverages for the long term.

Section 6.2 includes a description of different manufacturing processes used today and key manufacturing trends, as certain formats and materials are becoming more predominant due to a variety of market and economic factors. The emergence of new metal container formats and various can-shaping technologies for effective shelf differentiation is also covered.

Section 6.3 addresses the many diverse roles for coatings with metal packaging, including branding, informing the consumer, differentiation and providing long shelf life for the product. Current and emerging print methods are addressed in the context of consumer's and brand owner's needs. The discussion of internal coatings and sealing compounds includes the role of the market on material choice going forward and non-traditional materials such as polymer-coated cans.

Section 6.4 defines the importance of innovation across a range of metal packaging containers. This section has an emphasis on lids (the ends and closures used to seal containers after filling), as these innovations can be particularly important to the consumer. Developments in rigid and flexible food can ends, including lidding directly heat-sealed onto the can body, highlight the role of filler investment in driving future change. Metal closure developments include the emergence of composite closures and novel easy-to-open designs. Beverage end developments, driven by the need to reduce costs and/or improve consumer functionality, are addressed, as well as new metal packaging concepts responding to these same drivers.

Section 6.5 covers key packaging trends and the advent of new metal packaging technologies. The chapter closes with conclusions and commentary on the sustainability of metal packaging and its growing importance to society by reducing the resources needed to deliver safe, long-lasting, high-quality foods and beverages to the consumer.

6.2 Manufacturing technology developments

Cans date back to the nineteenth century when hand-made tinplate canisters were first used to preserve food. Food was literally cooked within the can by maintaining both can and contents at a high temperature to kill bacteria. The sterilized contents remained protected from further bacterial attack by the hermetic seal of the can. The principles of producing canned food are much the same today, although manufacturing quality and speed has been greatly increased as mass production techniques were developed by US companies such as American Can and Continental Can.

6.2.1 General manufacturing trends

Material trends

Metal packaging manufacturers have been iteratively reducing the weight of cans for many years. This makes good business sense, as metal is the largest part of the total package cost. Light-weighting has accelerated in recent years because of increasing interest amongst fillers, retailers and consumers in sustainability and minimization of packaging. Steel and aluminum producers have facilitated package weight reduction by finding ways to decrease metal thickness while increasing material strength and formability. With tinplate, techniques such as composition (steel alloy), additional cold rolling operations (double reduced) and continuous annealing have been developed to facilitate light-weighting. In similar ways, aluminum producers have improved the cost effectiveness of aluminum can stock through improved strength and technology to roll thinner metal. For both metals, thickness variation has been reduced and better uniformity of material properties achieved with higher quality, permitting metal packaging producers to remove material while maintaining performance.

Choice of metal (aluminum or steel)

Traditionally, tinplate was used for mass production of metal packaging because of its high strength, formability and the ability to make high integrity joints or 'seams' at high speeds using soldering. However, since the introduction of the drawn aluminum beverage can in the 1950s, this material has increasingly been preferred for beverage products. The clean and modern appearance, formability and superior corrosion resistance are the main reasons for this trend. Aluminum packaging is also now taking a growing share of the aerosol container market. However, steel is still the preferred material for food containers, as its inherently higher strength is beneficial during thermal processing and for the distribution of cans through the retail chain to the consumer.

Container construction

There are two primary construction methods for making cans, 2-piece and 3-piece manufacturing processes. The 2- and 3- refer to the number of separate components (or pieces) that make up the final container. Three-piece can construction was initially developed in the early 1900s. The body, formed from a rectangle of tinplate, is rolled into a cylinder and the edges joined together (Fig. 6.1). The joining technique for the side-seam was changed from soldering to welding years ago, as it is a more reliable and cleaner process (Turner, 2001). Some 3-piece cans, mainly produced in Japan, have adhesively bonded side seams instead. The ends are then joined to the cylinder by a folding/crimping process known as 'double-seaming', which creates a hermetic seal between the body and end via an elastomeric compound placed within the seam (double seaming will be described in more detail later in this chapter). With 2-piece cans, the can body and bottom are drawn from a single piece of metal; the top or end is the second part of the '2-piece' construction. Three-piece cans are almost exclusively made from tinplate, whereas 2-piece cans are made from both steel and aluminum.

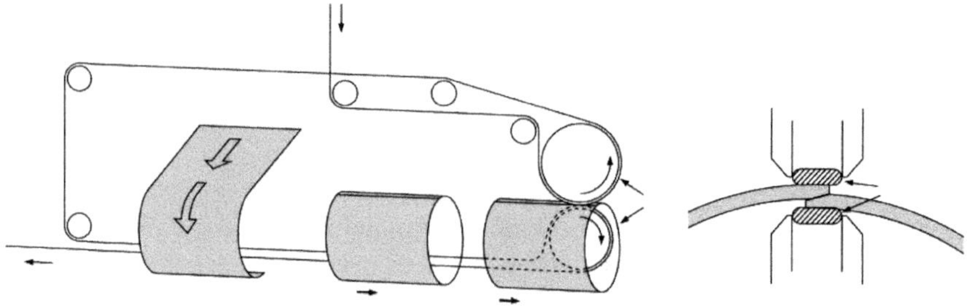

Fig. 6.1 Three-piece can body construction.

Two-piece cans are manufactured by two methods. Both start with a blank and draw stage to produce a shallow cup followed by one or more redrawing operations in which the diameter of the cup is progressively reduced and the height increased without thinning the container wall. This produces a Drawn/ReDrawn (DRD) container (Fig. 6.2). A further 'wall-ironing' operation may also be used to reduce the thickness of the can wall. This second method produces a Drawn & Wall Ironed (DWI or D&I) container (Fig. 6.3). The wall-ironing operation is normally done in a number of stages, as the material strength limits the amount of thinning possible in a single operation. The number of stages depends upon the overall thickness reduction required. Typically the wall is thinned from the original metal thickness by 50–75%. DWI cans are formed from plain metal in a flood lubricated cold working process. The can body is subsequently washed to remove the forming lubricant, with protective coatings and outside printed design added *after* forming. Most food and beverage metal packaging is now produced this way. In contrast, the coatings and printing on 2-piece DRDs and 3-piece cans are added to coil or flat sheet metal *before* the can is formed.

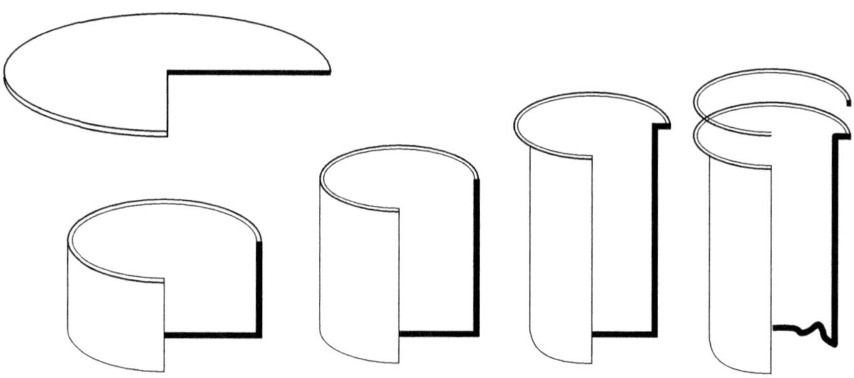

Fig. 6.2 Two-piece DRD can body construction.

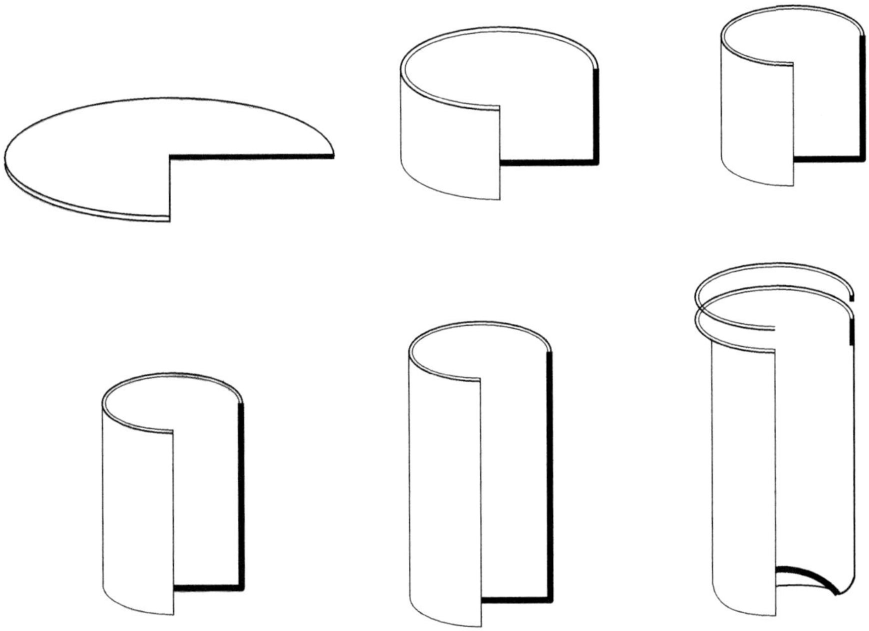

Fig. 6.3 Two-piece DWI can body construction.

Manufacturing trends

Continuous monitoring of critical processes, 100% inspection and careful attention to detail ensures that the defect rate for metal cans is extremely low. Manufacturing equipment for 2-piece containers is expensive, allowing high-speed production of typically 1500 cans per minute or more for a beverage can-making line. A large factory building is needed to contain such production as well as extensive infrastructure including sub-processes for water and effluent treatment, compressed air and vacuum generation, can conveying, inspection and packing. Two-piece manufacturing equipment is also relatively inflexible for changing can sizes. Therefore, 2-piece technology is normally most appropriate when very large can volumes (i.e. >500 million units) of the same size are to be produced at a single location. In contrast, 3-piece can-making equipment is smaller, less expensive and much more easily changed from one can size to another. The material weight in a 3-piece container is generally greater than an equivalent 2-piece can, but this additional metal cost is balanced against the cost of the can-making equipment and size change flexibility. The trend in recent years has been to convert cans from 3-piece to 2-piece manufacturing where the market size exists to support the extra investment for 2-piece can-making equipment. This trend has been most common in the food and aerosol can markets and has been driven by package image and cost reduction. New factories built to meet the growing global demand for beverage cans have almost exclusively utilized aluminum DWI manufacturing methods.

Emerging can formats

The beverage can market has become highly standardized for beer and soft drink cans and interchangeability of can and end between various producers has been achieved through harmonizing critical dimensions. Despite this and the fact that DWI production lines do not favor can size changes, there has been a growing demand from beverage brand owners for new can sizes. This has created a growing number of new can sizes (5–32 oz) in a wider range of diameters and heights, beyond the ubiquitous 12 oz can size. Can-makers have found more efficient ways to change can size while maintaining the very high output needed to cost-effectively produce metal packaging. Techniques such as 'flexible-manufacturing' and Single Minute Exchange of Dies (SMED) have evolved in response to this need.

The Monobloc process for aluminum extrusion has existed for many years and has been widely used to produce toothpaste tubes and aerosol containers from pellets or 'slugs' of very soft metal. With the increasing demand for aluminum containers in the aerosol market, this technology has become more common. The fact that the process uses such malleable material means that complex container shapes and neck forms can be made, albeit using complex manufacturing processes. The use of Monobloc technology has also been extended to produce metal bottles for the beer, energy and soft drinks markets. The metal bottle has provided alternative forms of opening, such as crown corks and screw caps, offering package re-closure, which has been desired by some brand owners. A number of other metal bottle formats have also emerged in the Japanese market and these are most commonly closed using a rolled-on aluminum screw cap. These new metal bottles are based on DWI technology and require relatively complex, expensive, low-speed manufacturing equipment. So far, the use of metal bottles has been restricted to niche and premium product sectors of the beverage market, but the market is growing.

6.2.2 Can shaping

Can-shaping is the term commonly used within the can-making industry for a process that takes a straight-walled, circular can section and reforms the body to produce a 'shaped can'. The process is commonly used to give either a more attractive and/or ergonomic shape to the can. Brand owners use cans with more attractive shapes to enhance shelf-presence and further brand differentiation on the retail shelf – the point of consumer purchase. A number of illustrative examples are shown in Fig. 6.4. Examples of shaped cans with ergonomic and economic advantages are rectangular cans for industrial and DIY products that are both easier to hold and boast a better internal volume to footprint ratio.

There are many different can-shaping technologies used within the can-making industry and the complexity of these processes varies immensely. The shaping process selected by the can manufacturer depends upon type of can required, the production rate, the market volume and the level of automation required. Table 6.1 provides a detailed technical comparison of several different metal-shaping

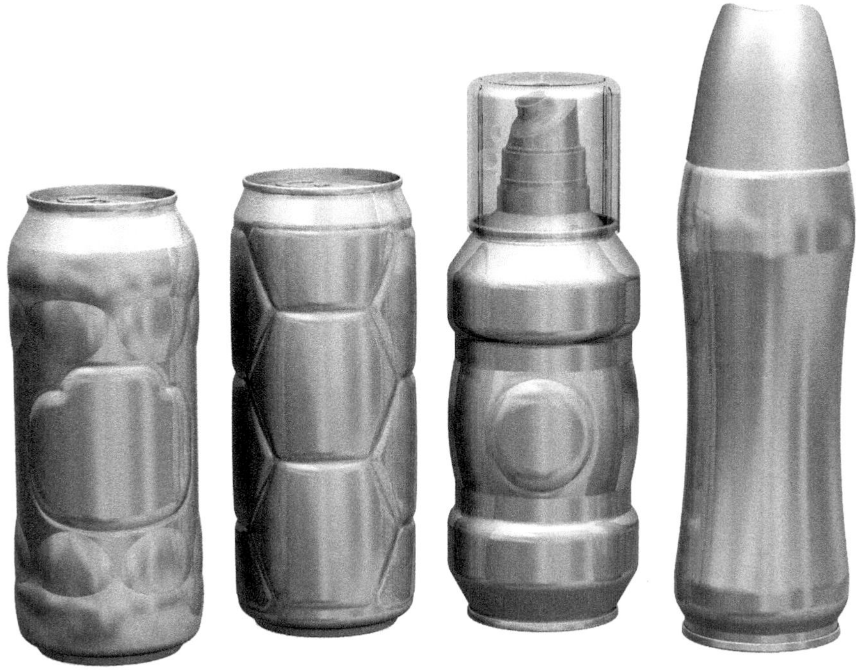

Fig. 6.4 Examples of shaped metal cans produced by blowforming.

processes. Some can-shaping technologies such as stretch forming have been available to the can-making industry for many decades. However, the drive from brand owners for more novel or attractive shapes has driven the industry to develop new can-shaping processes in more recent times. Some of the more common can-shaping processes used in the industry and their main uses and advantages are described below.

Blowforming

Blowforming is a relatively new process, which was developed and first commercialized in the 1990s. It is the most flexible of all the shaping processes and can be used to shape most forms of cans in both tinplate and aluminum at high speeds. The process uses high pressure air inside the unshaped can to expand the can walls into a shaped mold. Maximum expansion rates vary, depending on the material in the can wall. Typical maximum expansions are up to 8% in 2-piece aluminum beverage cans and up to 28% on 3-piece tinplate cans. The blowforming process offers the advantage of being fast and clean, with no physical contact to the inside of the can. The integrity of the internal coating can therefore be maintained, making this the preferred method of shaping cans that require good internal protection, such as beverage and aerosol cans. Shaping machines with outputs of up to 1200 cans per minute are used in commercial production.

Table 6.1 Comparison of metal can shaping techniques

Shaping process	Shaping medium	Can material	Can types				Advantages, disadvantages and limitations
			Two-piece beverage cans	Three-piece food cans	Aerosol cans	Specialist/ other cans	
Blowforming	High pressure air or inert gas	Aluminum	Max expansion 8%	NA	NA	NA	Most versatile fast, flexible, no internal contact, with or without ends and multi-material
		Tinplate	NA	Max expansion 25% SR plate 15% DR plate	Max expansion 25% SR plate 15% DR plate	Max expansion 25% SR plate 15% DR plate	
Hydroforming	High pressure water	Tinplate	NA	NA	Max expansion 25% SR plate 15% DR plate	Max expansion 25% SR plate 15% DR plate	Suffers from long cycle time, high energy costs and internal contamination
Rheoforming	High pressure water jets	Tinplate or aluminum	Max expansion 8%	NA	NA	NA	High energy, long cycle time and internal coating damage
Mandrel expansion	Internal expanding segments	Tinplate	NA	Max expansion 15% SR plate	NA	Max expansion 15% SR plate	Limited to open ended tinplate cans
Embossing	Matched male and female dies	Tinplate	NA	NA	0.012″ (0.3 mm) deep	Up to 0.060″ (1.5 mm)	Limited embossing depth given little or no visual shelf appeal
		Aluminum	0.012″ (0.3 mm) deep	NA	NA	NA	
Die necking and expansion	Metallic dies	Aluminum	NA	NA	NA	Diameter reduction up to 20%	High capital cost per unit cost

Note: SR = Single Reduced plate; DR = Double Reduced plate.

Stretch forming

The stretch forming process uses a number of shaped expanding segments inside the can to stretch the walls into a more attractive or ergonomic shape. As the segments expand inside the can, they stretch the can wall beyond its elastic limit, forcing the can to take the shape of the expanded segments. After the forming process, the can will have a segmented appearance caused by the can wall bridging the gaps between the segments. The larger the expansion, the wider the segmented areas appear. In some cases, brand owners have used this non-continuous or segmented appearance as a part of the overall can design. The process is only suitable for 3-piece welded cans where the integrity of the internal coating is not critical. Typical applications of mechanically expanded cans include rectangular or non-round cans for industrial products and non-processed or carbonated packages for the food and beverage industry. Automated stretch forming machines are commercially available with output speeds of 600 cans per minute.

Rheoforming

The rheoforming process uses high-pressure water jets projecting from a spinning wand to shape the can. The can is enclosed inside a mold and the wand is placed inside the can. As the wand rotates and traverses the length of the can, the water jets aimed directly at the wall of the can progressively stretch the material into the shape of the mold. Using this process, up to 10% expansion on 2-piece beverage cans can be achieved. The main drawbacks of this process are the long processing times and the damage to the internal coatings caused by the high-pressure water jets. Production machines with outputs of up to 20 cans per minute were developed; however, to date, the rheoforming process has not been successfully commercialized.

Hydroforming

As the name implies, the hydroforming process uses water to shape the can. The process is similar to the blowforming process in that the can is enclosed inside a shaped mold. Water is then introduced inside the can under high pressure, causing the can to expand and take the shape of the mold. It was developed for 3-piece tinplate aerosol cans in the early 1990s and an automated production machine was built and run commercially. The process suffers from long processing times, high energy requirements and potential contamination from the forming fluid. In an attempt to overcome these disadvantages, a variation of this process contains the forming fluid inside an elastomeric bag. While this option overcomes the high energy and contamination issues, the life of the elastomeric bag is limited.

Embossing

The embossing process makes use of the plastic properties of the can walls to raise or depress areas of the wall with respect to their mid-point. Embossing has been utilized by the can-making industry for many decades. However, prior to recent developments in production machinery, the process had to be carried out before the body of the can was formed, limiting its suitability solely to 3-piece cans. In more recent times, the development of high-speed rotary machines to emboss finished cans means that the process can now also be used for 2-piece aluminum beverage cans. The 2-piece embossing process uses matched and synchronized rotating mandrels inside and outside of the can to move the metal. As the mandrels rotate, they are cammed into the wall of the rotating can, forming an embossed pattern that can be matched to the can decoration. Machines with output speeds of up to 1500 cans per minute are commercially available. However, as the depth of embossing is limited to only 0.012″ (0.3 mm), the effect has little impact on shelf appearance and realistically can only be considered a tactile finish.

Multi-die necking with die expansion

The multi-die necking with die expansion process was originally developed by machinery suppliers for the shaping of Monobloc aerosol cans. The process differs from most other shaping technologies in that it does not expand the can beyond its original diameter. Despite this limitation, some very attractive shapes can be produced. The process works by using a series of necking dies to progressively reduce the diameter of the body wall. By varying the length and depth of the necking progression and following it with further expansion dies, effective symmetrical shapes can be produced with substantial shelf differentiation as well as ergonomic benefits. Typical commercial machine output speeds up to 200 cans per minute are available.

6.3 Developments in protection and decoration

This section addresses the many roles of coatings on metal packaging, including branding, informing the consumer, product differentiation, and protecting the container. Current and emerging printing methods are addressed in the context of consumers' and brand owners' needs. The discussion of internal coatings and sealing compounds includes the role of the market on material choice going forward and novel materials such as polymer-coated cans.

6.3.1 Printing and decoration

Printing directly onto tinplate containers dates back to about 1870, but the commercialization of branded products had its origins well before that time. Prior to this development, decorative fancy boxes were made by the use of transfers of lithographed pictures printed onto glassine paper. This was a slow process and not suitable for mass production. The lithographic (litho) printing process used for printing transfers was primarily developed for paper that is soft and absorbent, whereas direct printing onto tinplate required adaptation of the process. In 1875 a patent was granted to the Robert Barclay and John Fry (later co-founders of Metal Box) for 'off-set' lithography, where the printed image was transferred via glazed cardboard from the printing plate directly to the substrate (Dolphin, 2004). In the same year, another patented invention replaced the cardboard with a cylinder containing a rubber covering. By the time Mrs Beeton had published her famous *Beeton's Book of Household Management* (Beeton, 1895), some 60 examples of processed, branded food products in cans with high-quality printing were mentioned (Reader, 1976). This was the early beginnings of brand identity and differentiation, which remains one of the prime functions of printing onto tinplate and aluminum (metal) packaging products in the modern era.

Two main processes are used to print metal containers today, 'dry-offset' for beverage cans and 'sheet offset lithography' for fabricated food, aerosol, promotional containers and metal closures.

In flat sheet printing, the lithographic process uses 'planographic' (flat) printing plates with application of ink and water to separate the image and non-image areas. Successive colors are applied to the metal sheet via an 'off-set' blanket cylinder, one color at a time. The art of applying one color at a time without contamination was a skill for which the trade printer served a lengthy apprenticeship. With the invention and development of 'instantly dried' inks using high intensity ultraviolet light to dry (cure) each color before the application of subsequent colors, very high quality images are produced. By using sheets, multiple images can be printed simultaneously, enabling high output (in can-equivalents) at relatively modest print speeds up to 120 m/min. Most modern presses print at least 4 colors in one pass, but there are 6 and even 8-color examples in use today. This process is used to print a gamut of images from simple 1-color instructions to the 'fine art' images found on high-end promotional metal packaging.

Beer and carbonated beverage cans were originally fabricated from printed sheet into 3-piece containers but have now been replaced by drawn steel or aluminum 2-piece containers, which must be printed after forming due to the high degree of distortion required to print the metal before deep drawing. The offset litho printing process had to be adapted to enable each can to be decorated singly while being supported on a mandrel. Printing then became possible at high speeds, but with lower resolution than sheet lithography. Each color is printed individually onto a multi-segment blanket roller, analogous to the blanket cylinder of the sheet printing process. The ink is then printed onto the transfer blanket without being cured and then transferred still wet to the can as one image (Fig. 6.5), after which the inks are dried by passing the cans through a thermal oven. The printing plates used are a 'relief' type, where the image is raised and the background (non-image area) is recessed. In this way the image and non-image areas are separated, enabling the elimination of the water used for lithography (hence the name 'dry-offset'). Since the process prints only one container at a time, it has to run at much higher linear speeds (up to 1000 m/min) to be productive. This requires higher print pressures to transfer the ink, causing some image quality loss compared with sheet printing.

The printing process is only one element of the finished can appearance or 'decoration'. In response to the growing need for brand promotion and product differentiation, can-makers and their suppliers have also developed a wide range of special coatings and inks to provide new visual and surface effects. Tactile

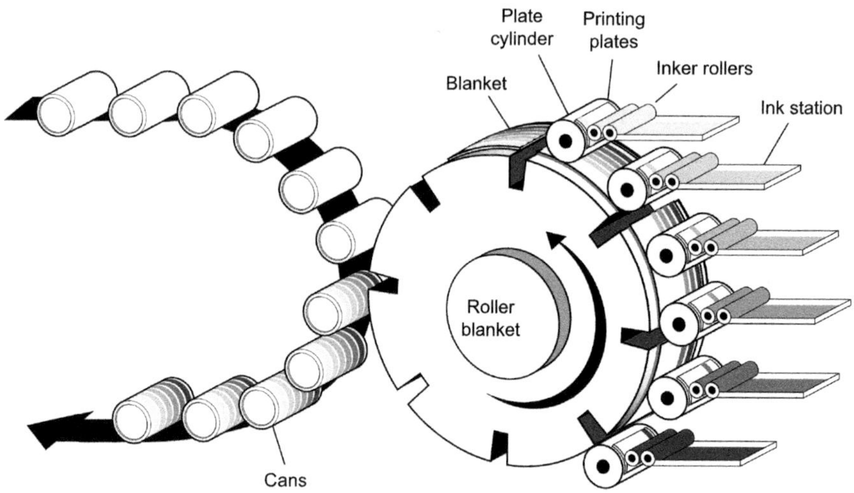

Fig. 6.5 Two-piece can printing (post-forming).

effects such as embossing have always been a popular choice, but embossing is slow and relatively expensive. The use of high build inks, printed in selected areas of the design decoration, has been developed to add a sensory dimension (Fig. 6.6). Other tactile finishes such as 'soft-touch' have also been created using special varnishes applied over the print. The use of special finishes for beer and beverage containers for promotions is very active and demand is growing. For example, 'thermochromic' inks, which change color with temperature, are used to inform the consumer when the beverage is suitably chilled. This trend for packaging to inform the consumer is likely to accelerate as electronics become cheaper and consumers' demand for information increases (Section 6.5).

Quality and consistency are key requirements of modern container decoration. The needs of global brands have driven digital color management and the automation of color control systems, using computers to reduce reliance on mechanical adjustment and the 'skill of the printer'. Computer-to-plate (CTP) systems have now largely replaced photographic film in the preparation of printing plates, again improving quality and consistency.

Deep drawn containers used for beer and soft drinks are now also used for food and aerosol containers. Production speeds and capabilities have increased significantly from first introduction in the 1970s, with modern decorating machines capable of 8-color printing and production speeds typically of 2000 cans per minute. The latest printing plates are far more capable than their predecessors, enabling higher image quality and finer detail to be printed even on the fastest lines, without distortion or loss of quality.

Today's challenging commercial environment is causing a reassessment of print production, planning and customer response. Smaller production batches, more frequent ordering, minimized inventory and shorter print runs are becoming

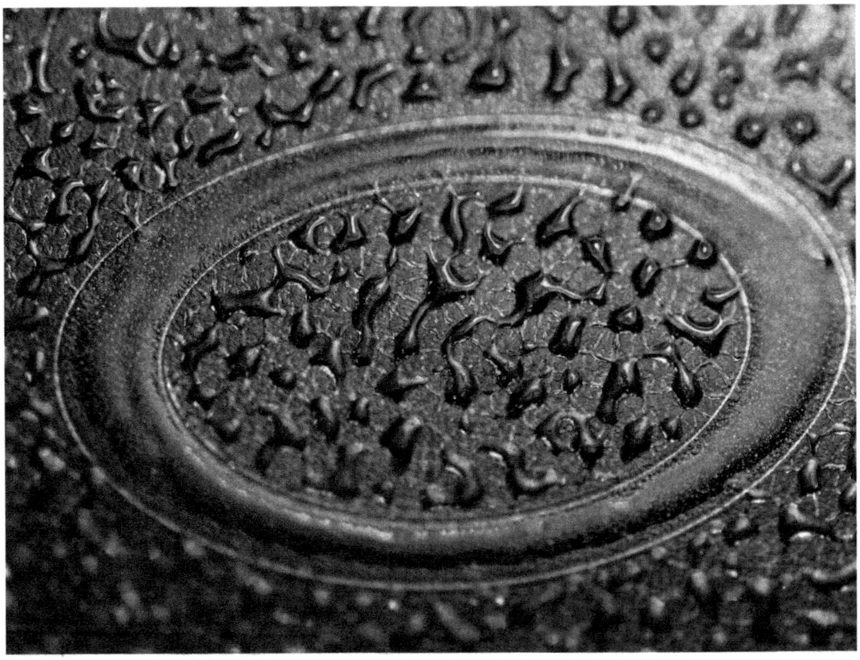

Fig. 6.6 Example of tactile finish.

the norm. For example, over the last 20 years, flat sheet run lengths have dropped by a factor of 10, production speed has trebled, and typical changeover times have dropped from 2 hours to 35 minutes. This is part process improvement and part procedural, but it is reaching the limit of what can be achieved with traditional equipment. Ultimately, as game-changing technology such as digital printing is adopted for metal printing, initially at reduced speeds, productivity will be improved since near-instantaneous design change will be possible. Coupled with lifestyle evolution, such as the use of 'Quick Response' (QR) codes and smart phones, engaging the consumer on a 1:1 basis with print and decoration will be realized.

6.3.2 Polymer coating

Manufacturing routes

Polymer coating for steel and aluminum packaging was developed in response to the desire for higher integrity coatings. Most commonly, polymer film is bonded to the metal while it is still in strip form. The polymer films are manufactured in a separate process by blowforming to make a bi-axially stretched material. The polymer film is brought into contact with a previously heated coil of metal and pressure is applied to form a strong and evenly bonded laminate. This process is called 'lamination' and one or both sides of the strip can be coated in a single operation. Polymers commonly applied to metal in this way include polyethylene terephthalate (PET), polypropylene (PP) and nylon. The applied coating is relatively simple in chemical make-up and the laminate is able to withstand severe and complex metal forming operations. Polymer coatings applied in this way are relatively thick compared to lacquer coatings (>3 times thicker than traditional metal coatings), so the technique is relatively expensive. To address the higher cost issue, at least in part, direct extrusion coating was developed. In this process a melted polymer coating is extruded in close proximity to the heated metal strip and then immediately brought into contact with it and, as before, pressure is applied. Direct extrusion eliminates the separate film production and handling cost incurred with lamination. Direct extrusion is most cost-effective when large batches are coated with the same or very similar resin, as the process is not well suited to small batch runs and changes in resin types and colors. Despite the film cost saving, direct extrusion coating is still generally significantly more expensive than the equivalent lacquering process.

Typical applications

Polymer coating of can-making material has been most widely adopted in Japan, where large numbers of steel and aluminum beverage cans, and also metal bottles, are produced using the technology. The advantages of more compact can-making plant, the absence of volatile solvents in the coating, and simpler coating formulations have been argued as the benefits that offset the higher cost of these coatings. Polymer coating has been used elsewhere in the world, when superior corrosion resistance or a more opaque white coating is needed than is possible with traditional lacquers. Examples of such applications include steel top and bottom components for aerosols and DRD food cans for high-value products like salmon. A number of suppliers in Japan and Europe are able to offer polymer coated steel.

6.3.3 Internal coatings and compounds

The role of internal coatings

The properties of metal are fundamental to the performance and perception of metal packaging by consumers. The metal itself forms the impermeable barrier, which prevents light and oxygen from degrading the quality of the food or beverage, allowing secure high temperature cooking of the product and providing long shelf life without the need for refrigeration. However, that same metal is highly susceptible to corrosion and for this reason almost all of today's metal packaging has some form of protective coating to isolate the packaged product from the container.

Generally internal coatings are applied to the metal surface as liquids made up of a dispersion or solution of polymer in an organic- or water-based solvent mixture. These coatings are applied using traditional roller coating or spray coating methods, either before or after component fabrication. The wet films are then subject to a 'curing' treatment, where the application of heat causes the solvent to evaporate and the film coating to develop as the polymer constituents cross-link on the surface of the metal. This process achieves dry film coatings with excellent adhesion to the surface of the metal, good chemical barrier performance and good corrosion resistance.

A small number of alternative film forming materials are also in use today, using thermoset or thermoplastic polymers in either powder or film form, which are fused with heat onto the metal surface. These materials are 100% solids and avoid the need for solvent or polymerization chemistry during curing, but their construction means they produce significantly thicker finished films, with excellent performance but at a much higher cost.

Once formed, the finished coating has to achieve several demanding technical requirements:

- The coating must be capable of withstanding the metal forming processes for component manufacture. Typically, the coating isolates the forming tooling from direct contact with the metal and must stay intact through, e.g. drawing, necking, beading and flanging processes used to make metal packaging (Section 6.2).
- The coating must be capable of withstanding the processes used in the filling and closing of metal packaging. Food and beverage processing normally involve filling of product at high speed, attaching a closing component such as an end by double seaming or a closure by capping, followed by a sterilization or pasteurization of the finished container at high temperatures, usually involving steam and pressure cycles.
- In food applications, the coating must perform to the required specifications to ensure that food safety is not compromised with respect to food-borne illness.
- The coating must perform to the customer's specifications for shelf life, appearance and taste of the packaged product. This is a very complex spectrum of requirements varying between countries, customers, markets and product types.

The role of sealing compounds

When joining metal ends to metal bodies or metal closure caps to glass jars, it is vital to create a hermetic seal between the components such that food safety is maintained throughout the product shelf life. Sealing compounds help to ensure an airtight seal in both applications.

Sealing can ends: Joining can ends to can bodies is achieved through double seaming, a process by which the end and the can are folded together in a high speed process either by the can manufacturer or the can filler. A compound material is used within this seam formation to fill up the naturally occurring voids in the folded structure. This structure provides a hermetic seal capable of withstanding the temperatures through filling and processing and ongoing stability to the seam during the transportation, retail and storage phase of the container.

The compound material is principally an elastomer with added ingredients to achieve the required performance from a manufacturing or product resistance need. There are many different materials to support seaming of containers used for industrial substances through to food and beverage products.

Compounds are material dispersions, which can be modified with water- or solvent-based diluents to achieve the required application and performance requirements. Typically, they are applied hot to the 'curl' of an end using a viscosity controlled system followed by a period of drying either in air or using a drying oven. Like coatings, material migration into the product has to be managed carefully with respect to consumer safety and product spoilage over the very lengthy shelf life expected from metal packaging.

Sealing closures: Joining a metal closure to a glass jar requires a flexible material to provide the hermetic seal needed for the multi-year shelf life expected from these containers. This gasket not only provides the secure seal but also has to facilitate the easy opening of the jar and the capability to withstand the temperatures through filling and processing and ongoing package security during the transportation, retail and storage phase of the jar. The sealing gasket used today is primarily a plasticized PVC thermoplastic polymer modified with waxes, blowing agents and other additives to achieve the required performance. The gasket is jetted into the metal cap, molded to achieve a consistent sealing surface and cured to fuse the mixture into a blown elastic compound.

Food safety is of critical concern for closure gaskets as they are often used on jars used for infant nutrition products. The raw materials selected and the associated migration into foodstuffs is highly regulated and change in this area is not undertaken lightly.

Influences on materials of choice

With such a wide range of product requirements, it is no surprise that today there are hundreds of coating specifications and many compound specifications, each designed to meet a specific need. The metal packaging industry has always sought to use the most appropriate system for each application, whether that is through regulation of allowed materials, regulation of volatile emissions (VOCs) during the application process or through customer preferences for coating colors or types.

When any materials are in contact with food, it is inevitable that some small amount of that material will migrate into the product. To achieve the level of safety required, the metal packaging industry relies upon the regulatory experts in global authoritative agencies (US FDA, European Food Standards Agency, Health Canada, etc.) to provide guidance about approved materials in contact with food for consumption.

There is constant and increasing pressure to minimize or remove types of coatings and compounds or their constituents from contact with food. This can be driven by regulators (e.g. REACH regulations in Europe), politicians (e.g. banning of polycarbonate materials for infant feeding bottles in Canada) or customers (e.g. desire to eliminate PVC materials from the packaging chain). In each case, making safe changes to find alternatives is both complex and significantly lengthy, but always starts with the need for the safety of the consumer.

Future trends in coating and compound development

The majority of coatings in use today are epoxy resin-based materials, cured with hot air ovens. The majority of closure gasket materials are plasticized PVC-based materials, cured with hot air ovens. Materials selection for the future will be influenced to an even greater extent by regulation (based on sound science) and legislation (based on political need). Based upon history, it would be difficult to predict which polymer species will be favored in the future. The only certainty is that the choice will be limited, given the demanding requirements such materials endure. Environmental needs dictate that control of volatile solvents released during the material curing phase will become increasingly stringent and while this can be controlled by incineration, it is an expensive solution. For the future we can expect to see a significant shift towards higher solids materials, which require different application and curing approaches.

Following trends in coil coating technologies for other industries, it is clear that radiation curing (e.g. electron beam, UV and infra-red) will find increasing use with high solids coatings in metal packaging, replacing thermal curing systems. This will depend on appropriate investment strategies. Using a metal packaging manufacturers' coating line to chemically polymerize coatings in their ovens also seems less than ideal when trying to produce the perfect coating. Powder coating and polymer film approaches, which simply fuse the coating onto the surface, are likely to be the direction for the future, with the potential to simultaneously reduce or eliminate solvents, complex curing and migration issues.

In the metal closures sector there is considerable pressure to replace PVC-based gaskets and it seems likely that a new, as yet unproven, materials class will be needed to satisfy the legislators and consumer groups. Injection-molded or compression-molded gaskets using thermoplastic elastomer polymer mixtures are already offering new opportunities in the market, albeit on a limited scale.

6.4 New product developments

New product developments are critical to any industry and this is especially true in packaging, where innovations can be highly valued by brand owners and consumers. Developments across a range of metal packaging containers are included, with particular emphasis on lids (the ends and closures used to seal containers after filling). Developments in rigid and flexible food can ends, including lidding directly heat-sealed onto the can body, highlight the role of filler investment in driving future change. Metal closure developments include the emergence of composite closures and novel easy-to-open designs. Beverage end developments, driven by the need to reduce costs and/or improve consumer interaction, are addressed, as well as new metal packaging concepts responding to these same drivers.

6.4.1 Easy-open food can ends

Easy-open ends for processable foods

The 'full aperture easy-open end' (FAEOE) used on processed food cans is by far the predominant application of easy-open ends within the food packaging market. For several decades these ends have provided consumers with a 'tool free' opening, which has been significantly improved over the years. Key milestones in this improvement include Carnaud's (now Crown's) EOLE™ end, which first achieved realistic consistency of performance and subsequently led developments with the use of special double reduced steel. However, the manufacturing process is complex, requiring well considered operating practices and quality regimes, and the quality produced by different manufacturers remains variable.

It is difficult to foresee significant further development of such ends, as the fundamentals of easy-open ends for processed food applications are now predominantly limited by the food sterilization process used by the can filler. Traditional continuous retort equipment requires that the metal container act as a pressure cooker, with both positive and negative pressure (ΔP) existing outside the can during the process cycle. Differential pressures (ΔP) can achieve more than +40 to −15 psi and this requirement puts extreme limitations on the material types, thicknesses and end designs that can be used for processable FAEOE and competes directly with the needs of the seaming process and consumer opening performance.

FAEOE for processed foods are therefore generally limited to steel as the only material that can meet this extreme performance with acceptable economics. However, aluminum FAEOE can be used in circumstances where the can aspect ratio is below 1:1 height to diameter, since the variations in ΔP are reduced, putting less extreme requirements on the end. Aluminum in this situation imparts a noticeable ease of opening benefit without economic disadvantage.

With the requirements of traditional canning processes, particularly retorting, it is now difficult to foresee other than minor evolutionary progress on processable FAEOE. Anything beyond evolution will be defined primarily by retort development and can filler investment that enables less severe performance requirements. Alternative retort processes such as 'balanced over-pressure batch retorts' are used with glass jars and plastic packages. These systems significantly reduce the constraints on FAEOE design and specification, permitting a more flexible approach including wider use of aluminum, since the ΔP resulting from this process would normally not exceed +7.5 to −7.5 psi. However, such retorts are unfortunately rarely utilized by today's can fillers, except within the fish canning industry.

Improvements to the consumer opening is generally either in direct competition to the fundamental end performance or to the relatively low cost of the FAEOE, although advances have been achieved through the use of harder double reduced steels and through compromising manufacturing efficiency to provide the consumer with enhanced 'tab access' (i.e. the ability to access the tab to lift with the tip of the finger rather than the finger nail). It is unlikely that these improvements will advance significantly further and suggestions of 'two-finger tabs' (larger tabs) are generally unacceptable on cost grounds.

Indeed, the extreme pressure on packaging cost and weight reduction, together with consumer expectations for convenience, may trigger developments in so-called 'tool opening ends' (i.e. ends where a tool rather than an integral tab on the end is used to open the end). The classic 'non-easy open' end already provides this facility via the traditional can opener, but more convenient, safe and easy to open options are conceivable.

Peelable foil ends

The most significant development in food can openability within the last decade has been the design of peelable foil ends that involve the removal of a central panel from a conventional metal end with the resultant aperture then covered with a heat-sealed foil (Fig. 6.7). Advantages of this design include improved openability and the ability to be applied to the can after filling via the conventional double seaming process. Such peelable ends are often polymer coated with PP, which also serves as the inner layer of the laminated foil structure. The laminated foil structure also includes aluminum to provide gas and light barrier equivalent to traditional FAEOE.

These ends provide easier opening for the consumer but are significantly more expensive than conventional FAEOE and require the balanced over-pressure retorting referred above. This frequently requires significant investment by the can filler, considerably limiting market penetration. While such ends are substantially less robust than traditional ends, there is significant evidence that the market can adapt its methods to handle the reduced performance. However, more traditional consumers can be negative about the loss of robustness.

Ends for dry products

Easy-open ends for non-processed 'dry food' products such as milk powder are more frequently manufactured in aluminum, since the performance requirements

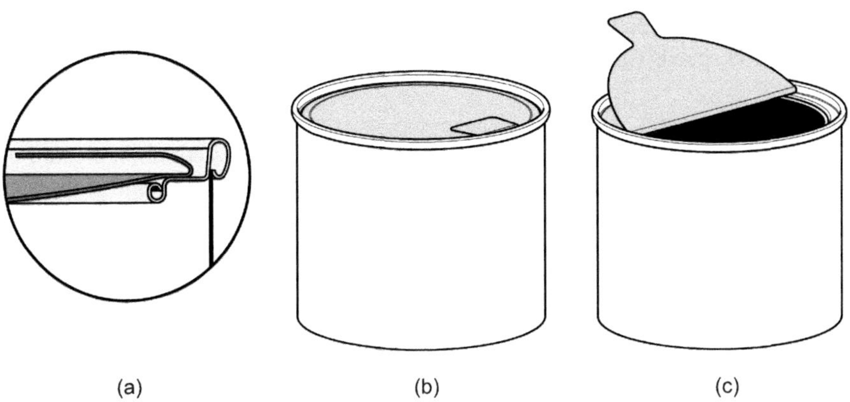

Fig. 6.7 Peelable foil food end: (a) enlarged section showing end double seamed onto can body, (b) closed can, and (c) opened can.

for such applications are neither as extensive nor as extreme. Again, aluminum generally provides better ease of opening than steel, although obtaining controlled ease of opening on large diameter ends does require careful design. Larger ends for this purpose can take advantage of design features such as 'two-finger tabs', which provide a significant aid to opening, as the additional cost of the tab relative to the overall cost of the package and product is small.

The significant development in ends for dried foods is likely to be (again) peelable foil. The most cost-effective solution is to heat seal the peelable foil directly onto the can body. Such products have been available from a range of suppliers for many years but have largely been restricted to non-critical products, as the seal quality has often been inadequate for wider use. However, new developments in both manufacturing and materials enable such 'direct heat seal' solutions to be exploited for more critical products, including such high value products as milk powder.

Future innovations with easy-open food can ends

The most significant factor influencing easy-open end development in future years will be the development of food sterilization processes and the investment by can fillers in such developments, as this is the only way to liberate radical improvements to can opening. Based upon recent history, this will be a slow process, both because of the necessary conservatism of the food industry and because of cost pressure. Metal packaging for food will achieve a new lease of life and relevance once such development and investment is progressed.

6.4.2 Metal vacuum closure developments

The four most important metal vacuum closure designs in use today are shown in Fig. 6.8, with cross-sections detailing key differences in the sealing mechanisms. These designs include traditional single-part closures (Fig. 6.8(a,b)) and more recent two-part designs (Fig. 6.8(c,d)). A comparison of these four designs can be found in Table 6.2, which summarizes typical uses, application mechanisms, application speeds (in cpm or caps per minute) and other attributes.

Twist-open closures

Twist-open closures or 'lug caps' (Fig. 6.8(a)) are the most common closure on glass jars. They are used on containers for a wide range of products and processing conditions such as cold fill, hot fill, pasteurized and fully sterilized with shelf life up to three years. The resulting pack has achieved a strong reputation for integrity and reliability built over the last century. The seal to the jar is usually made using a flowed-in gasket material that is applied via a nozzle into the 'channel' and then thermally cured. A vacuum is generated during 'steam-flow closing' on the capping machine that excludes oxygen from the package and also compresses the gasket to create a robust seal. The use of vacuum enables a 'vacuum-indicating button' to be incorporated, providing assurance of package integrity and hence product safety. Additional tamper evidence may be provided with plastic shrink

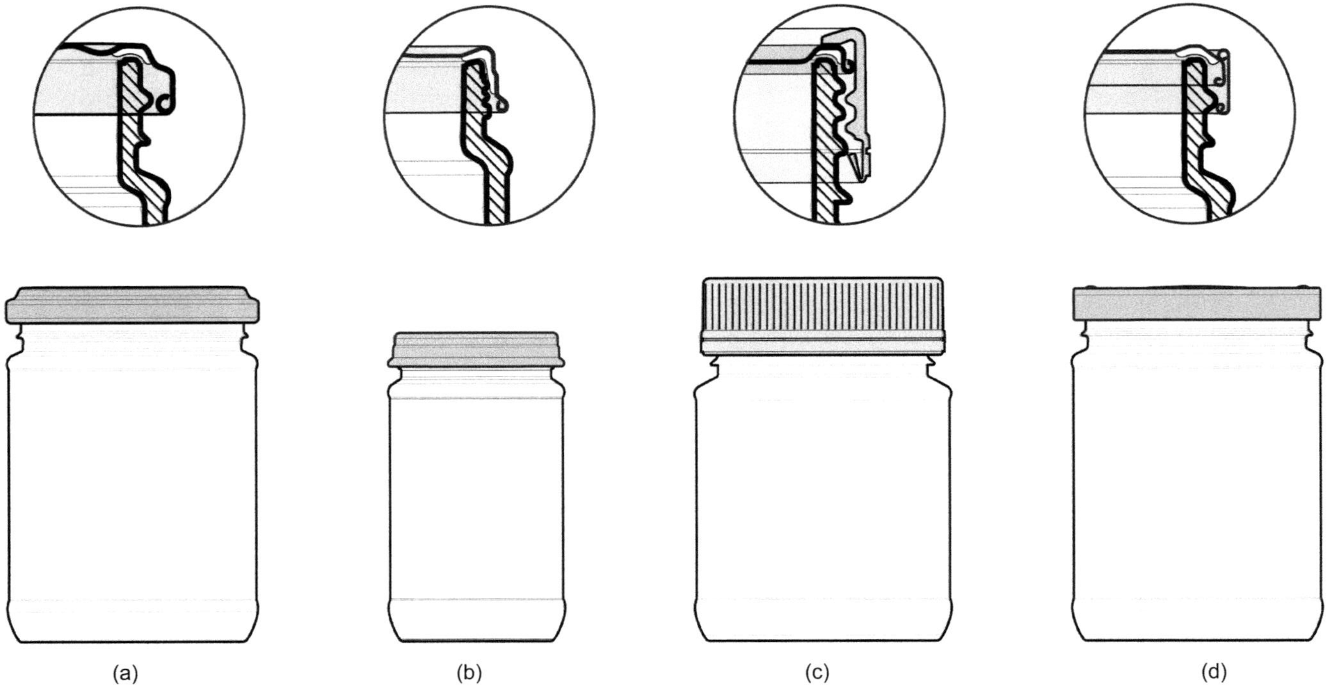

Fig. 6.8 Comparison of metal vacuum closure designs: (a) twist-open or lug closures, (b) press-on twist-open closures, (c) metal–plastic composite closures, and (d) all metal 2-part closures.

Table 6.2 Metal vacuum closures: typical uses, application and benefits

Closure type	Twist-open (or lug cap)	Press-on twist-open	Metal–plastic composite	All-metal 2-part
Type of container	Glass	Glass	Plastic or glass	Glass
Heat sterilized products	Yes	Yes	Yes	Yes
Pasteurized products	Yes	Yes	Yes	Yes
Hot filled products	Yes	Yes	Yes	Yes
Application machine type and typical speed	In-line 400–800 cpm	In-line up to 1500 cpm	Rotary up to 300 cpm	In-line 400–800 cpm
Vacuum-indicating button	Yes	Yes	No	Yes
Tamper evidence	With shrink-sleeve or tag	Integral plastic drop band, shrink-sleeve or tag	Integral plastic drop band	With shrink-sleeve or tag
Sequential opening to reduce removal torque	No	No	Yes	Yes

sleeve or tag, if needed. The disadvantage of this closure for the consumer is that both the vacuum and the friction between the lugs and the glass threads must be overcome simultaneously when the pack is opened, requiring high removal torque (a particular issue with larger diameter closures).

Press-on twist-open closures

These closures are most typically used on small glass jars for baby-food packs (Fig. 6.8(b)). They were developed decades ago to prevent particles that may be present between the jar threads and the closure entering the pack during opening, and to enable cap application at higher speeds via the press-on (rather than twist) motion. The gasket is applied as a flowed-in liquid and in a second operation a 'hot compound molding punch' squeezes the gasket up the closure wall where it 'pre-sets'. During capping, the sealant adopts the shape of the glass threads to allow twist opening. A tamper evident feature may be adopted using an injection molded plastic 'drop band' that separates from the metal cap when the consumer opens the pack (Collins, 1994), or by the same means as for the aforementioned twist-open closure.

Metal–plastic composite closures

'Composite' closures combine a plastic outer component (or band) with molded threads and a second 'floating' metal component (panel) that contains sealant in a channel. The plastic band typically incorporates an integral tamper evident feature ('drop-band') that detaches during opening (Fig. 6.8(c)). This closure is principally used on threaded heat-stable plastic bottles and on glass containers for a wide range of products and processing conditions, such as hot fill, pasteurized and fully sterilized. The advantage of the floating metal panel is that the panel remains in place (held by vacuum) when the plastic band is twisted to open the closure. As the plastic band is rotated further, the metal disk is levered off by the plastic threads to break the vacuum. Hence, the forces required to open the pack are sequential and lower, offering a large benefit to the consumer. The metal disk in this closure provides a complete gas and moisture barrier. Moreover, when used for thermally processed foodstuffs in plastic containers, it enables optimum heat transfer into the product, minimizing processing time and optimizing product quality. Single-piece plastic closures can also be used on plastic containers for thermally processed foodstuffs. These require a secondary metal foil component (pre-heat sealed to the rim of the container) to provide the required gas tightness.

All metal 2-part closures

In a recent development, the principle of a floating metal panel used in the aforementioned composite closure is adopted in an all-metal version. This creates a significantly easier-to-open version of the metal twist open (lug) closure that has met with critical acclaim. To enable this closure to be cost-effective, patented technology enables the metal parts to be manufactured from the same metal shell (Ramsey and Dunwoody, 2011). The shell is cut into two parts, one part to provide material for the outer ring and the second for the floating panel. With the same sealing principle as the metal twist-open closure, this version maintains the same integrity and versatility as its longstanding predecessor, while offering much improved openability (Fig. 6.8(d)).

6.4.3 Beverage end developments

Beverage ends are produced on a huge scale (over 200 billion units annually) and their design has been refined and optimized over many years. Some of the key design features for a typical end shell (before the tab and the score are formed) and the corresponding nomenclature are represented as a cross-section in Fig. 6.9(a). The fully formed end in its two most common configurations (Stolle- and DRT-style designs) and corresponding generic nomenclature are shown in Fig. 6.9(b).

Cost reduction – trends on end gauges and diameters

There have been four major developments of ends for beverage cans since World War II. The first stage was the transition from steel to aluminum ends in the 1960s. The softer material enabled easier opening with a tool as well as the development of easy open ends (EOE). However, the high cost of the magnesium alloys required for these ends put immediate pressure on costs and everything that has followed is a consequence of this change. The second stage was the development

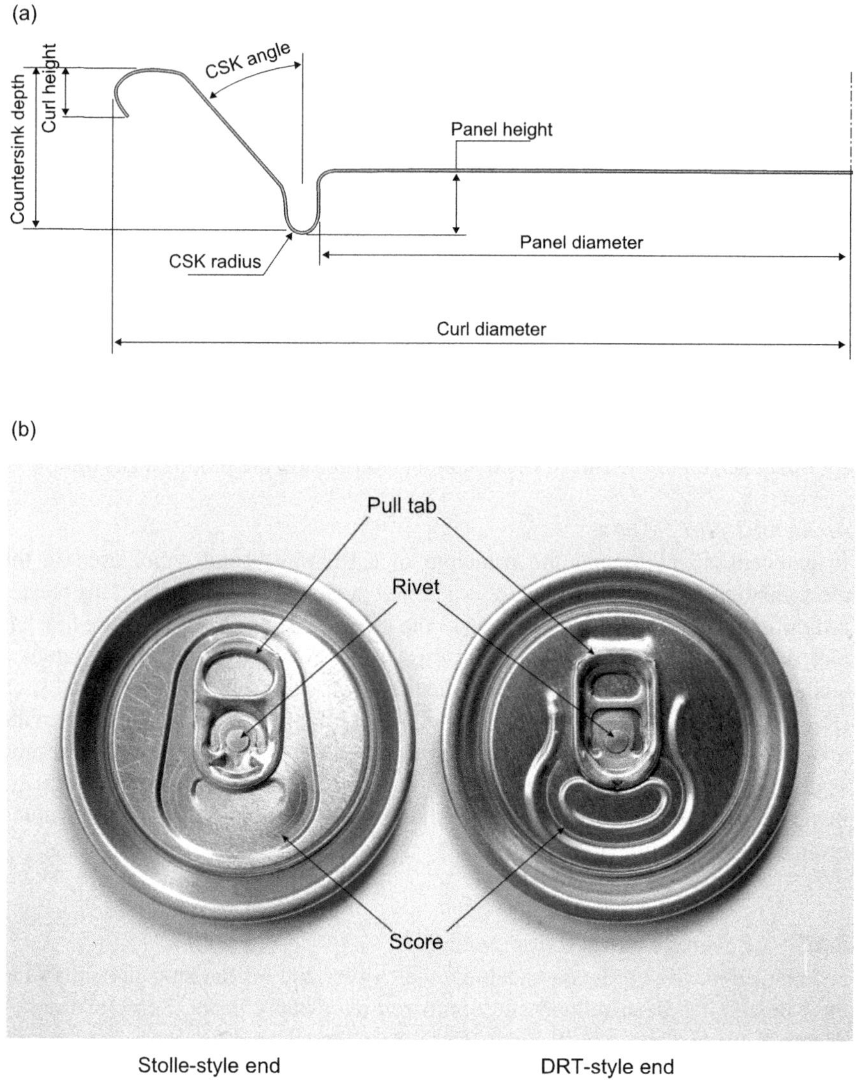

Fig. 6.9 Beverage end nomenclature: (a) typical shell design terminology, and (b) typical end terminology.

of can body necking that enabled smaller diameter and thus thinner gauge ends to be used on cans. This reduction of total metal volume started with the reduction from 211 (2 11/16″) to 209 (2 9/16″) and though 207.5, 206, 204 to the present day dominance of the 202 diameter end. Smaller diameters generally have disadvantages that offset the additional saving.

The third stage was the invention of the reform method of making ends. Prior to the 1980s, all ends were made by drawing metal over fixed tooling geometry.

The major disadvantage to this approach was that the countersink bead radius, which enhances the pressure or buckle performance of ends, could not be made smaller than approximately 0.028″ without severe metal thinning. The 'reform' method, developed by Alcoa (Shultz, 1978) and MetalBox/Crown (Taube and Roberts, 1986), used tooling with differential movements, enabling countersink radii as small as 0.018″. This then enabled further gauge reduction. The fourth and latest stage was the invention and subsequent commercialization (in 2001) of Crown's 'SuperEnd®' (Brifcani *et al.*, 2005). This end provided a similar step change in metal volume reduction that had previously only been possible by can neck and thus end diameter reduction.

Prior to SuperEnd® the countersink wall angle on beverage ends was nearly vertical to provide the necessary strength against the nearly 100 psi internal pressure requirements for many beverage can applications. However, this fundamental invention demonstrated that an increased countersink wall angle of approximately 30–60 degrees enabled both substantial strengthening and a large reduction in the volume of the metal disk used to manufacture the end. As a result, significant metal volume reductions of about 10% were achieved. This invention ushered in a new era in beverage end manufacturing and today SuperEnd-like ends with increased countersink wall angles are produced on a scale of nearly 100 billion a year.

Performance requirements

The critical performance requirements of EOE beverage ends include 90 to 100 psi buckle resistance (depending upon the product being filled), a robust seal, the ability to make an integral rivet without fracturing, a scored area that will open under the action of a pull tab, and finally the ability to deliver the end to the customer, intact and undamaged. The gauge reduction developments described above have been possible because the development of seam roll technology enables folding the curl geometry to achieve spur free double seams. The challenge of producing low gauge ends, which will open consistently and not leak at the rivet, has not been as difficult as would have been imagined 20 years ago. This is mainly due to improvements in the quality of aluminum coil manufacture and the development of self-lubricating coatings. The final performance need, that of making very low gauge ends that can be delivered undamaged to the can assembly, is probably the next great challenge limiting future significant gauge reductions. Handling equipment, from track work in the end manufacturing plant to the double seaming equipment in the customer's plant, have not fundamentally changed in 30 years, and in some respects have not kept up as manufacturing speeds have increased almost three-fold.

New beverage ends – reseal-ability

The advent of the PET bottle for soft drinks put pressure on the beverage can to achieve the same ability to reseal the container. Numerous attempts have been made in this area, notably Bound2B's marrying of plastic with metal to create an end that can be rotated to reseal (Perra, 2010), as well as screw tops on metal bottles (commercialized by Ball). However, no end maker has yet managed to commercialize a price competitive re-sealable end. The future may do away with the EOE altogether and move towards a closure style top on a necked in can.

Drinkability/pourability

Over the past 20 years, consumers have consistently asked for enhanced drinking and pouring utility in their cans. This has largely been addressed by the introduction of larger aperture score profiles. These ends create enhanced flow by maximizing the vent area in the aperture, above the liquid, as it is being poured, reducing glugging. In addition, there has been the recent Alcoa development of a venting feature in the score aperture (Mathabel *et al.*, 2011). This feature creates a vent 'tube' in the flow of liquid, enabling trapped air to be vented, even when the liquid is obstructing the aperture. Further developments in the area of improved venting are likely to be seen in the near future.

Improved branding/loyalty

The most significant recent development in branding is Crown's full aperture end for beverage cans. This has been used successfully in South Africa and South America. The intent of this new product is to marry the can to the glass, by making the can into the glass. Branding with the tab has been achieved by use of colored coatings, cutout shapes and logos on the tab and laser incising of similar details on the tabs. The push for enhanced brand loyalty has largely been led by making promotional ends, enhancing the collectability of the cans themselves. One of the first attempts with EOE ends involved Crown's Proof of Purchase (POP) ends in the 1980s (Pavely *et al.*, 1993). These ends had plastic tokens inserted into the tabs, which could be collected for a prize, or had printed codes on the underside awarding major vacation getaways to the winners. More recent development of this idea include the use of inkjet or laser coding on the tabs and/or ends coupled with an internet campaign, where the consumer is encouraged to log the numbers to receive prizes. A number of these innovations are shown in Fig. 6.10.

6.4.4 New metal packaging concepts

Metal packaging has the widest range of manufacturing methods and applications of any packaging material. This feature leads to a constant stream of new innovations that aim to satisfy new market needs. So what makes a successful new concept and what lessons can we learn from past launches? Are successful new products designed to take advantage of emerging technologies or simply meet shifting consumer trends? The range of outcomes will be represented through a number of current examples shown in Fig. 6.11.

A commonly known example is the in-can 'widget' that miraculously produces the perfect 'head' on a glass of beer poured from the can. The effect is a result of the sudden drop in pressure within the can upon opening, causing a jet of nitrogen gas to leave the widget and nucleate carbon dioxide bubbles from within the beer (Fig. 6.11(a)). Here the key to success is that the improved drinking experience

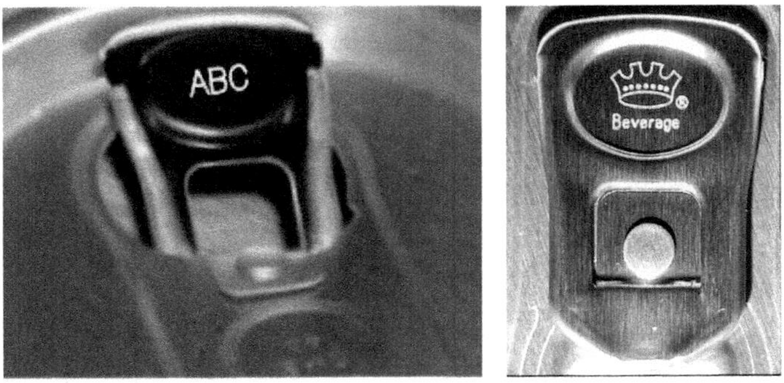

Laser etched TABs

Full aperture beverage end POP end

Fig. 6.10 Examples of beverage end innovations.

outweighs the relatively small added cost and complexity. Therefore this innovation has become a mass market success.

Another product-driven innovation is the self-heating can. A wide range of technologies have been proposed mainly based upon exothermic chemical reactions of which the slaked lime and water reaction is the most common. Here the package is typically inverted and a bi-stable button pressed releasing water into the lime compartment. Heat from the reaction then transfers over several minutes into the product (Fig. 6.11(b)). The main issues are the manufacturing complexity and the volume of the lime and water compartments, which require approximately 50% of the container volume. Therefore, the self-heating can has so far only been successful in niche markets, where the improved convenience justifies the extra cost.

Self-chilling cans using endothermic chemical reactions have also been launched as niche products, but in general these reactions are less efficient than the exothermic reactions and the time to cool is too long. The alluring quest for an instantly cooled

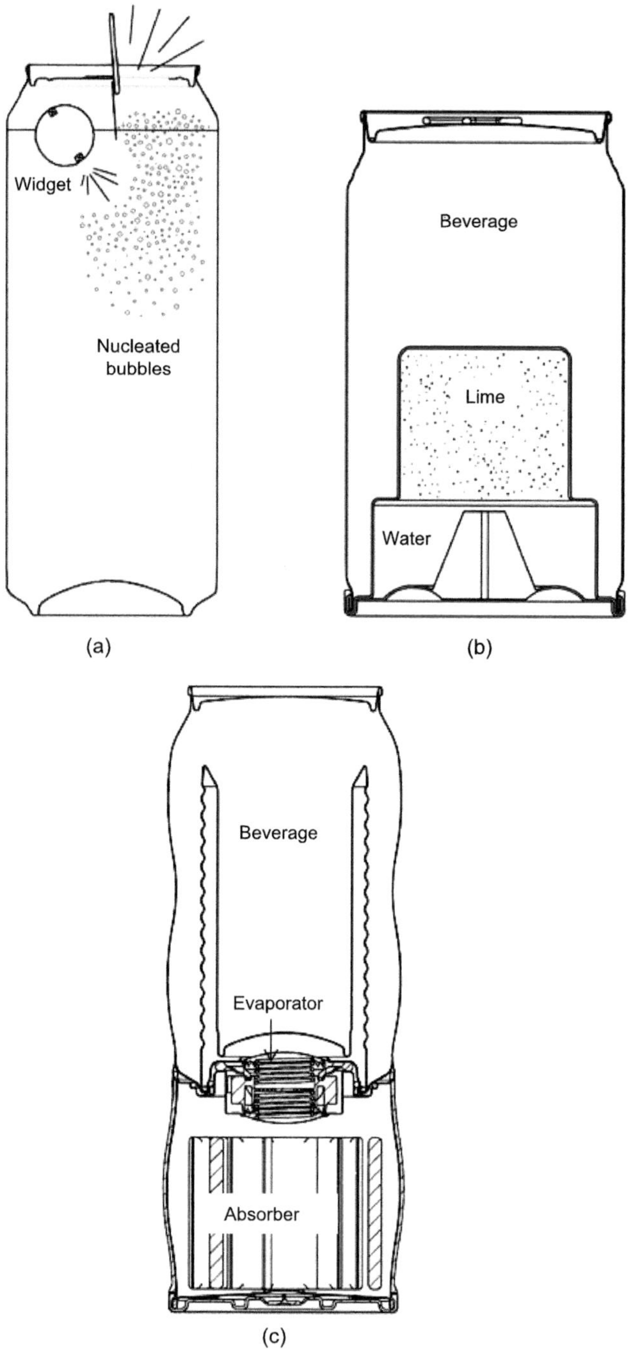

Fig. 6.11 New metal packaging concepts: (a) floating widget, (b) self-heating can, and (c) self-chilling can.

drink remains strong and many alternative mechanisms have been developed over the years. The most advanced mechanism was based upon heat pump technology where water is evaporated from a chamber within the can and condenses in a desiccator unit attached to the base of the can (Fig. 6.11(c)). The evaporation and adsorption of the water creates very effective heat transfer to rapidly chill the beverage. However, the complex technology resulted in excessive manufacturing costs versus the consumer benefit and this product has yet to be commercialized.

Often a more effective means of providing innovation is to design the package to interact with existing technology in new ways, rather than to develop wholly new technology for the package. A good example is microwavable metal bowls that provide even product heating, even though metal is a barrier to microwaves. These bowls are designed with a shallow aspect ratio and large opening to let the microwaves penetrate from above. Package interaction is also demonstrated by metal packaging for DIY tools, such as hand drills, where the metal packaging has a secondary use after purchase to store the drill, charger and accessories.

Decoration can also be used to provide interactivity. For example, thermochromic pigments may be added to beverage can decoration to give a color change when the container has been adequately chilled. Thermochromic pigments have also been used as content level indicators on packaging such as metal drums. Here the strip is warmed by touching and changes color in the headspace region of the drum where there is no liquid product to act as a heat sink.

Over recent years, innovation has also been strongly influenced by environmental drivers. For example, aerosol propellants have switched from CFCs to hydrocarbons, and are now moving toward compressed gasses. The move from a volatile propellant to compressed gas brings the challenge of how to maintain dispensing performance during the reduction in package pressure with use (typical reductions are from ~8 to 2 bar). Novel valves have been developed with mechanisms to either attenuate the pressure or to balance the flow rate as the pressure drops.

Although metals such as aluminum and steel are recyclable without loss of performance, there is also environmental pressure to reduce the amount of material used. This provides benefits beyond material savings such as the energy used for manufacture, transportation and recycling. Therefore, much innovation has centered on new ways to either reduce ingoing gauge thickness or to distribute the material more effectively so as to reduce the blank size (area of material) used per container. For example, a typical 400 g food can uses around 30% less material than 15 years ago. This has been achieved by using stronger, higher temper materials and improved manufacturing processes, allowing gauge reduction of the body and ends.

Further material reduction has been achieved by developing the methods of construction so as to reduce the number of components in the container or to limit the amount of material used in the bonding mechanisms. In some cases, joining by mechanical double seaming has been replaced by heat sealing.

An example of a metal package light-weighting progression demonstrating this trend follows:

- 3-piece can with EOE;
- 3-piece can with peelable foil closure on ring;
- 2-piece can with peelable foil closure on ring; and
- 2-piece can with direct sealed foil closure.

Foil pouches use a minimal amount of material in the primary package. However, these also require unique processing conditions and rigid secondary packaging (e.g. carton box) for stacking and to provide abuse resistance for distribution.

Material developments have been driven not only by sustainability or cost reduction. Advances in formability have also enabled new shaping processes that enhance the package appearance and branding. For example, in the syrup market, shaped cans are now produced with 20% expansion, and new high formability steels have been produced allowing up to 40% expansion.

In summary, we can group the success criteria for new metal packaging into three key factors, an example of which is shown in Fig. 6.12. A successful new product needs to provide a valued consumer benefit, environmental sustainability and cost effectiveness. In the illustrative diagram, the size of the circles provides an indication of how well the factor is satisfied and the size of the overlap region determines the potential market opportunity. Given the complexity of addressing these competing criteria, finding the overlap region can be an exciting challenge.

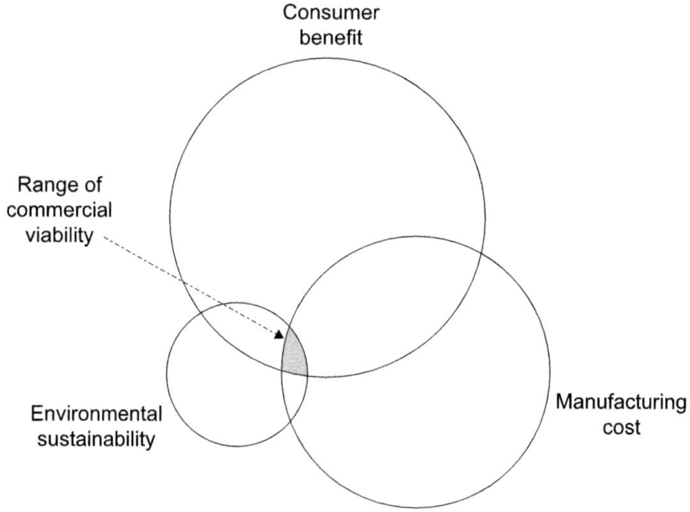

Fig. 6.12 Packaging success diagram.

6.5 Future trends

6.5.1 Appearance – shelf image/shelf presence and point of sale

Printing and decoration have been critical to both the presentation and attractiveness of products to the consumer (shelf image) and are largely responsible for establishing brand recognition (Section 6.3.1). When it was widely introduced to supermarkets in the 1980s, the Universal Product Coding system (UPC bar coding) of products generated a step change in the accuracy of container decoration required to enable accurate laser reading of the code. Bar coding drove a major change in the way retail stores and supermarkets deliver and manage product inventory and the resultant check-out data is now essential marketing data.

Container decoration will play a similarly critical role in the promotion of consumer products in the future. Another significant change in consumer interaction with packaging is expected with the introduction of low-cost printed electronics. In combination with smart phones, electronic codes printed or embedded into container decoration and retail store shelves and displays will add another dimension to the ability of brand owners to interact personally with the consumer and retail outlets will gain even greater functionality in managing their product inventory. Consumers are increasingly using smart phones to obtain product information, using Quick Reference (QR) codes present on packaging that enable the consumer to connect directly to the internet to obtain more information about the product. This is also an invaluable resource for brand owners to communicate with the consumer.

Animated decoration will also be possible with technology currently under development, using OLED technology together with self-contained miniature power sources. Increased product management security and warehouse management will be accommodated by printed RFID tags, making it possible for manufacturers to track even individual product items through manufacture and distribution to retailers with unparalleled accuracy.

6.5.2 Manufacturing flexibility

More people are expected to be living alone in the future and the demand for more personalized cuisine, with greater variety, catering for broader cultures and in smaller portion sizes, will drive a change in consumer expectations. With the increasing demand for convenience, consumers will expect more freedom, convenience and choice in the way they prepare and consume food and beverages. An example of how consumers' growing demand for convenience will drive container design is the increasing popularity of coffee and food pods enabling single servings, in a wide variety of flavors, on demand. It is expected that more food containers will be designed to integrate with domestic equipment for preparing food.

Much of the manufacturing equipment installed for metal packaging has been designed for mass production, enabling the production of high volumes of standardized containers at consistently high levels of quality and efficiency.

Container size standardization has evolved from past consumer lifestyle requirements, which will inevitably change. As marketers drive brand differentiation and consumers increasingly come to expect customized and even personalized products, container manufacturers will be required to accommodate frequent product changes including smaller batch volumes. 'Mass customization' is commonplace for fashion products and electronic consumer goods. The challenge for metal container manufacturers will be to similarly accommodate this increased demand for manufacturing flexibility, while retaining production efficiency and competitiveness. The traditional size and container volume standardization, central to the canning industry, will inevitably give way to the lifestyle changes expected in the future.

6.5.3 Effect of sustainability on packaging fabrication

Recycling of raw materials from manufactured products has long been a feature of modern industrial infrastructure, but with increasing demand due to growing consumer wealth in the West, the development of emerging economies, and rapidly increasing global population, more attention is being paid to the overall impact on the environment and the sustainability of raw materials and utilities used for production. 'Cradle to Grave' (Life Cycle) analysis is used to fully understand the total impact of manufactured products such as the consumption of raw materials and the availability of utility resources (energy, water), including potential environmental impact through water and atmospheric discharges.

Modern industrial production is subject to more comprehensive scrutiny of the total impact of product manufacture and this 'good husbandry' of the world's resources will increasingly become a feature of product marketing as the consumer is increasingly being made aware of the sustainability of the product choices they make. Global standards for the assessment and definition of sustainability have not yet been harmonized, but we can expect in future the consumer will increasingly take sustainability into account in their selection of products. Legislation will also increase the control of resources used by manufacturing companies; however, well informed consumers are likely to have a greater impact in the long term.

6.5.4 Emerging manufacturing and material technologies

Section 6.2.1 describes existing metal fabrication processes. However, several new technologies originally developed for applications outside of metal packaging will be incorporated into future container manufacturing and will potentially offer greater capability and flexibility, such as laser welding and cutting, robotic handling, modern adhesives, three-dimensional printing, low cost printed electronics and distributed electronic control.

New materials are constantly being introduced into the packaging industry, but the drive for recyclability, process water reduction, and energy reduction will accelerate the introduction of 'natural materials' such as bio-polymers and waxes.

Traditional protective and decorative coatings for metal packaging will be replaced, under pressure from legislation and consumer preference, to reduce migration into food and drinks. There will also be new technologies introduced and developed to reduce the energy required to dry and cure these metal packaging coatings. Radiation cured decoration is now commonplace, but currently limited to the external metal container surface due to concerns over migration of any residual catalyst. However, there is also the potential for 'electronically cured polymers' (being developed for electronic visual displays) that may overcome these concerns, where the chemical catalyst is replaced with an electronic mechanism.

The introduction of new materials and manufacturing processes has traditionally been relatively slow within the metal packaging industry, due to a combination of factors such as cost, quality, efficiency, mass production and the capital implications of making obsolete existing capital equipment. As discussed above, future pressure from consumers for new products, increasing competition between brands, the inevitable competition from alternative packaging (i.e. glass, plastic, carton, laminates) and the global market will drive considerable diversification.

6.6 Conclusions

Innovation has driven the metal packaging industry since its inception and this trend will continue. Indeed, the key properties of metal, including strength, flexibility, barrier and recyclability, have fuelled these innovations and enabled a remarkable range of manufacturing process and resultant products that is still expanding today (i.e. metal bottles and novel closures). These same properties provide metal packaging with environmental credentials that are second to none. Unlike nearly every other packaging material, metal is infinitely recyclable (as aluminum and steel suffer no loss of properties through recycling). Moreover, metal is 100% recyclable and is recycled by consumers around the world at high rates. As a result, these metals represent permanent resources that can be used continually, effectively embodying McDonough and Braungart's (2002) 'Cradle to Cradle' vision.

Food and beverage cans are hermetically sealed, preventing microbes, light and oxygen from damaging the product inside the container. When combined with its strength and robustness, metal packaging offers the longest shelf life and superior security for foods and beverages. As a result, metal packaging plays a vital role in society by reducing wastage and therefore the environmental footprint of these products. By ensuring the substantial resources (raw materials, water, energy, etc.) invested by society to produce foods and beverages are not lost, metal packaging plays a larger role in enhancing sustainability than is apparent from the environmental credentials of the package itself. Indeed, metal packaging enables the consumption of more food and beverage with fewer resources, reducing world hunger and improving the quality of life for billions of individuals, an impressive resumé for a package entering its third century. The versatility of metal packaging combined with its capacity for innovation and impressive environmental credentials ensure it will continue to play an important role for years to come.

6.7 References

BEETON I. (1895) *Beeton's Book of Household Management*, SO, Beeton Publishing.

BRIFCANI M., HINTON P. and KYSH M. (2005) *Can end and method for fixing the same to a can body*, US 6,848,875.

COLLINS M. (1994) *Screw closures for packaging containers*, EP 0450959 B1.

DOLPHIN T. (2004) 'Biscuit Tins – The Tin Pages – Tin Manufacturers'. Available from: http://www.britishbiscuittins.co.uk/index.html

MATHABEL R., CHASTEEN H., JACOBER M., SANTAMARIA A. and CLARK A. (2011) *Vent tube for liquid container*, US 7,975,884.

MCDONOUGH W. and BRAUNGART M. (2002) *Cradle to Cradle*, New York, North Point Press.

PAVELY A., MARR R., FOSKETT P. and WHITE P. (1993) *Method of retaining a token in a ring pull or lever*, US 5,191,695.

PERRA A. (2010) *Device for sealing food product containers and food product container provided with such a device*, US 7,819,280.

RAMSEY C. and DUNWOODY P. (2011) *Method for producing a twist metal closure with separated disc and ring from a single closure blank*, EP 2086700 B1.

READER W. J. (1976) *Metal Box: A History*, London, Heinemann.

SHULTZ F. (1978) *Method of forming a pressure resistant end shell for a container*, US 4,109,599.

TAUBE W. and ROBERTS D. (1986), *Method of and apparatus for forming a reinforced can end*, US 4,571,978.

TURNER T. (2001) *Can-Making for Can Fillers*, Oxford UK, Sheffield Academic Press.

7

Paper and paperboard innovations and developments for the packaging of food, beverages and other fast-moving consumer goods

R. Coles, RichColes Packaging Associates Limited, UK

DOI: 10.1533/9780857098979.187

Abstract: This paper has five main parts: drivers of innovation and trends in the paper and paperboard packaging market with sustainability recognised as an increasingly important guiding influence; packaging design optimisation which considers the need to balance the requirements of consumers, the supply chain and the environment; brand communication through packaging in the retail environment; the application of sophisticated packaging technologies that promote consumer well-being, package integrity, brand authenticity and ecological solutions – these include developments in smart or intelligent packaging, active packaging, modified atmosphere packaging, biodegradable and compostable packaging; and an overview of innovations in main packaging technologies made from, or utilising, paper and paperboard materials.

Key words: paper and paperboard packaging, markets, sustainability, brand communication, innovation and design, technology.

7.1 Introduction

The paper and paper-based packaging industries are well placed to meet future commercial requirements by responding to evolving marketing, economic, supply chain and sustainability goals. They are also in tune with demographic trends, and people's changing lifestyles and values. This chapter will discuss innovations occurring in the manufacture of paper and paperboard and within the wide variety of pack formats, whilst highlighting the versatility of designs available.

Paper and paperboard materials can be printed, glued, cut, creased and shaped. However, they are permeable to gases, moisture, oils and fats. Consequently, these materials often require treatments such as coatings and laminations. These processes may involve specific materials, plastics and barrier materials, such as aluminium foil, to extend their packaging applications and the shelf life of the products they contain.

7.1.1 Drivers of innovation

The paper and paperboard industries are developing packaging materials, processes and systems, which are flexible and responsive to the increasingly complex mix of marketing, technological and logistical requirements of their customers. Key drivers of innovation, which are illustrated with examples throughout this chapter, include:

- Adverse world economic climate stimulating market demand for packaging solutions, which can increase operational efficiency in the supply chain and save costs, thereby improving profitability. These solutions involve the development of more efficient manufacturing processes and innovations in technology/materials.
- Intense competition amongst branded goods manufacturers and retailers who seek to gain a competitive edge through using packaging, which better engages consumers and meets their needs.
- Growing concerns of consumers, brand owners and governments to address supply chain concerns regarding pack integrity, brand authenticity/product provenance, food safety, waste, product quality, counterfeiting, product tampering and theft.
- Fast-growing global market demand for products which promote consumer well-being and are ecologically responsible.
- Brand owners seeking to enhance their environmental or corporate social responsibility profile and gain a share in the green consumer market through, for example:
  - lowering the carbon and water footprints of their packaged product and the services used to deliver them;
  - use of sustainably sourced packaging materials verified by traceability systems and supporting certification;
  - reducing their reliance on non-renewable fossil fuel-based materials;
  - adopting light-weight, recyclable and compostable packaging to avoid or minimise waste disposal to landfill and incineration.
- Regulatory compliance and responding to increasing sustainability pressures from a range of stakeholders, which include governments, retailers, financial institutions/ethical investors and environmental NGOs. This is in order to address strategic issues, meet standards and fulfil voluntary agreements, especially with regard to environmental/waste management, labelling and food safety concerns. For example, these include:

- minimising packaging and eliminating waste/pollution throughout the supply chain;
- recovering and utilising used packaging materials in biological and/or industrial cradle-to-cradle cycles;
- reducing carbon emissions and associated financial impact of carbon levies or taxes;
- achieving renewable energy targets and carbon reduction commitments;
- meeting product safety and traceability requirements;
- responding to governmental labelling regulations.

- Increasing international competition between paper and paperboard packaging suppliers and converters who are investing in new materials, energy and water technologies to:

  - mitigate the rising cost of resource inputs;
  - address growing resource scarcity;
  - meet growing volume of market demand;
  - enhance quality standards.

7.1.2 Sustainability – a commercial opportunity to add brand value

In addition to the basic commercial drivers of profitability, minimising costs and marketing spurring innovation for competitive advantage, a key focus of attention for the future of paper and paperboard packaging suppliers and their customers will be *sustainability*. As environmental, social and economic concerns continue to grow, packaging will play an increasingly important role in the sustainability agenda of industry. The development of sustainable packaging is one of the main challenges facing the packaging industry (Nampak, 2010).

The drive for sustainability – interlinking social, environmental and economic factors – is spurring the adoption of lower carbon biobased packaging materials. This is especially relevant in the case of packaging materials constructed using paper and paperboard for the food and drink sector. The aim is to deliver packaging solutions that add brand value, through better meeting consumer needs and offering lifestyle benefits in a sustainable way. The sustainability credentials of a product and its packaging can serve as a point of differentiation for a brand and add value for consumers through assurance of ecological and social responsibility. However, it is important that benefits of innovation for sustainability are clearly communicated, such that consumers can understand and fully appreciate them (refer to Section 7.2.1 for further discussion of this topic).

Being a natural, annually renewable resource supporting valuable eco-services, such as carbon/water cycling and biodiversity, paper and paperboard packaging materials offer distinct environmental sustainability credentials. These materials, therefore, present a commercial opportunity for retailers, brand owners and packaging suppliers to enhance their brand image and demonstrate corporate social and environmental responsibility. The mission for paper and paperboard suppliers also involves supporting their customers' efforts in seeking a competitive edge whilst contributing to a more sustainable society.

7.1.3 Trends in production

The worldwide production of paper and paperboard, including packaging, continues to rise, as reported by the Bureau of International Recycling (BIR, 2011) and indicated in Table 7.1. Corrugated and solid fibreboard account for approximately 30% of the total tonnage. According to a Confederation of European Paper Industries (CEPI; www.cepi.org) survey of its membership, paper and paperboard used for packaging accounted for 43% of output in Europe during 2009. As can be seen from Table 7.1, the overall production in 2010 had recovered to what it was in 2007 before the 2008 economic crisis.

The expansion in production of paper and paperboard is shifting towards rapidly industrialising countries in Asia and Latin America. As can be seen from Table 7.2, production in Asia has significantly increased by over 10% between

Table 7.1 World production of paper and paperboard

Year	Total tonnage (million tonnes)
1980	171
1990	238
2000	324
2005	367
2006	382
2007	394
2008	391
2009	371
2010	394

Source: BIR, 2011.

Table 7.2 World production percentage of paper and paperboard by region between 2002 and 2010

Region	World production by region (%)		
	2002	2009	2010
Europe	30.7	27.4	27.2
Latin America	4.8	5.2	5.1
North America	29.6	22.8	22.5
Africa	1.1	1.1	1.1
Asia	32.7	42.4	43.1

Source: BIR, 2011.

2002 and 2010. This follows massive investment, most notably in China, in large and fast machines as well as new mills.

7.1.4 Global trends in consumption and market demand

The global economic crisis in 2008 and the financial uncertainty since then have depressed market demand for goods and their packaging in developed countries. This, together with an accelerating shift over recent decades in the production of manufactured goods to emerging economies, largely accounts for the changes in the levels of comparative regional consumption, or relative affluence. This shift is reflected in Table 7.3.

A rapidly rising population, urbanisation and growing affluence in many developing countries and emerging economies are underpinning expansion in market demand for paper and paperboard, including packaging. However, this is, at the same time, exerting increasing pressure on resources, particularly energy and water.

About 50% of all paper produced is used for packaging (Datamonitor, 2008). The largest share of global packaging is accounted for by paper and board packaging, with sales of $165 billion in 2003, equating to 39% of the market (World Packaging Organisation, 2008). Also, paper and paperboard represents the largest proportion by weight of packaging material used. The food and beverage industry is the largest user of packaging generally.

By 2050, global food supply will need to increase by around 70%, which is disproportionately higher than the population growth of approximately 30% predicted by the United Nations from 7 billion people currently (FAO, 2009). Thus, packaging innovations, which serve to better protect product, reduce wastage and extend product shelf life – whilst meeting the trend for using more sustainably sourced materials – have an important role to play in contributing to the development of a more resource efficient and sustainable future.

Table 7.3 Apparent per capita consumption of all types of paper and paperboard in 2000 and 2010

Location	Apparent per capita consumption (kg) in 2000	Per capita consumption (kg) in 2010
North America	303.3	234.8
Europe	201.0	142.0
Australasia	147.6	135.0
Latin America	34.8	45.5
Asia	28.2	40.0
Africa	6.1	7.8

Source: BIR, 2011

7.1.5 Eco-innovation for a sustainable future

By combining innovation with ecology, eco-innovation can meet sustainability goals and enable companies to stay competitive. A key challenge for brand owners, the paper and paperboard industry, and industry in general, is how to adopt sustainable principles and goals whilst addressing cost, performance and market pressures. Eco-innovation or sustainability strategies in relation to packaging involve developing new materials, technologies and systems that consume less energy, water and other raw materials. There are numerous examples of companies investing in eco-innovation strategies to generate significant cost savings, both in the middle and long term, thereby improving turnover and increasing margins.

Plastics packaging prices are closely tied to volatile oil and resin markets. Petrochemical-based raw materials are not renewable or sustainable, although sizeable fossil fuel reserves are currently available. Unless recycled or reused, their use in packaging may possibly incur a high carbon footprint, particularly when used as an energy source.

By contrast, paper and paperboard materials are annually renewable, readily recyclable, biodegradable and compostable, and can be sourced from sustainably managed forests. In addition, growing trees absorb carbon dioxide from the atmosphere thereby alleviating climate change and, at the same time, generating oxygen that is vital for life. Also, sustainably managed forests promote biodiversity and provide a range of important eco-services.

A mill needs steam-heated drying rolls plus electricity for drying paper and board. Modern mills are investing in the use of combined heat and power plants (CHP), which are significantly more efficient in producing steam and electricity from either biomass, natural gas or oil. They are also investing heavily in biomass energy, which usually involves wood chips/wood waste. The paper and board industry is the largest producer and consumer of biomass-based energy – 25% of the EU total (ProCarton, 2012a). For example, Iggesund Paperboard (www.iggesund.com), a leading European manufacturer of folding box board and solid bleached board, has invested in an advanced biomass CHP and a wind turbine at its integrated mill (i.e. manufacturing both pulp and board) in Cumbria, UK. Iggesund plans to export any surplus electricity to the grid. Thus, packaging materials, which utilise biomass or other forms of renewable energy in their manufacture, can also assist governments in meeting their renewable energy and carbon reduction targets. Over the next decade, with higher fossil fuel prices and ongoing climate change concerns, further market demand for packaging, which helps to reduce carbon emissions and limit fossil fuel dependency, is likely to be stimulated.

7.2 Package design optimisation

Companies continually seek to:

- Deliver a better user experience to differentiate their brands and enhance consumer appeal.

- Minimise costs and enhance supply chain performance.
- Improve the environmental credentials of their products and services.

A main challenge for industry is to optimise design of the packaging system. This involves striving to create packaging that is balanced in terms of providing product protection and preservation, is cost-effective, creates maximum consumer appeal and at the same time takes into account environmental responsibility (Nampak, 2010).

7.2.1 Design for the consumer

Increasingly, brand owners are adopting a more 'user-centred design', or 'consumer-led' approach, to developing products and their packaging, which are well aligned to the needs and values of a clearly defined target user group. For example, smaller pack sizes that meet the needs of smaller households and more price sensitive consumers. Packaging solutions, which combine convenience and enhanced aesthetic appeal with minimal or reduced ecological impact, are increasingly valued by consumers and represent an important future trend. Paper provides an excellent printing substrate with potential for high-quality graphics to effectively communicate brand values and sustainability credentials (Section 7.1.2).

Pack usability

Convenience features, which enhance pack usability, particularly with regard to ergonomics and openability, represent another important trend. With the demographic shift to an ageing population in many advanced economies, packaging designs that facilitate quick identification of product, provide good legibility, offer ease of opening and use will gain favour with discriminating older consumers. Pack usability is also important to many of today's younger consumers who may have higher expectations regarding convenience. Consumers are less tolerant of brands that do not fully consider their needs and, increasingly, will switch to brands that do.

Ease of opening of food packaging is a high priority for many consumers, particularly the elderly, the visually impaired and those with disabling conditions such as arthritis, who often have difficulty opening cans, bottles and plastic packs. The reader is referred to the *Inclusive Design Toolkit* developed by the University of Cambridge (2011).

Paper and paperboard-based packaging with ease-of-tear open or pull-open features may offer more convenient solutions. For example, paper-based composite cans and retortable paperboard-based cartons with their easy pack opening features present an alternative option to the metal can (Sections 7.5.6 and 7.5.10).

An example of a disruptive innovation offering ease of opening and resealability of packages for dried product is the Zipbox® (www.zipbox.net) from the US. This is a linerless paperboard carton combined with a heat seal attached flexible film header containing a DoubleZip zipper profile. Zipbox® claims other benefits for consumers include product freshness and improved ease of pouring. As Zipbox®

is linerless, the package can contain between 10% and 40% additional content in comparison to bag-in-box packaging. Thereby, both source reduction (ratio of product to packaging) and cube efficiency in distribution are maximised.

Innovative closure or seal designs can be used to enhance the consumer's brand experience or engagement with food and non-food products through the process of opening the pack, dispensing product and reclosing the pack. This valued pack characteristic is well recognised at the premium end of the cosmetics and perfumery markets (Chapter 3).

Trend towards 'green' consumerism

Packaging and its sustainability credentials are integral to a brand's market performance. A Pro Carton (www.procarton.com) report on consumer attitudes, entitled *Is Sustainability in Packaging Important?* showed:

> . . . unequivocally that sustainability in packaging is important and consumers are interested in, and concerned about this issue. Over 64% of respondents said that packaging should consist of environmentally friendly materials and over 55% felt that packaging should contain as little plastic as possible (ProCarton, 2012b).

The market trend to greener products is being spurred by the growing group of health-conscious and environmentally aware consumers who have an emotional connection, or 'biophilic' preference, for biobased materials and 'natural looking' packages. This group of consumers may be characterised as one that aspires to a 'Lifestyle of Health and Sustainability' (LOHAS), as conceived by the Natural Marketing Institute (NMI, 2009). For example, there is packaging for a branded cheese stick employing a laminated paper with an ultra-thin plastic layer on the inner surface to provide both oxygen barrier and sealability. The outer brown unbleached kraft paper layer provides a substrate for printing as well as presenting a 'natural look or feel' to the pack.

Increasingly, retailers and brand owners are responding to the growing demand for more 'natural' eco-designed packaging by adopting paper and paperboard-based constructions. For example, GreenBottle Limited (www.greenbottle.com) in the UK has launched its radically innovative bottle consisting of a moulded pulp outer and a thin inner flexible plastic lining. This weighs significantly less than an equivalent capacity glass bottle. Also, it is claimed to offer lower carbon footprint and transportation costs. The GreenBottles can be moulded to almost any shape for use in most liquid markets including wine, milk and drinks, liquid food, liquid detergents/fabric softeners and motor oils. After use, the plastic liner and paper components can be separated by hand for the purpose of recycling or composting the paper, and recycling the plastic liner.

7.2.2 Design for the supply chain

Packaging innovation can contribute to developing more robust and resource efficient supply chains, particularly for the fast-moving consumer goods (FMCG) markets. For example, these include:

- Ensuring that product is sold well before deterioration in quality approaches an unacceptable level (or expiry date is reached).
- Minimising use of resources, whilst preventing wastage or product damage.
- Securing brand authenticity or product provenance.
- Monitoring package integrity.
- Preventing theft.
- Reducing costs and enhancing logistics performance.

The focus is on improving the supply chain process to minimise costs, increase profitability, optimise product quality, conserve resources and prevent product damage, counterfeiting, tampering and theft. With regard to the latter, the National Association for Shoplifting Preventions (NASP, www.shopliftingprevention.org) claims that more than $13 billion worth of goods are stolen from US retailers each year (NASP, 2012).

In response to supply chain challenges, paper and paperboard packaging material suppliers and package converters are continuing to focus their efforts on providing innovative material solutions and pack designs, which yield cost savings, functional and environmental benefits. For example, there are new specifications of corrugated fibreboard being launched with innovative designs of fluting, which have smaller flute heights and more compact flute arrangements. These can provide cost-effective alternatives to commonly fluted boards, such as B-flute, to deliver improved printing, better appearance, machine line efficiencies and significant savings in logistics for FMCG.

Through such innovations, the corrugated fibreboard manufacturer can deliver more packaging per pallet to the packer, who in turn can deliver more packs per pallet to the retail customer. Thus, both cost savings and environmental benefits can be derived in terms of reduced fuel usage, truck trips, carbon emissions and other pollutants. These innovations are stimulating market interest, particularly with regard to high-quality print primary packaging, for example retail display cases of wine and retail ready packaging (RRP). However, they also have potential application as transit packaging for products such as glass jars and bottles, where good flat-crush resistance properties are required.

Another main challenge for the supply chain is continued industry action to alleviate climate change, conserve fossil fuels and water resources, and implement other sustainability initiatives.

7.2.3 Design for the environment

It is important that more sustainable and lower carbon or carbon neutral packaging solutions using biobased materials, including paper and paperboard, are sought. One way this can be achieved is through eco-design, which aims at reducing the environmental impact of products, including their energy consumption (and, by association, carbon footprint if derived from fossil fuel resource) throughout their entire life cycle.

With continued growth in market demand for biobased packaging materials in general, the prospects for paper and paperboard packaging appear positive. Although the use of paper and paperboard may, in certain applications, currently incur more cost than their conventional petroleum-based plastic counterparts, it is important to also consider the full lifecycle from manufacture to end-of-life, including the cost of disposal to landfill or incineration. This can make the use of biobased packaging a more favourable commercial proposition (Sections 7.1.1, 7.1.5, 7.2.2 and 7.4.4).

7.3 Brand communication through packaging

In addition to packaging's prime design consideration of functionality (containment, protection and preservation), there is its strategically important communication's role as the 'silent salesperson'. Packaging serves to persuade consumers to purchase the branded product at the 'First Moment of Truth' (FMoT) at the retail point of sale. This is particularly important for FMCG brands competing for the consumer's attention in the busy shopping aisles. However, this FMoT is increasingly being complimented by the 'Mobile Moment of Truth' (MMoT), as consumers use their smart phones to scan Quick Response (QR) codes printed on packs within the retail environment and in their homes. A radically innovative brand communicative packaging technique, using a recognisable marker not requiring a specific bar code or QR code, is Augmented Reality (AR). AR was recently adopted by Lego (www.lego.com) to provide in-store three-dimensional (3D) animation for its boxes of children's model toys (AIPIA, 2012).

7.3.1 Packaging design – brand perception

Packaging needs to catch the consumer's eye whilst scanning the supermarket shelves through the use of striking or aesthetically attractive graphics and distinctive design elements, shapes, logos and promotional messaging. For example, at the luxury end of the food market, there is a globally branded premium chocolate maker, Godiva® (www.godiva.com), offering its praline chocolates in a gold coloured carton. The cartons are finished with Henkel Corporation's (www.henkelna.com) Mirafoil® technology. This uses a novel gold-shimmering non-volatile and UV-curable liquid coating incorporating highly reflective micro-thin metallic platelets formed by physical vapour deposition (PVD). MiraFoil® liquid coating technology presents a cost-effective and environmentally sound alternative to speciality papers and foil board laminating or hot foil stamping, which are commonly employed in the carton industry (*Packaging World*, 2009). Unlike foil laminates, MiraFoil® is 100% recyclable and reduces waste in the printing process.

In addition to attractive or distinctive appearance, packaging needs to communicate product attributes and characteristics valued by the modern consumer such as integrity, safety, authenticity, convenience, health, naturalness, freshness, purity, tradition, comfort, well-being, ethical and minimal ecological impact. Consumers are prepared to pay a premium for packaging, which delivers distinct benefits as well as novelty. In this regard, paper and paperboard materials are well positioned to provide packaging solutions that can satisfy consumer perception and effectively communicate desired brand values, particularly with respect to sustainability credentials.

In its report entitled *The Power of Limbic Packaging*, the Association of European Cartonboard and Carton Makers (Procarton, www.procarton.com) provides insights to the evolving area of brain research regarding brand perception (ProCarton, 2011). The report emphasises the capability of paper and paperboard to communicate brand attributes through the creative use of their distinct material properties, use of colour and shape. These packaging characteristics provide a powerful on-shelf marketing tool, which can reflect brand presence and convey brand attributes. Paper and paperboard permit a wide variety of designs and pack formats. For example, a major innovation has been the growth of several types of in-mould labelling (IML).

In-mould labelling (IML)

In-mould labelling (IML) enables the use of labels with high-quality digitally printed graphics with high resolution in a range of colours, to produce attractive designs and maximise brand impact at the point of sale. These labels are applied when shaping plastic containers using blow moulding, injection moulding and thermoforming processes.

There are also packages where a paperboard-based complete side wall is heat sealed during injection moulding to a plastic framework of minimal weight. A similar package has been developed where the sidewall with a glued side seam is attached to the rim such that it is easily removed from the plastic frame for separate recycling after the container has been used.

Innovative IML based on renewable raw materials can provide more sustainable solutions whilst delivering high brand differentiation capability and shelf impact. The paper label can be inserted using a reel fed system into the mould, prior to combining with the plastic melt to form the finished container. By using paper as an alternative to plastic film as the pre-printed label, a significant saving in the share of non-renewable fossil fuel resources can be derived. In addition, the paper label combines with the plastic melt to form an integral unit that is much stiffer than packaging made using a conventional film label. Polypropylene is commonly used for the production of containers employing IML (e.g. yoghurt pots). Although the saving per container may be small in terms of the amount of plastic displaced by using a paper label, the collective saving in resources for billions of containers manufactured on a global basis is significant. For example, in the case of its recently launched InnoMould® injection-moulded pot, Stora Enso (www.storaenso.com) claims that:

> The higher bearing pressure of the pot permits higher stacking and the walls of the pots can be produced with up to 50% less material. That means brand owners can easily achieve a significant reduction in the percentage of non-renewable materials in their packaging (Stora Enso, 2012).

InnoMould® is also promoted on its ability to reduce the carbon footprint of packaging, achieve premium quality printing, increase container rigidity and reduce costs.

Pressure-sensitive self-adhesive (PSA) labelling

Pressure-sensitive self-adhesive (PSA) labelling represents another example of a technology enhancing the shelf impact of branded products. Conventionally printed labels are widely used, but their rate of sales growth has been superseded by pressure sensitive self-adhesive (PSA) labels that have been digitally printed. The growth of PSA labels is expected to continue with expanding investment in digital presses and changes in governmental labelling regulations, encouraging the drive to shorter production runs. They are used for shorter production runs at slower speeds in labelling products such as beverages, food, cosmetics and pharmaceuticals. In recent times, the expansion in the use of plastic containers has led to a further expansion in PSA label use. A wider range of label shapes are now available and some containers have front and back labels. Release papers are also used in the manufacture of self-adhesive materials and components for an ever-increasing number of applications. Significant growth potential exists for the production of high-grade pressure-sensitive labels (Ahlstrom, 2012).

The production of PSA labels involves the label stock comprising the label with the pressure-sensitive adhesive being applied to the reverse side and laminated to a siliconised paper base. The die-cutting of the label has to be very precisely controlled so that the siliconised backing paper is not damaged. The labelling operation requires the removal of the label from the backing paper and winding both the backing paper and the label paper waste, which was left surrounding the label after die-cutting.

There is, therefore, interest in the development of a linerless self-adhesive label. This requires a release coating over the print and micro-perforations surrounding the label, to maintain its position in the reel to the point where the label is removed for application on the labelling machine (Catchpoint, 2012). Another application innovation is the ability to have front and back labels on the same backing liner.

7.3.2 Certification schemes

There are global trends of increasing ‘green consumerism’ or ‘ethical consumerism’; and ‘ethical’ or ‘cause-related marketing’, demonstrating engagement with ecological and social issues. They are well exemplified by the worldwide success of schemes such as FairTrade™ (www.Fairtrade.org.uk) and the Rainforest Alliance™ (www.rainforest-alliance.org), with their logos now common on packages within the retail marketplace.

Increasingly, both private sector FMCG companies and public sector organisations are expressing preference for packaging using label certification schemes offering high brand recognition and certified by internationally recognised bodies, for example the Forestry Stewardship Council™ (FSC™, www.fsc.org)

and the Programme for the Endorsement of Forest Certification™ (PEFC™, www.pefc.org). For example, the FSC™ chain of custody scheme provides internationally recognised standards setting, trademark assurance and accreditation services to companies, organisations and communities. Interest in such certification schemes, including those promoting materials with natural and recycled fibre content, is likely to increase as more pulp is sourced from new forest areas.

These schemes provide assurance to users and consumers that materials are sourced from forests that are managed sustainably. They are often supported by global organisations such as the World Wildlife Fund (WFF, www.wwf.org), the Rainforest Alliance™ and Greenpeace (www.greenpeace.org.uk). Many governments have issued standards for forest management. For example, the comprehensive UK Forestry Standard issued in 1998, sets out guidelines for sustainable forest management (www.forestry.gov.uk).

In the case of paper and paperboard packaging, the use of labels or symbols certifying schemes, such as recycled content, biobased and sustainable forestry, serve to support a brand's sustainability credentials. The development of such schemes responds to the growing need for brand owners to develop transparent sustainability policies. For example, through its policy devised in 2010, for sourcing paper and board, Unilever plans to secure 75% of these materials, either from sustainably managed forests or recycled materials by 2015 (Canadean, 2012).

In 2011 the United States government's Department of Agriculture (USDA, www.usda.gov) launched its 'biobased' certification standard label (www.biopreferred.gov). The USDA's definition of a biobased material is:

> Commercial or industrial products, other than food or feed, which are composed in whole, or in significant part, of biological products or renewable agricultural materials (including plant, animal and aquatic materials), or forestry materials (USDA, 2011).

The adoption of this label is being promoted by a 'BioPreferred® program', which encourages users and manufacturers to source products based in total, or significant part, on bio-derived sources. It is likely that a similar biobased certification standard or programme will be adopted by the EU in the near future.

7.3.3 Retail ready packaging (RRP)

Retail ready packaging (RRP) is a system of secondary packaging for merchandising that allows goods to be moved direct to the point of sale with minimal handling, enabling products to be displayed effectively within the secondary packaging medium. When empty, RRP can be easily collapsed and disposed of ready for recycling.

There is a higher rate of brand innovation to meet an increasingly diverse range of consumer needs in an expanded array of retail formats in both existing and new global markets. Consequently, this brand fragmentation is expected to lead to more frequent use of RRP, especially corrugated fibreboard merchandising displays, for a wide range of FMCG products and pack promotions in both existing and new retail contexts.

Industry trends, such as increased emphasis on the quality of presentation, particularly for value-added high margin products, will promote adoption of RRP for an increasing number of products and drive innovation in RRP. For example, MirriNor (www.mirrinor.co.uk), a joint venture between Smurfit Kappa (www.smufitkappa.co.uk) and Mirri (www.mirri.co.uk), offers a biodegradable metallised film laminated to T-Flute micro-flute corrugated board, which can be used for RRP. This can produce impressive graphics to deliver a premium pack appearance for a wide range of FMCG products, whilst reducing packaging weight and yet providing high strength.

RRP facilitates faster filling of supermarket shelves, thereby reducing in-store labour costs, which represent a significant portion of a retailer's Direct Product Cost (DPC). Retailers' valued characteristics of RRP also include ease-of-opening, ease-of-consumer access to products and attractive or engaging display using high quality graphics.

An important requirement in self-service retail operations is to meet the need for ease of handling and transfer of packed product from the back of the store to the shelves. Also, for the restocking at the point of sale to be as efficient as possible and for the presentation of the product on the shelf to remain orderly as the stock on the shelf is sold.

Packaging's point of sale presentation can be enhanced through easy-to-assemble (as well as easy-to-dismantle for reuse or recycling) corrugated fibreboard merchandising displays, which can provide stability and facilitate handling of packs. Free-standing merchandising displays, rapidly erected utilising the transit pack and easily dismantled, is an important example of RRP. For example, innovative modular corrugated board displays are available using a few standard components, which reduce the number of parts required for fast assembly. Merchandising displays are usually temporary structures and strategically located at the end of store aisles to promote sales of new products. Modular displays are becoming increasingly popular with retailers.

Corrugated board is a highly versatile material and its creative application can deliver striking merchandising effects and inspiring retail environments to attract and engage the consumers' interest. Attractive merchandising displays are particularly effective when launching new products in innovative promotional package designs, which communicate well with the consumer. Another term to describe this form of packaging is 'display-ready packaging'.

For further discussion of RRP, please refer to Section 7.5.7 and the UK's Institute of Grocery Distribution (IGD, www.igd.com).

7.4 Consumer well-being, package integrity, brand authenticity and ecological packaging

Strong growth is anticipated for innovations that meet the needs of modern consumers, particularly those aspiring to a LOHAS, who value products and their packaging that:

- Ensure brand authenticity or product provenance and which are perceived as perfectly safe.
- Prioritise convenience through extending shelf life when purchasing product in bulk.
- Provide an assurance of quality and maintain product integrity.
- Promote unadulterated taste and offer fresher products free of (or contain minimal) preservatives and additives.
- Possess sound ecological attributes.
- Provide value for money.

This growth will be encouraged by improvements in technical performance and influenced by a number of factors such as the economic climate and urbanisation. There are a range of innovations that can enhance performance with regard to consumers' valued product characteristics and packaging attributes. These include security packaging, smart or intelligent packaging, active packaging (AP) and modified atmosphere packaging (MAP).

7.4.1 Security packaging

There is a wide range of tamper evident packaging features available for paper and paperboard packaging, ranging from simple traditional paper security seals to holographic labels and other more sophisticated designs and technologies. Security packaging is an on-going development area and there are several possible methods to achieve anti-counterfeit and tamper evidence. This can be achieved by using a paper or paperboard with a characteristic feature, such as the presence of special fibres, and similar techniques are the subject of continuous development.

Other methods are achieved by using special printing techniques, such as using inks that are only readable under specific lighting or using colours that change under specific conditions, for example UV light identifiable QR codes and printed security marks, together with forensic and track-and-trace systems.

Features can be printed that are difficult to copy. For example, there are several types of Optically Variable Devices (OVDs) and Diffractive Optically Variable Devices (DOVDs), of which holograms are typical examples. The reader is referred to the International Hologram Manufacturers Association (IHMA, www.ihma.org) and Chapter 3.

7.4.2 Smart or intelligent packaging

Intelligent or smart packaging refers to packaging that senses and informs (Summers, 1992; Day, 2001). It is anticipated to play an increasingly important role in delivering added brand value through providing assurance of pack integrity, tamper evidence, product safety and quality. Development of intelligent/smart packaging is being accelerated by rapid innovation in enabling technologies such as nano-coatings, nano-taggants and nano-codes. Relatively low cost and disposable intelligent indicator systems can now be directly printed onto paper and paperboard packaging materials (please refer to Chapters 3, 9 and 10).

Colour change labels

Smart or Intelligent Packaging can serve to monitor quality and prevent product wastage, thereby saving resources. Conventional 'use by' and 'best before' date labelling are relatively crude measures, which often result in products being thrown away unnecessarily. It is expected that labels incorporating devices – such as combined temperature and time indicators (TTIs), freshness indicators, chillness and produce ripeness indicators – for use with paper and paperboard packaging, will become a widespread feature within the retail marketplace. These labels are sensitive to changes in the internal condition of the pack and will effect a visible colour change to alert or inform the consumer or user.

This technique may also be employed to indicate package tampering. For example, tamper proof labels are available for hermetically sealed gas flushed or vacuum packs which, if exposed to the atmosphere, either though accident or malicious intent, can effect an oxidative colour change of the label.

Colour change labels can also serve to monitor the condition of product along the supply chain. However, more sophisticated techniques, such as radio frequency identification (RFID) labelling technology, are available.

Radio frequency identification (RFID) labels and tags

RFID micro-chips can provide a 'track-and-trace' facility when used with pallets of packed product or as sensors applied to paper and paperboard packaging. An example is Mondi's (www.mondigroup.com) Intelligent Box, which is a RFID-enabled corrugated case. RFID may be integrated into packaging using very thin labels or laminates, which can significantly lower the cost of this technology. In addition to its application for anti-counterfeiting and anti-theft technologies, RFID is also being used to automate the management of stock and also enable remote monitoring of an individual product or package unit's condition, thereby providing full transparency of the supply chain. For example, an alarm can be raised if parameters change outside of set limits, such as in the case of temperature- and moisture-sensitive products.

As the technology rapidly advances and scale of production increases, the cost of the disposable integrated RFID sensors will become a more affordable commercial proposition. It is anticipated, therefore, that an accelerating number of market opportunities is likely in the near future. However, alternative technology solutions, such as QR codes, are starting to make significant inroads in the labels market.

In Germany, the Organic Electronics Association (OE-A, www.oe-a.org) is developing low-cost, printable and thin flexible electronics on polyester substrates, which can be easily integrated into packaging. RFID technology offers strong commercial potential for a range of industries and many companies. Paper and paperboard manufacturers are also actively engaged in its development (please refer to Chapter 3).

7.4.3 Active packaging (AP)

There is a range of active packaging (AP) techniques, which can be used in combination with, or incorporated into, paper and paperboard-based packaging. A few of these are discussed below, although there are other techniques such as odour and flavour absorbers or releasers. AP may also be used in conjunction with other food preservation technologies, such as MAP. Strong growth in the use of AP solutions is anticipated over the next decade.

Anti-microbial coatings and treatments

Paper-based AP packaging containing anti-microbial coatings and treatments are increasingly being used to extend the shelf life of a wide range of perishable food, including baby food, fresh produce, cheese, snacks and cold meats. For example, odourless and taste neutral anti-microbial agents can be released onto the exposed surfaces of food where microbial contamination is most likely to occur. The preservative effect of the agent results in significant reductions in food spoilage and infections caused by microbial growth. In addition, savings in both cost and environmental impact can be achieved by reducing wastage throughout the supply chain. AP technology can render the use of artificial preservatives in the product unnecessary or minimal. This caters to the desire of a growing number of health conscious consumers for 'fresher' or more 'natural' products, which promote a healthier lifestyle. An example of antimicrobial packaging is Sanocoat® from Mondi (www.modigroup.com/gogreen).

A recent development from Israel is paper coated with silver nano-particles (killer paper) which, reportedly, could provide an alternative to common food-preservation methods such as radiation, heat treatment and low-temperature storage (*Packaging News*, 2011) (please refer to Chapter 4).

Insect repellents

Insect infestation of both stored commodities and packaged food is an issue worldwide and use of insect repellents for paper and paperboard packaging has been employed for many years.

An example of a recent innovation is an Environment Protection Agency (EPA) approved water-based insect repellent coating called BugBan 9000™, available from Michelman (www.michelman.com) based in the US. This can be applied to the surface of folding cartons, paper and corrugated fibreboard, for example paper-based pallets used in the transportation of food. Michelman claims that its insect repellent is 100% effective in eradicating a wide variety of insects – including yellow jacket wasps, fire ants, German Cockroaches and Indian Mealmoths – within 24 hours (*FoodProductionDaily.com*, 2006). Michelman also states that insect infestation in the sugar, cereal, flour and pet food markets (US) totalled $226 million annually. According to Michelman, BugBan 9000™ is specifically designed to not only repel insects, but to effectively eliminate the insects before they can penetrate the package – for the entire life of the package (Michelman, 2012).

Ethylene absorbers

There is also growing interest in ethylene absorbers for produce packaging. Ethylene absorbers can be particularly useful in produce warehouses where ethylene gas, a natural plant growth regulator, can rapidly accumulate. During storage, high concentrations of ethylene can lead to accelerating respiration rate and subsequent senescence. Use of ethylene absorbers can significantly extend shelf life, thereby reducing waste in the supply chain and deriving economic benefit for suppliers and retailers. These absorbers can be embedded into paper bags or corrugated fibreboard transit packaging used for produce storage and transport. They can also be incorporated as sachets into retail packs of produce to extend produce shelf life for the benefit of the retailer and, potentially, the consumer. For example, Fresh Food Technology (FFT) Limited's (www.foodfreshnesstechnology.com) range of 'Its Fresh!'™ absorbers contained in paper sachets are being used by major UK retailers to extend the optimal quality shelf life of a range of produce including strawberries, tomatoes and avocados (FFT, 2012).

7.4.4 Modified atmosphere packaging (MAP)

Modified atmosphere packaging (MAP) utilising paper-based materials can be applied in combination with barrier plastics in a variety of lightweight package formats. These include press-formed containers, moulded trays, paper-based composite cans, thermo-formed paper-based containers (e.g. Fibreform®), plastic/paper laminated pouches or sachets, paperboard cartons and sleeves. Hermetically-sealed MAP packaging is being used to extend the shelf life of a wide range of food products, by offering barrier properties against the permeation of water, oxygen, aromas and MAP gases. It can also protect against the possible effects of light.

For example, MAP for ready meals and 'cook chill' food products can utilise paper-based packaging with appropriate plastic film barriers and lidding materials. The shelf life of ready meals and cook chill products can be extended from between 2 and 5 days to between 5 and 10 days if MAP is used (MAP, 2012). MAP has also been shown to delay the onset of a characteristic phenomenon related to pre-cooked meat and poultry known as 'oxidative warmed over flavour'– but its development may be retarded by using low oxygen level (MAP, 2012).

The product quality, convenience attributes and reduced wastage offered by such lightweight, high performance MAP solutions are valued by consumers, brand owners and retailers. Retailers and brand owners will continue to drive forward the adoption of MAP technology as their operations expand, particularly in emerging and transitional economies, which are experiencing rapid urbanisation and population growth.

Case study: Alexipack®

In 2008, UK-based Alexir Partnership (www.alexir.co.uk) launched its lightweight, fully printable and press formed Alexipack® cartonboard tray system which can be used for chilled, frozen and dual-ovenable food products. The PET coated cartonboard trays, developed in conjunction with Stora Enso (www.storaenso.com), are designed to provide high moisture and grease resistance. If MAP is required for chilled ready meals, a *deluxe* tray is available featuring a reinforced PET rim to provide high seal integrity. The design of the pack allows full direct branding whilst removing the cost of labels, u-cards, printed films and printed sleeves (Alexir, 2012).

A unique feature of the Alexipack® tray is that it has flat flanges to easily accommodate any lid type, for example standard film (single or laminate) and ventable film (for improved cooking performance). This enables high-volume in-line heat sealing to be achieved using board or plastic film lidding options, which are compatible with common sealing equipment. In addition, the trays offer good dimensional stability and heat distribution if the product requires reheating in a microwave or conventional oven.

Alexipack® cartonboard trays are lighter than CPET (plastic) trays. This assisted a UK retailer, which adopted this pack format in 2008 for its 'Healthy Living' ready meal range, to achieve a 35% weight reduction on the competing pack format used previously. It also served to meet this retailer's weight reduction programme and deliver environmental benefits. Over 75% of the Alexipack® tray is made from direct food contact approved paperboard, which is derived from FSC™/PEFC™ certified sustainable sources. The Alexipack® tray has attracted interest from both the food retail market and airline catering industry. Please refer to Chapter 2 for further discussion of MAP.

7.4.5 Biodegradable and compostable packaging

There are increasing concerns globally about non-biodegradable plastics, or plastics-based, packaging pollution. This is particularly the case with regard to flexible packaging and degraded plastics fragments or particulates, which wildlife may become entangled with or ingest. Hence, a number of governments have imposed or are planning bans, taxes and recycling obligations for plastic shopping bags. In addition, more retailers and brand owners in advanced economies are banning the use of shopping plastics bags altogether in favour of paper bags. Packaging waste pollution, or littering, is not only aesthetically undesirable but may create a health risk if food contaminated waste accumulates. Increasingly, the adoption of biodegradable packaging is being encouraged where this may be more appropriate ecologically, for example in the foodservice sector.

Litter is not only an issue in advanced economies but a growing one in other regions of the world where population growth is rapid and demand for packaged products is rising sharply. It presents a particular challenge for brand owners and their packaging suppliers to develop appropriate packaging solutions.

In South Africa, for example, Sappi Paper and Packaging (www.sappi.com) has, in recent years, been encouraging retailers to adopt its biodegradable and fully recyclable 'Green Paper Bag' (Sappi, 2009). This shopping bag is made from Kraft paper using locally sourced and manufactured paper in a process that uses a high level of biomass-based energy. It is offered as a 'greener' alternative to the non-biodegradable plastic carrier bag. The printing inks used are water-based, fully recyclable and do not release toxic chemicals when decomposing. The heavy duty Green Paper Bag with reinforced handles enables shopping loads of up to 10 kg to be carried, often involving long distances by public transport.

Paper and paper-based packaging, which is fully compostable (conforming to at least EN 13432 standards), offers an alternative to recycling, disposal to landfill or incineration. Compostable packaging can play a valuable role in producing a mulch, or soil improver, which can enhance water retention, thereby increasing water availability and providing fertiliser for plants.

In EU member states, disposal to landfill is becoming less possible due to the implementation of the EU Directive on Landfill (99/31/EC as amended in 2003 and 2008), with the consequence that biodegradable and compostable packaging options (in accordance with the EN 13432 standard) are increasingly being adopted by companies whose post-consumer used packaging is not readily recyclable. For example food contaminated single-use disposable plastics packaging in the takeaway foodservice sector.

There is a growing market demand for paper and paperboard packaging incorporating plastic materials which, whilst providing barrier and other functional needs, are also biodegradable and compostable. This is reflected in a wide range of both food and non-food product available in such packaging. For example, a 100% biodegradable and compostable tea and coffee bag tissue is available with fibre based on the Ingeo™ brand of polylactic acid (PLA) produced by NatureWorks LLC (Ahlstrom, 2009) (please refer to Section 7.5.5 and Chapter 4).

7.5 Other innovations in paper and paperboard packaging

Modern paper and paperboard packaging material suppliers and converters continue their efforts to improve the economic and environmental performance of their operations. This is being achieved through technological innovation and by maintaining environmental management systems certified to ISO14001. The industry is making significant investments to:

- Improve resource efficiency of raw material (pulp) processing.
- Use biomass energy and CHPs.
- Deliver cleaner production methods.
- Create new packaging designs using recycled and/or lightweight materials.

There are a range of manufacturing innovations involving micro-crêping, mineral surface coating, on-machine measurement, computer-based operational controls and finishing operations. Manufacturing innovations, together with associated developments in conversion (i.e. laminating, printing, coating and package manufacture) and supply chain processes are delivering increases in the consumption of recovered fibre. They are also deriving significant reductions in electricity usage, water consumption and carbon emissions. Consequently, more cost-effective and environmentally responsible production is being achieved without compromise on performance and quality.

7.5.1 Micro-crêping

Micro-crêping is a process that involves compacting the paper web in the machine direction (MD) and imparting a high degree of stretchability by passing the web between a roll and, for example, an endless rubber blanket (IP3, 2012). When applied to a thicker paper, the resulting product is capable of 3D forming into trays and other shapes (Billerud, 2012). An advantage of forming paper into containers is that existing packing lines originally designed for thermoforming plastics can be used. This means that no large investment is required and switching from plastic can be cost-effective. Innovative soft round natural-shaped containers with a range of surface textures and patterns can be formed to create distinctive brands with high shelf impact. The lightweight containers can also be embossed, coated with barrier material, printed and made sealable. For example, in 2010 a well-known UK retailer launched its sliced meat in Flextrus Paperlite® packaging based on Billerud FibreForm® (www.billerud.com/fibreform) FSC™ sourced paper (Billerud, 2010).

7.5.2 Nanotechnology – micro-fibrillated cellulose (MFC) technology

There is growing industry interest and research investment in the area of micro-fibrillated cellulose (MFC) technology. This can be used in existing paper and board products as well as to provide novel fibre-based products and barrier material solutions. MFC technology enables the development of lighter, stronger paper and paperboard packaging materials using less resource input, particularly in terms of energy and water use. For example, Stora Enso (www.storaenso.com) has invested heavily in the development of a pre-commercial plant at Imatra in Finland. With likely price rises for fossil-based plastics and aluminium, together with sustainability concerns, innovation in renewable MFC packaging materials offers a promising future in a range of product sectors (please refer to Chapter 9).

7.5.3 Innovative barrier solutions against mineral oil hydrocarbon (MOH) residues

A range of process treatments and barrier coatings against mineral oil hydrocarbon (MOH) migration are now available for every type of food packaging incorporating paper and paperboard. This development is a result of the alert raised in 2011 following studies by the Food Safety Authority (FSA) of the Canton of Zurich in Switzerland. In 2011, this FSA reported the possible health risks posed by high levels of MOH residues detected in a range of food products packaged in carton board containing recycled fibres (Grob *et al.*, 2011).

Mineral oils are widely used to make printing inks and chemical substances used for newspaper and packaging. They may be retained in recovered paper and paperboard produced from these sources. At ambient temperatures, MOH residues can evaporate or migrate from cartons to contaminate dried foods, for example pasta, semolina, rice and breakfast cereals. The Swiss researchers identified solvent-based ink components used in offset printing as the main source of the mineral oils. Most of the mineral oils are derived from newsprint that represents the main supply of recycled paper for carton manufacture. A smaller fraction was found to come from the printing inks used on the food packaging. The scientists also indicated that most conventional plastic bag liners in cartons in use do not provide sufficient protection from the transmission of mineral oils.

However, it still remains to be determined whether any specific dangers are posed by mineral oils at levels usually detected in food. The industry has argued that there is no proven danger of the mineral oils causing any harm to the public. In response to the mineral oil issue and in the absence of clear scientific and regulatory guidance, leading paperboard carton suppliers are adapting their manufacturing operations to produce innovative mineral oil barrier boards and removing mineral oils from their printing processes and products. It has been reported that the industry has already made significant progress with reductions in some cases of up to 90% in mineral oil over the 24 months up to the end of 2011 (CITPA/CEPI, 2011).

In December 2011, a decision was taken by the Confederation of European Paper Industries (CEPI) and the European converting body, the International Confederation of Paper and Board Converters in Europe (CITPA, http://www.citpa-europe.org), that all paper and paperboard packaging should:

- phase out use of printing inks based on mineral oils, and
- phase out use of mineral oil-based process chemicals used for food contact paper and paperboard packaging material.

The reported industry consensus is that this should be achievable, probably within the next five years. Meanwhile, investigations are in progress by the European Food Safety Authority (EFSA, www.efsa.europa.eu) and the German Federal Risk Assessment Authority (BfR, www.bfr.bund.de).

7.5.4 Digital printing

In recent years, digital printing has become competitive, particularly for short-run and medium-sized packaging printing for a range of paper-based formats, such as in-mould labels, self-adhesive labels, folding cartons, laminated tubes, pouches and sachets. Increasingly, the adoption of digital printing solutions is becoming more widespread globally for leading labellers and, to a lesser extent, converters of paper-based flexible packaging and folding cartons.

Digital solutions are compatible with the need to quickly respond to opportunities arising from fast-evolving retail market trends, with brand owners constantly seeking to reduce costs and reduce lead times. These market trends also require the creation of innovative and high-quality packaging alongside demand for more frequent, shorter and fast production runs.

Digital printing offers equivalent or superior print quality to flexographic processes, whilst enabling the immediate adaptation of packaging and labels to meet rapidly changing market demands. These demands may include, for example, seasonal promotions on paperboard carton wraps for multipacks of food or beverage containers.

Digital printing also enables rapid prototyping of innovative packaging that cost-effectively facilitates market trials of new designs, promotions, etc. Through its capability of using variable data, together with high accuracy and reliable colour matching, digital printing technology can provide customised print that can be personalised for individual consumers or users. Digital printing is also being applied for the purpose of providing information for improved stock control, product tracking and security.

Leading digital providers include Xeikon (www.xeikon.com), which is a world leader in dry toner digital printing, HP Speciality Printing (http://h10088.www1.hp.com) and Eastman Kodak (www.kodak.com). Unlike the dry toner process employed by Xeikon, the HP Indigo liquid toner process requires that substrates are treated with a special top/primer coating. Consequently, papermakers are developing innovative material solutions appropriate to each specific digital printing process. Although the dry toner and inkjet process is suitable for most current types of label stock, there may be a requirement for a top or primer coating treatment for certain paper applications.

A potentially disruptive innovation is nano-graphic digital printing technology from the Landa Corporation (www.landalabs.com), which uses nano-sized pigment particles. Landa claims its digital press uses less ink (water-based) and produces images of unprecedented sharpness. Fine-image detail is particularly valued for packaging of premium-quality branded products such as cartons of chocolates or cosmetics. Several major printing press companies, including Heidelberg (www.heidelberg.com), Komori (www.komori.com) and Manroland (www.manroland.co.uk) have signed strategic partnership deals with Landa (*Packaging News*, 2012).

7.5.5 Biobased materials used with paper-based packaging

There is a wide range of developments in the fields of biodegradable and compostable biobased plastics, printing inks, fibres and coatings. These can be used in combination with paper and paperboard materials in tray, pot, carton, bag and other pack formats for a variety of product applications. Such developments respond in part to brand owners' requirements for a totally compostable pack, which is aligned to their eco-innovation or sustainability strategies. In addition, there have also been significant developments in the commercial scale production of non-biodegradable biobased plastics, which meet brand owners' need for a biobased pack.

To date, the potential of biobased plastics, used in combination with paper and paperboard, appears promising with increasing market demand evident, albeit from a relatively small base. Although the scale of this growth remains uncertain, there are a number of key legislative drivers in Europe, including the EU Directive on Packaging & Packaging Waste and the EU Directive on Landfill (99/31/EC). These, together with the EU Lead Market Initiative (LMI), will encourage development of the biobased materials market. The EU LMI (http://ec.europa.eu), like the USDA BioPreferred® Program (www.biopreferred.gov), aims to promote the development and adoption of more biobased product.

Main bioplastics attracting interest for potential application with paper and paperboard include those based on or incorporating:

- Starch complexes (e.g. Novamont's Mater-bi®, www.materbi.com).
- Cellulose derivatives (e.g. Innovia Films' Natureflex™, www.innoviafilms.com).
- Polylactic acid or polylactide (PLA) (e.g. Ingeo™ PLA from NatureWorksLLC, www.natureworksllc.com).
- Polycaprolactone (PCL) (e.g. *CaPA*® from Solvay Belgium, www.solvaypress.com and Tone® from Union Carbide, www.unioncarbide.com).
- Polyethylene from bio-ethanol (Bio-PE) (e.g. from Brazil's Braskem www.braskem.com.br, which is due to start production of Bio-PP).

There is also growing interest in the use of other bioplastics, such as polyhydroxyalkanoate (PHA) and polybutylene succinate (PBS).

Paper and paperboard packaging incorporating PLA is being used in a variety of mainly fresh food packaging and short shelf-life product applications that do not require sophisticated barriers. However, it should be noted that although PLA can be commercially composted (certified according to Din Certo or OK Compost standards), PLA is not suitable for home composting (Song *et al.*, 2009).

Examples include PLA-coated paper wraps for bread and other baked products, sandwich cartons with PLA film for the windows and heat-resistant grade PLA-coated paperboard cups for hot beverages and liquid food, such as soup. Heat-resistant grades of PLA polymer, which can withstand temperatures of up to 175°C, extends its use in more widespread applications, such as microwaveable trays. In the US, for example, International Paper (www.internationalpaper.com) markets its PLA-coated Ecotainer® paperboard cup and other foodservice containers.

7.5.6 Paperboard-based carton systems – retortable, aseptic and chill preservation treatments

Cartons can offer FMCG brand owners lower production and distribution costs, whilst also meeting the needs of more price-sensitive consumer for an affordable high-quality product in a convenient and attractive package. Cost considerations have become more important in the prevailing economic climate, with distribution costs representing up to 50% or more of the price for many 'commodity brand status' products, such as milk and juice drinks. With changing lifestyles linked to rapid urbanisation, high global population growth, rising world food and beverages prices and new market opportunities, continued strong demand for food and beverage cartons appears assured.

Cartons provide an excellent medium for high-quality printing to project brand image and enhance market appeal. They also represent an excellent example of lightweight and energy efficient packaging, typically, 96% product and 4% packaging by weight. In the case of aseptic cartons, the product's shelf life can be extended by up to six months or more without the need for refrigeration. Refrigeration is an intensive user of fossil fuel derived energy resources that are associated with the generation of emissions, including greenhouse gases.

Several large manufacturers, with Tetra Pak being the world leader, market complete processing, packaging and distribution systems delivering solutions for beverages and a wide variety of other food. These include milk and soy drinks, juices and nectars, still drinks, wine, ice cream, cheese, dry foods, fruits, vegetables and pet food.

Plastic-coated cartons can be used for fresh/chilled liquid food and beverages. These cartons have paperboard as their main component (at least 75%), with plastic as an additional component. In the case of laminate constructions, as used for retortable and aseptic product applications, there is also aluminium foil.

Another innovation, which is available for aseptic cartons, is to replace aluminium foil by ethylene vinyl alcohol (EVOH), an excellent oxygen barrier and more easily managed in the waste stream.

Retortable paperboard-based carton systems

Both Tetra Pak and SIG Combiloc have introduced retortable cartons for food products (up to 500 ml capacities) with the respective brand names, Tetra Recart® and CombiSafe™. These combined processing and packaging systems are designed for in-container sterilisation of food products, particularly those containing chunky and highly viscous components, for example chopped tomato in juice, soup with particulates, sauces, peas, beans, pet food, semolina, fruit and rice pudding. In the case of Tetra Recart®, the carton comprises a six-layer laminate including paperboard, aluminium foil and polypropylene, with heat seals adequate to withstand retorting up to temperatures of around 130°C.

These cartons are ergonomically shaped for firmer grip and are available with complete laser-perforated tear-open tops. Added convenience value for consumers is provided by a re-close feature and ease of disposal. A key valued packaging characteristic claimed for these cartons is that they can be opened relatively easily without the need for opening tools.

As in the case of aseptically packed drinks cartons, retortable cartons can project a fresh natural image with excellent graphics potential and shelf impact capability. The lightweight cartons also improve shelf space efficiency by up to 40%, the speed of shelf replenishment and unit product fill per pallet. In addition, these carton formats assist retailers with reducing their Direct Product Costs (DPC). Both the empty Tetra Recart® and CombiSafe™ cartons are produced in the flat, thus deriving economic and environmental benefits in transportation. Environmental, social and economic considerations will play an increasingly important role in the direct competition between retortable cartons and can packaging systems.

The packaging material used for a Tetra Recart® carton weighs 64% less than that of a can for a 400 g package weight equivalent (Tetra Pak, 2012). Thereby, the lighter weight and efficiency in space utilisation of the Tetra Recart® carton package derives significant saving in resources, carbon emissions and transportation. Tetra Pak also reports that 32 000 empty flat Tetra Recart® cartons (for a 390 g package) can fit on a 1150 × 800 mm pallet, compared to only 3120 cans (for a 400 g package). Tetra Recart® cartons can be filled at either rates of 24 000 packages per hour with the Tetra Pak R1 line or 6000 packages per hour with the Tetra Pak R2 line. Market reaction to these innovative carton formats appears positive since their launch. Tetra Pak reports that over 100 brands have adopted this innovation since the launch of Tetra Recart® in 2001 (Tetra Pak, 2012).

Paperboard-based carton packaging systems for chilled and aseptic liquid food and drink

Tetra Pak and SIG Combibloc dominate the global market for cartons of chilled fresh and aseptically packaged drinks. Elopak is also a main player with its range of gable-top Pure-Pak® and other carton types.

Like retortable cartons for food, aseptic carton systems offer similar features and benefits including long shelf-life product, high-quality graphics capability, lightness, low fragility, low cost and space efficiency in distribution. In addition, there are associated savings in transport fuel and emissions, including greenhouse gases.

Although a limited range of standard carton sizes are offered for both aseptic and chilled fresh drinks, a variety of designs are available, including a number of ergonomic pack shapes and a wide variety of convenience in-use design features. In the case of pack sizes ranging from 500 ml capacity, there is a wide range of plastic screw-action closures with tamper evident features and peelable foil-based tabs. For small portion size drinking-on-the-move cartons, there are ease-of-pack entry features for use with plastic straws.

A recent AP innovation is Australia-based Unistraw International Limited's (http://unistraw.com) patented on-pack straw drinking system, which can deliver probiotics, flavours, vitamins, minerals, etc. stored in the straws. Following an agreement with Tetra Pak, these 'sensing straws' can be used with small-sized Tetra Brik Aseptic® (TBA) cartons of milk, juice and other dairy products.

In addition, there is a growing range of ergonomically-shaped and distinctive cartons on offer. For example, in addition to its distinctive Tetra Pisma® Aseptic carton, used for products such as juice drinks and wines, Tetra Pak offers a new ergonomically-shaped container called the Tetra Evero® Aseptic one-litre bottle for milk, launched in 2011. This package facilitates handling and ease of pouring by a broader range of consumers. Also, Tetra Pak recently launched its shaped Tetra Brik Aseptic Edge® carton, with its distinctive sloping top panel, for small portion and one-litre sizes. The Tetra Brik Aseptic 1000 Edge® for the one-litre carton features a lightweight ergonomic closure design, the LightCap 30 screw cap. SIG Combibloc markets its combishape™ series of aseptic cartons for brand owners to distinctively position their brands. Elopak also markets a range of shaped cartons for chilled fresh liquid product, for example Pure-Pak Curve.

In addition to its standard combibloc™ and combifit™ aseptic cartons, SIG Combibloc markets its combibloc EcoPlus™ aseptic carton pack for liquid foods. The latter uses a special new composite structure cardboard which, according to SIG Combibloc, enables the generation of 28% less CO_2 compared with the same-format one-litre carton pack. The CO_2 reductions are based on product lifecycle from the production of raw materials to the finished carton pack (SIG Combibloc, 2012).

Future innovations in drinks cartons will include a focus on technologies, such as direct-injection closures, new types of material made from biobased plastics and new carton shapes. Carton manufacturers are continuing to 'green' their operations and are developing innovative packaging such as lighter weight cartons and lightweight closures which are user-friendly; and foil-less aseptic cartons using EVOH as an alternative oxygen barrier material. There is also the option of using a 100% biobased carton using bioplastics, such as bio-polyethylene (bio-PE) derived from sugar cane that is certified according to sustainability criteria, for example Bonsucro™ (www/bonsucro.com).

7.5.7 Corrugated fibreboard packaging

Corrugated fibreboard packaging represents the largest use of paper and paperboard for packaging. Also, being the main form of transit packaging, it plays a significant role in logistics. It is mainly used as secondary packaging in the form of trays and cases. It is also extensively used for RRP (Section 7.3.3). In addition, it can be used for primary packaging which can be defined as all packaging intended for purchase and take-away by the consumer.

For cases and trays, a number of individual flute specifications are possible and new specifications are regularly being launched. With corrugated fibreboard, the designer of the pack has to balance good stackability and compression strength with print quality. There is a continuing and growing trend to introduce trays and other designs of packaging that use less corrugated fibreboard. These packs, which are shrink- or stretch-wrapped with plastic film, allow the product to sustain handling in distribution and storage, as well as be seen by consumers.

In the retail packaging sector, the trend continues for converters and distributors to use grades of premium quality, high brightness and whiteness top liner made from bleached kraft. This material provides an excellent substrate for accurate colour reproduction and a high strength material suitable for heavy duty protective packaging applications. These liners are valued for brand stand-out in packaging forms, such as corrugated fibreboard merchandising displays, on-shelf trays and boxes.

The industry is seeking to exploit the potential for lower grammage liners in the corrugated fibreboard sector. However, a reduction in grammage has implications for compression strength properties, though this may be addressed through modern handling systems. The fine, or lower height/amplitude, flutes – commonly known as micro-flute – enable good print presentation suitable for primary packaging applications. However, they are not the strongest for compression strength in distribution and storage. Micro-flute corrugated fibreboard is commonly used as primary packaging for retail shelf display of fragile goods. For example, it may be used for products such as canned drinks where a degree of shock or impact protection, puncture resistance and high-quality graphics for brand impact are required (please refer to Section 7.2.2).

7.5.8 Folding cartons

Cartons are small- to medium-sized paperboard boxes and comprise a significant proportion of the packaging found in the retail sector. They are used for a wide range of products, including food such as those involving storage in frozen and chilled conditions and where the product is intended to be reheated in a microwave, radiant or convection heated oven.

Examples of innovations include:

- *Those introduced in the pre-press field.* Artwork production is now digital, with numerous benefits such as electronic communication and sharing information – packaging and graphic design companies can now work 24/7 around the world. Pre-press proofing makes the approval and sharing of information more convenient.
- *Sheet fed litho printing* has increased significantly. Press changeovers are much faster as a result of being able to record and save settings and having on-machine push button support for both colour density and register control. There are also improvements in both front and side sheet lay control. Settings can be stored so that preparing the next print run of the same design is a faster process.
- *Sheet fed machines, which cut and crease paperboard* for folding cartons have been continuously developed to improve speed, reduce set-up time, record settings for repeat orders and achieve more consistency with increase performance.
- *Machines which glue carton side seams and, for some designs, bottom flaps,* have also been developed to improve speed and efficiency. Additional features have been added at the pre-feeding end resulting in similar improvements. Code readers can be used to prevent the mixing of carton designs, which have different print but the same designs. At the delivery end, automated counting, packing into transit packaging and palletising are available.
- *Use of UV curing of inks and varnishes has improved rub and product resistance in sheet fed litho printing.* This has made it possible to carry out subsequent operations without it being necessary to wait for oil-based inks to dry.
- *Machines are now available, which meet the production demands of both high and low folding carton volumes.* For high carton volumes, high output wide machines are used with in-line gravure printing and rotary cutting and creasing.

Where volumes are low, narrow web machines with in-line flexo printing and flat bed cutting and creasing are available.

- *Use of water-based inks and surface coatings* has increased in response to environmental demands for reducing the amounts of organic solvents in the atmosphere.
- *Innovative adhesive formulations* are being developed to enhance efficiency in the production of cartons or paperboard packages. For example, Henkel's (www.henkel.com) Aerbond® system features a laminating adhesive and a foaming device, which optimally disperses air into the adhesive, maximising the volume used but minimising the amount of adhesive applied, thereby saving cost and resources used while enhancing bond security. Conditioning times are significantly reduced (maximum of 2 hours) compared with conventional adhesive blends, which typically possess a high moisture content requiring conditioning times of up to 24 hours. This technology is being used globally in sheet-to-sheet and for in-line lamination.
- *Carton packing – involving carton erection, product loading, closing and packing in transit packaging by end-users* – maybe by manual, partly manual and mechanically assisted, or fully automatic means. Many innovations have been introduced in the machinery used. Typical fully automatic speeds are from 60 per minute upwards, though not many machines operate above 500 per minute. Machines that run a range of cartons have automatic setting recording to reduce down time at changeovers – they also have on-line fault detection and support.
- *Anti-counterfeiting techniques* (Sections 7.4.1 and 7.4.2).
- *Insect repellent techniques* (Section 7.4.3).
- *'Scratch and sniff'* can be used in a surface printed design.
- *Use of Braille*. This is a form of surface design using embossed letters and numbers, which enables the visually impaired to read and identify printing.
- Many examples of folding cartons being used as formed, or erected corner sealed, trays where they are combined with plastics for food products. These products are distributed in frozen or chilled systems and reheated in microwave, conventional radiant or convection ovens. Cartonboard trays are also used in fast food outlets in place of plastic trays.
- *Use of innovative biobased polymers films and coatings* (Section 7.5.5).

7.5.9 Paper-based flexible packaging

Paper-based flexible packaging is widely used for a variety of product applications in retail, catering and other market areas. There is a wide range of flexible packaging incorporating paper comprising sachets, pouches, strip packs and bags – made on form-fill-seal (FFS) equipment – and overwrapping materials. It is also used as a lidding material to provide product protection and tamper evidence when used as a membrane or diaphragm in rigid metal, glass, composite and plastic containers. Paper-based flexible materials are also used for bag-in-box lining material, multi-wall and single-wall bags.

In food products, flexible packaging may use paper-based constructions incorporating:

- Moisture, aroma and gas barrier materials, e.g. aluminium foil and barrier plastic film or coatings.
- Light barrier coatings and materials, e.g. metallised paper and aluminium foil.
- Coating materials/bonding agents/adhesives, which enable cold seal or heat seal.
- Paper and plastic materials resistant to grease or fat contained in the product.

For several decades, paper has been used as part of a laminate construction, often combined with high barrier substrates and a heat seal layer. The foil laminate pouch for dehydrated soup is a classic example. There is a large range of laminations and coatings, which can be water-based, solvent-based, hot melt or plastic extrusion coated.

A wide range of paper-based flexible packaging structures printable by a number of printing processes are available. Paper has properties of good printability, crease-ability, fold-ability, run-ability and glue-ability on packaging machines. However, paper is permeable to water, including aqueous solutions and emulsions, moisture vapour and common gases. With the exception of greaseproof paper, glassine and vegetable parchment, it is permeable to fat and fat-containing products. Innovations drawing together technology from several areas have resulted in a wide range of packaging applications that address this issue. These areas include plastics, inks and printing processes, converting machinery such as extrusion coating and lamination, packaging machinery and aluminium foil.

Barriers are now achievable at various levels for moisture vapour, gases, fatty substances, for flavour and aroma retention and for protection from the ingress of odours from the surrounding environment. A wide range of plastics are used to provide such barriers, along with aluminium foil. EVOH polymer offers potential as an alternative gas barrier to aluminium foil. Paper can also be metallised by forming a surface barrier coating of aluminium several microns thick. Metallisation offers the potential for a more sustainable alternative to aluminium foil-laminated papers. However, there are a number of developments underway for innovative high-performance plastic barrier materials. The Fraunhofer Institute for Electron Beam and Plasma Technology (www.fep.fraunhofer.de/en.html) is actively engaged in research in this area.

The range of other materials now combined with paper in flexible packaging to provide a wide range of barrier properties include:

- Polyethylene (PE).
- Polypropylene (PP).
- Polyethylene terephthalate (PET or PETE).
- Ionomer Surlyn®.
- Ethylene vinyl alcohol (EVOH).
- Polymethyl pentene (PMP).

- Lamination with plastic films or aluminium foil.
- By treatment with wax, silicone or fluorocarbon.

The range of plastic materials has been extended with the addition of biopolymers, which are biodegradable and may be sourced sustainably. The plastics used also frequently provide heat sealability, though for heat-sensitive products, cold sealing coatings are available.

In packaging machinery, innovation is involved in developing form-fill-seal (FFS) machinery for a wide variety of packs such as tea bags and sachets, strips, pouches and bags for powders, granules, solid items and liquids. Other machines include over-wrappers and machines such as those for lidding trays. These provide product protection and tamper evidence when used as a membrane (diaphragm) in rigid plastic, metal and glass containers. Paper-based flexible materials are also used for bag-in-box lining material, multi-wall and single-wall bags. Packaging has also involved the development of machines that vacuumise and gas-flush with nitrogen, or mixtures of gases including carbon dioxide, to extend the shelf life of food contents.

7.5.10 Paper-based composite cans

Increasingly, paper-based composite cans are being adopted by major brand and private label companies around the world in response to the demand for lower cost, safer and easier to use packs with good graphics capabilities. They are available as individualised pack designs, such as oval shapes, for high impact on-shelf brand differentiation or in rectangular/square shape designs for optimal shelf space utilisation and efficiency in distribution. Both conventional round and distinctive ergonomically-shaped designs are available together with designs using spiral-wound 100% recycled paperboard. Composite cans are valued for their light weight, low cost transport and resource efficiency, so conserving energy and reducing ecological footprint.

They represent an alternative to glass containers and metal cans for a wide variety of dried food products, for example breakfast cereals, milk powder, freeze-dried coffee and pet food. Composite cans are also used for drinks, frozen concentrates and non-food products. The global leader in the composite can market is US-based Sonoco (www.sonoco.com).

Innovations were introduced in material selection, design of the cans including both the sidewalls and ends, methods of can manufacture and in the packing operation. The sidewall materials which have been developed are either unbleached kraft or recycled paper. The kraft may also contain specified recycled fibre content. The outer ply/label is usually a bleached (white), kraft paper.

For gas/aroma/liquid/moisture barrier performance, a range of paper laminations and extrusion coatings are available to provide the level of performance required. The barrier materials used include aluminium foil, low and high density polyethylene (including white pigmented grades), polypropylene, metallised PET (or PETE) and Surlyn®. Gas flushing with nitrogen is used to provide long shelf life for products that are oxygen sensitive such as nuts and potato crisps (chips).

A wide variety of can ends, which are seamed onto the can bodies, have been developed using paperboard, metal or plastic materials. Easy-opening ends have also been introduced. Some cans have a simpler end applied after the product has been filled that uses a heat sealed membrane and a simple design of plastic cap, which also becomes a reclosure after opening the package.

7.5.11 Moulded pulp containers

The most widely used examples of moulded pulp packaging are the clam shell type of egg boxes, trays of all kinds for fruit and vegetables, and top and bottom protectors for bottles and jars.

Moulded pulp packaging is based on recycling fibre recovered from waste paper, dispersed in water, formed and dried. Conventional moulding involves forming conducted in moulds by vacuum suction or by pressure. The moulds are perforated to assist water removal, which may be by suction or pressurised hot air. Drying may either be by oven or using thermoformed 'Cure-In-The-Mould' technology. The latter involves heated forming moulds, which press and densify the moulded product. Rotary press-to-dry pulp moulding processes produce thin-walled containers, which are strong with well-defined smooth surfaces suitable for high end quality packaging.

Whilst recovered waste paper is the main source of fibre raw material, there have also been innovations in the source of fibre used to make moulded pulp containers. Examples include the use of sugar beet co-products, palm fibre, bagasse (sugar cane residue), coconut husk and indigenous grasses, such as bamboo and reeds.

7.5.12 Paper bags

Paper bags are one of the oldest forms of packaging. Traditionally, there have been several well established designs including flat bags, satchel bags (with gussets) and several designs of the self-opening satchel (SOS) bag, which has a rectangular base that enables the bag to be free-standing.

Innovations have included:

- A range of carrier bags with handles made from twisted paper, rope, cotton twill, ribbon, cord, polypropylene and die cut paper. Carrier bags are either basic brown shopping bags or more luxurious with high-quality printing and gloss.
- Bags with strip polypropylene windows, which can be micro-perforated to maintain the correct product moisture balance.
- Barrier properties constructed using paper laminated with PE, PP or PET to ensure the right moisture vapour transmission rate for the product. Bags may be heat sealed after the product is inserted if a leak-proof performance is required. Bags can be made using greaseproof or grease resistant paper if the nature of the product requires grease resistance.
- A wicketting system that joins the bags at the end of the bag machine in counted quantities. The sets of wicketted bags are necessary for the in-feed of bags into a bread packing machine.

- Stronger carrier bags that are packed by the product producer in weights up to 12.5 kg. In the case of potatoes, the bag can have a window made from netting.
- Higher grammage brown kraft paper for the larger supermarket check-out bag, also referred to as a 'shopper' or 'chuck' bag, which is particularly popular in North America.
- Metal strips that can be applied to bags to act as a reclosure feature.

7.6 References

AHLSTROM (2009) 'Biodegradable Solutions for Infusion Products (Tea and Coffee)', press release. Available from: http://www.ahlstrom.com/en/sustainability/newsletter/Documents/Sustainability%20matters%20newsletter%201_2009.pdf

AHLSTROM (2012) Ahlstrom speciality papers, products: release papers. Available from: http://www.ahlstrom.com/en/products/specialtyPapers/Pages/Releasepapers.aspx [Accessed October 2012].

AIPIA (2012) 'Augmented Reality – See in 3D Before You Buy', producer features, Active and Intelligent Packaging Association. Available from: http://www.aip-worldcongress.org/ [Accessed October 2012].

ALEXIR (2012) 'Press Form Trays'. Available from: http://www.alexir.co.uk/packaging/international/products/press-form-trays [Accessed October 2012].

BILLERUD (2010) 'Marks & Spencer Launches Sliced Meat in Paper Packaging Based on Billerud FibreForm®', Fibreform® News article, Issue: 6 October 2010. Available from: http://www.billerud.se/en/Media/Press-releases/2010/Marks--Spencer-launches-sliced-meat-in-paper-packaging-based-on-Billerud-FibreForm/ [Accessed October 2012].

BILLERUD (2012) 'Fibreform®'. Available from: http://www.billerud.se/en/Our-Offer/Materials/Billerud-FibreForm/ [Accessed October 2012].

BIR (2011) 'The World Recovered Paper Market in 2010', prepared by Magnaghi, G., Bureau of International Recycling (BIR) report. Available from: http://www.bir.org/industry/paper/ [Accessed October 2012].

CAMBRIDGE UNIVERSITY (2011) 'Inclusive Design Toolkit', Engineering Design Centre, Cambridge. Available from: www.inclusivedesigntoolkit.com/ [Accessed October 2012].

CANADEAN (2012) 'Global Packaging Industry Survey 2012: Trends and Opportunities in Packaging, Budget Allocation, Procurement and NPD', report. Available from: http://www.canadean.com [Accessed October 2012].

CATCHPOINT (2012) 'Catchpoint Linerless Labels'. Available from: www.catchpointlabels.com/ [Accessed October 2012].

CITPA/CEPI (2011) 'Paper-based Packaging Industry Takes a Commitment on Mineral Oils in Food Packaging', issued: 8 December 2011, press release, Brussels. Available from: http://www.citpa-europe.org/index [Accessed October 2012].

DATAMONITOR (2008) 'Global Paper and Paperboard – Industry Profile', June 2008. Available from: http://www.datamonitor.com/ [Accessed October 2012].

DAY, B. P. F. (2001) 'Active packaging – a fresh approach *Brand* ©', *The Journal of Brand Technology*, 1(1), 32–41.

FAO (2009) 'How to Feed the World in 2050', Executive Summary, Food and Agricultural Organization. Available from: http://www.fao.org/ [Accessed October 2012].

FFT (2012) 'It's Fresh™', press releases. Available from: http://www.foodfreshnesstechnology.com/?page_id=27 (Accessed October 2012).

FOODPRODUCTIONDAILY.COM (2006) 'Packaging Keeps Bugs Away', *Food Production Daily*, Issue: 30, May 2006. Available from: http://www.foodproductiondaily.com/Packaging/Packaging-coating-keeps-bugs-away

GROB, K. *et al.* (2011) 'Migration of mineral oil from printed paperboard into dry foods: survey of the German market', *European Food Research and Technology*, Issue March.

IP3 (2012) 'Micro-crêping', Institute of Paper, Printing and Publishing', glossary. Available from: http://www.ip3.org.uk/glossary.htm [Accessed October 2012].

MAP (2012) 'Modified Atmosphere Packaging of Prepared Food & Ready Meals'. Available from: http://www.modifiedatmospherepackaging.com/Applications/modified-atmosphere-packaging-ready-meals.aspx [Accessed October 2012].

MICHELMAN (2012) 'Michelman Introduces BugBan™ 9000 as the Solution for Eliminating Insect Infestation on Packaging'. Available from: http://www.michelman.com/News-%26-Information/Releases/Michelman-Introduces-BugBan-9000 [Accessed October 2012].

NAMPAK (2010) 'Sustainable Packaging, Annual Report 2010', p. 46, sustainability report. Available from: http://www.nampak.com/DynamicData/AnnualReport/Current/Sustainability%20report.pdf [Accessed October 2012].

NASP (2012) 'Shoplifting Statistics, National Association for Shoplifting Preventions (NASP)'. Available from: http://www.shopliftingprevention.org/ [Accessed October 2012].

NMI (2009) *The LOHAS Report: 2009*, Harleysville, PA, Natural Marketing Institute.

PACKAGING NEWS (2011) 'Scientists Develop Silver-Coated "Killer Paper" for Packaging Applications', Issue: 20, January 2011, article. Available from: http://www.packagingnews.co.uk/news [Accessed October 2012].

PACKAGING NEWS (2012) 'The five lessons of Drupa 2012', *Packaging News*, June, 25.

PACKAGING WORLD (2009) 'Godiva goes green while printing gold', *Packaging World*, 30 October. Available from: http://www.packworld.com/package-design/graphic/godiva- goes-green-while-printing-gold [Accessed October 2012].

PROCARTON (2011) 'The Power of Limbic Packaging', report. Available from: http://www.procarton.com/files/publications_item/brainpacks_english1.pdf [Accessed October 2012].

PROCARTON (2012a) 'Sustainability and Cartonboard Packaging', report. Available from: http://sustainability.procarton.com/files/file_manager/sustainability_files/sust_folder_en.pdf [Accessed October 2012].

PROCARTON (2012b) 'Is Sustainability in Packaging Important?' Available from: http://www.procarton.com/files/file_manager/press_0210/procartonmarktforschung_engl.pdf [Accessed October 2012].

SAPPI (2009) 'Case Study: Paper Bags the Way to Go', Sappi Sustainability Report. Available from: http://sappi.investoreports.com/sappi_sdr_2009/paper-bags-the-way-to-go/ [Accessed October 2012].

SIG COMBIBLOC (2012) 'Less CO_2, Higher Content of Renewable Resource – Better for the Environment', Combibloc EcoPlus. Available from: http://www.sig.biz/s

SONG, J.H., MURPHY, R.J., NARAYAN, R. and DAVIES, G. (2009) 'Biodegradable and compostable alternatives to conventional plastics', *Philosophical Transactions of the Royal Society*, B1, 1–12.

STORA ENSO (2012) 'Innomould – Paper for In-mould Labelling'. Available from: www.storaenso/innomould [Accessed October 2012].

SUMMERS, L. (1992) *Intelligent Packaging*, London, Centre for Exploitation of Science and Technology.

TETRA PAK (2012) 'Tetra Recart®: From A to Z. In 6000 or 24000 p/H'. Available from: http://campaign.tetrapak.com/tetra-recart/ [Accessed October 2012].

USDA (2011) 'BioPreferred® Program'. Available from: http://www.biopreferred.gov/Default.aspx [Accessed October 2012].

WORLD PACKAGING ORGANISATION (2008) 'Market Statistics and Future Trends in Global Packaging', p. 11. Available from: http://www.worldpackaging.org/publications/documents/market-statistics.pdf [Accessed October 2012].

8

International environmental and sustainability regulatory and legislative frameworks for the packaging of food, beverages and other fast-moving consumer goods

G. R. Inns, PEC Partnership Limited, UK

DOI: 10.1533/9780857098979.221

Abstract: Legislation regarding legislation in respect of the environment and sustainability has focused exclusively on the former. Even within the environmental aspect, there is an overwhelming focus on recycling, although some important jurisdictions do seek minimisation. This chapter is not an exhaustive list of all legislation within this category but rather examines the most important pieces of legislation in Europe, North America, Australasia and Asia. It also seeks to look to the future, in particular the potential influence on packaging regulation of the shift in policy emphasis towards reducing product waste. If that were to take place in a truly sustainable fashion, then the positive influence of packaging in reducing waste would have to be taken into consideration in legislation. Should this happen, it will represent an important step forward in the fight to create a truly sustainable use of global resources.

Key words: legislation, sustainability, Europe, North America, Australasia, Asia.

8.1 Introduction

The regulatory framework covering environmental and sustainability issues regarding packaging has a long and complex history of development starting from concerns regarding packaging in litter. 'Keep Britain Tidy'[1] came into being in the UK as long ago as 1954. The 'Tidyman' started to appear on packaging in 1969, by which time the incidence of packaging in litter was regarded by many as a

Published by Woodhead Publishing Limited, 2013

serious issue. Similar trends were appearing at the same time across Europe and around the developed world.

In 1974, not long after the Tidyman initiative, INCPEN[2] was formed by the packaging industry, one of its first priorities being to address the litter problem. Littering is still a cause of concern today and as recently as May 2011, CEOs from Europe and North America came together in an event organised by Europe's convenience food packaging association, Pack2Go, and North America's Foodservice Packaging Institute to discuss means to address the issue.

Litter is not just a problem for the developed world. In the developing world, poor consumers can often only afford to buy unit doses sufficient for a single use. Whilst the rag pickers that constitute the main means of waste collection in poor areas will pick up and sell most large pieces of packaging, they tend to ignore these small packs as having little or no resale value. In countries such as India, this has led to pressure for legislative action to ban such packs.

This original concern about litter inspired a much wider debate about the role of and need for packaging that continues today, becoming more complex and antagonistic. One regrettable thing that early debate did do was to fix in the minds of the public and politicians the idea that packaging was only about its end-of-life phase, so the positive roles of packaging and its benefits were mostly overlooked. That idea that packaging is all about waste and not about waste prevention still influences the manner in which legislation is conceived and its focus on recovery and reduction.

This core debate and its origins in concerns about litter have spawned a range of side issues concerning packaging and its use, notably with regard to packaging for alcoholic beverages and soft drinks. Famously, in 1982, Denmark introduced a bill prohibiting the use of cans for soft drinks and beers, a law that was only repealed under pressure from the EU 20 years later. Ironically, it was arguments about barriers to imports of foreign beverages in cans rather than waste prevention that brought about this change.

Still in The Netherlands most glass and plastic beverage containers carry a refundable deposit that is credited with helping to achieve high levels of recycling. However, for trade reasons this may again be lifted.

It was something of a logical progression over time to move from concerns over litter, through questions about the waste that litter represented, to an eventual wish to understand the environmental impact of that waste. That wish for understanding was a major reason why by the 1970s lifecycle thinking and Life Cycle Analysis (LCA) came to the fore as a means of evaluating and comparing environmental impacts of products. Whilst the technique was intended to be widely applied in response to these concerns, packaging became its focus.

There have been many attempts to include LCA analysis in legislation. The Swiss have got the nearest thus far by making LCA studies fundamental to decisions about environmentally-based legislation. LCA does, of course, produce an increasing range of different factors for the analyst to compare and it is not unusual that different factors support different courses of action. Determining the best course of action from an LCA thus requires a balanced judgement to be made and analysts will not always agree.

Legislators have on occasion proposed that an algorithm be composed that would automate the LCA balance process and produce a single figure, such as the Packaging Environmental Indicator (PEI) proposed some years ago by the EU. Fortunately, they were persuaded that to utilise such a scientifically unsound approach would be misleading and help nobody.

In any event, measuring packaging alone without including it within the product impact is largely meaningless. In addition, LCA calculations are designed to examine multiple impacts, emissions to air and water and the consumption of energy being the three commonest. The magnitudes of these different impacts are often seen in different rank order when comparing alternate systems. Legislating in a meaningful way under these circumstances is almost impossible.

It may still be that in future we see attempts to utilise other apparently simpler measures, such as the carbon footprint of packaging in legislation. However, the two fundamental problems will remain; assessing packaging in isolation from product tells you nothing and carbon footprint on its own, whilst important, is only part of the story of environmental impact. As will be seen later, more than 50 factors have been put forward as important in valuating environmental impact. Carbon footprint is one, and a very important one, but by no means does it represent the total picture.

From the early 1980s, when specific packaging legislation started to be introduced, there has been concern that an obsessive fixation on packaging was distracting from the fundamental issue of preventing waste overall. Unfortunately, that bigger debate became obscured by arguments about growing landfills, misconceptions about the relative contributions made by industrial as opposed to domestic waste and within domestic waste, and debates about what contribution came from each of the different constituent contents of domestic waste bins. All of which tended to exaggerate the contribution from packaging.

That focus on landfills has turned attention to recovery and recycling and that, because of the visibility of packaging as an apparently large constituent of domestic waste, kept the focus there. Accordingly, whilst there had been a Waste Framework Directive in place since 1975, it was not until 2005 that a thematic Strategy on the Prevention and Recycling of Waste was put in place.

As a further consequence of this approach, the concept of 'prevention' of waste has up to now remained synonymous in legislative thinking, with 'reduction' in the quantity of raw materials used in the manufacture of an article. The concept has not yet come to embrace waste prevention in its most fundamental form, that is to say, including materials that are consumed but never usefully applied.

However, in one potentially positive move in 2011, the EU published their *Roadmap to a Resource Efficient Europe*,[3] containing a specific goal for a 50% reduction in edible food waste by 2020. That target has the potential to become a tipping point in the way that packaging is addressed in legislation. Although it misses altogether the importance of prevention of transport damage, and hence a key role of packaging, the roadmap does set out that:

- The manufacturing sector will have to increase use of its waste as by-products.
- The retail sector will need to improve its stock management and selling methods to prevent products reaching their end-of-life before being sold.
- End-of-life dates alone appears to be responsible for about 20% of the edible food wasted.
- The restaurant and catering sectors will need to manage their systems and services, so as to reduce drastically their food waste losses.
- Consumers will have to be encouraged to find new ways of using edible food, which they would otherwise throw away.

For consumers, the reasons why people waste edible food are diverse and include cultural attitudes, lack of knowledge of good conservation and storage practices, outsize portions, and lack of awareness about how to re-use leftovers.

Unfortunately, according to the EU Environment Commissioner, the Roadmap still characterises packaging in the following way: '. . . the packaging used to wrap the food is also costly and needs managing as waste'.[4] So, as yet, legislators are still failing to recognise the vital role played by packaging in reducing and eliminating waste, even though achieving such ambitious goals as are set out in the Roadmap must involve more effective use of packaging.

However, the debate on preventing product waste still has a long way to go. As it progresses, there is some hope at least that legislation will evolve to embrace the positive aspects of packaging in that field.

It is worth noting that in some places, voluntary agreements have been found to be at least as effective as legislation and in some places have taken its place. For example, packaging covenants exist in The Netherlands and in Australia, aimed in particular at reducing the quantities of packaging to be used. Whilst compliance is by definition voluntary, failure to sign up or follow the rules can lead to other measures being applied. For example, in Australia, a company not complying becomes subject instead to the National Environmental Protection (Used Packaging Materials) Measure.[5]

This voluntary approach has proved to be popular in a number of regions, not least in South America. In Brazil in particular, an organisation called CEMPRE[6] has had a great deal of success in working with industry to create an entirely voluntary scheme for collecting and recycling packaging waste. This voluntary scheme has achieved results comparable to the legislative approach in Europe and has been copied by a number of neighbouring countries as well as spreading to Asia.

8.2 The environmental and sustainability regulatory and legislative framework in Europe

There are no less than 33 environmental or sustainability-based EU directives, regulations and programmes that can be seen as impacting on packaging or on packing operations. Only those directly aimed at packaging will be covered here, but the following is a list of some to be aware of:

- The Packaging and Packaging Waste Directive 94/62/EC.[7]
- EU Directive Future Policy on Industrial Emissions (IPPC) 2008/1/EC.[8]
- 'EU Timber Regulation' (Regulation (EU) No 995/2010 of 20 October 2010, laying down the obligations of operators who place timber and timber products on the market),[9] also related to FLEGT (Forest Law Enforcement, Governance and Trade)[10] EU Forestry Strategy.
- Directive 2010/75/EU of 24 November 2010, on industrial emissions[11] (waste incineration).
- Directive 1999/31/EC of 26 April 1999, on the landfill of waste.[12]
- Regulation (EC) No 1013/2006 of 14 June 2006, on shipments of waste.[13]
- Sustainable Consumption and Production and Sustainable Industrial Policy Action Plan COM/2008/0397 final.[14]
- Regulation (EC) No 1221/2009 of 25 November 2009, on the voluntary participation by organisations in a community eco-management and audit scheme (EMAS).[15]
- Regulation (EC) No 1013/2006 of 14 June 2006, on shipments of waste (Basel Convention).[13]
- Council Directive 1999/13/EC of 11 March 1999, on the limitation of emissions of volatile organic compounds due to the use of organic solvents in certain activities and installations.[16]
- Regulation (EC) No 1005/2009 of 16 September 2009, on substances that deplete the ozone layer.[17]
- Thematic Strategy on the Sustainable Use of Natural Resources (COM670 2005).[18]
- 'Taking sustainable use of resources forward: A Thematic Strategy on the prevention and recycling of waste', COM (2005) 666.[19]
- Regulation (EC) No 66/2010, on the EU Ecolabel.[20]
- 'Implementing the Partnership for Growth and Jobs: Making Europe a Pole of Excellence on Corporate Social Responsibility', COM (2006) 136 final.[21]
- Consumer Information Unfair Commercial Practices Directive (UCP) 2005/29/EC.[22]
- Directive 2004/35/EC of the European Parliament and of the Council of 21 April 2004, on environmental liability with regard to the prevention and remedying of environmental damage.[23]
- Directive 2008/98/EC of 19 November 2008, on waste.[24]

NOTE: the references above were correct at the time of going to press but Directives and strategies are subject to frequent modification and checks should always be made via the EU web site: http://ec.europa.eu for the latest versions and for new legislative initiatives.

Many European countries within and outside of the EU have chosen to focus national legislation on the recovery and recycling of waste, with a particular focus on packaging. Indeed there have been frequent complaints that in many countries this approach has been substituted for the implementation of the Packaging and Packaging Waste Directive and, in particular, the enforcement of its 'Essential

Requirements'. The EU has taken action to try to correct this situation, but is not yet clear to what degree it will succeed.

Undoubtedly, the best known piece of national legislation regarding packaging is the German Packaging Ordinance and the 'Green Dot' operated by the DSD.[25] This has served as a model for many other countries in Europe and around the world. There are currently 24 countries in Europe using the mark, whilst internationally many countries have given consideration to the scheme in the course of developing their own legislation.

Pro Europe s.p.r.l. Packaging Recovery Organisation[26] is the umbrella organisation of 34 European national producer responsibility systems engaged in the selective collection and recycling of packaging waste. Detailed country-by-country information can be found on their web site. That information includes national rules and provides links to national sites giving details of charges and statistics on the collection performance. The coverage of the site is extensive, going well beyond 'Green Dot' countries and EU members.

In an attempt to facilitate the recovery operations called for by the Packaging Waste Directive and national collection schemes, a number of European countries, as listed below, now require specific types of packaging to be marked according to constituent material type. The EU has established a system of marking, Commission Decision 97/129/EC,[27] but different systems are also permitted within individual country regulations:

- *Austria*: plastic material type.
- *Croatia*: plastic and composite packaging to be marked with defined numbers and symbols.
- *Czech Republic*: type marking is mandatory for plastics, metals and composites, whilst voluntary for paper, glass, wood and textiles. Certain types of packaging components, such as shrink film, closures and dosing devices are also exempt.
- *Latvia*: requires material marking based on the EU material coding system and an indication of recycled content (if used) on metal, plastic, glass, paper, paperboard, wood, textiles and composite material packaging.
- *Lithuania*: packaging must be marked according to the materials used.
- *Poland*: aluminium, with the exception of multilayer containers and those of less than 5 g and plastic packaging must be marked according to material content and recyclability.
- *Romania*: requires packaging material marking.
- *Slovakia*: requires material coding following the EU coding system and the use of the Tidyman symbol. Multi-trip packaging must be clearly and legibly marked with the words 'Návratny obal' (returnable packaging). There are exemptions for small packages.
- *Turkey*: all packaging must be marked with the appropriate EU material code and the Möbius loop. Companies who recover their own waste must label packaging with a Ministry-assigned code. All companies who are part of a recovery system must label packaging with the relevant symbol of the organisation.

Looking towards future potential improvements in their results, some countries such as Belgium, The Netherlands (Packaging Covenant), Greece, Slovakia and Spain all require companies to submit detailed long-term packaging reduction plans. Belgium, The Netherlands and Greece only require these plans for products actually manufactured in those countries; Slovakia and Spain require them for all products on the market.

8.2.1 Switzerland

EU regulations do not, of course, apply in Switzerland, but for obvious reasons they do influence the thinking of Swiss legislators and companies. Switzerland has an intention to align its packaging regulations with the EU over time, but has not yet set out a timetable to do so.

Swiss academic institutions were amongst the first to become involved in the development of LCA methodologies mentioned in the 'Introduction'. The first official Swiss LCA studies on packaging were published in 1984[28] and work has continued to ensure that an official list has been kept up to date. The results of these studies, which consider every stage of the package life through recovery, reuse, recycling or disposal, are used to guide all new environmentally focused legislative measures.

The key piece of Swiss packaging legislation, the Beverage Container Ordinance,[29] requires 75% of glass, PET and aluminium containers to be recycled. Manufacturers, importers, exporters and dealers must notify the authorities by the end of February of each year of the volume of beverages produced and imported and the weight of beverage containers returned or recycled. This also applies to companies that export beverage containers for recycling. In addition:

- Packaging must be compatible with existing recycling systems for glass, PET bottles, aluminium cans and tin cans.
- Reusable packaging is subject to a deposit and marking.
- Disposable packaging in PET or metal (aluminium and tin) incurs either a fee to an existing recycling organisation or must be taken back.
- Disposable packaging in PVC is subject to an obligatory deposit.

8.3 The environmental and sustainability regulatory and legislative framework in North America

Canada and the USA approach packaging in distinctly different ways. The US Environmental Protection Agency addresses packaging as a sub-set of the overall approach to waste and recycling, an approach that manifests itself only as an umbrella approach to 'Product Stewardship' focused on recovery. It recommends an integrated, three-tier hierarchical approach to MSW management composed of four components: source reduction, recycling, combustion (with energy recovery) and landfilling. The incorporation of 'incineration' in this hierarchy is in marked contrast with the European approach, where its employment remains controversial.

In the US, managing recovery and recycling and interpreting the 'Hierarchy' headings are delegated to individual states. This has led to the creation of a patchwork of differing state actions. Partly in response to this diversity of action and not uncommon inaction and partly to facilitate delivery on their commitments to sustainability as good corporate citizens, an increasing number of corporations have set up their own recovery and recycling schemes nationally. Given the patchwork of regulators, their different levels of apparent commitment to action and their different priorities, many have come to see the corporate approach as the best hope for a uniform national scheme.

8.3.1 Regulation in Canada

Although the Canadian environment agency Environment Canada[30] also delegates regulation on waste and recycling to provincial governments, there has been much more of a concerted effort to tackle sustainability and packaging. In October 2009, the Canadian Council of Ministers of the Environment[31] published *Canada Wide Strategy for Sustainable Packaging*. This Strategy is based on an Extended Producer Responsibility approach and the development of a Sustainable Packaging Framework, together with supporting measures. Thus, whilst Canada currently relies on a similar umbrella approach to the US, in future its approach will be more uniform.

8.3.2 Regulation in the USA

The Resource Conservation and Recovery Law 42 USC 6901[32] was primarily intended to enforce the safe management of hazardous waste, but also contains a protocol for the management of non-hazardous solid waste. Its stated aims are to:

> . . . protect us [US citizens) from the hazards of waste disposal; conserve energy and natural resources by recycling and recovery; reduce or eliminate waste; and clean up waste that may have spilled, leaked or been improperly disposed of.

The Act permits the use of the recycling symbol (Möbius loop) and the printing of the terms 'Recycled' and 'Recyclable', but their use is regulated by the Federal Trade Commission[33] via their Environmental Marketing Guide. The Sustainable Packaging Coalition (SPC),[34] a collaborative organisation bringing together major companies with interests in packaging across the US, has launched a Packaging Recovery Label system based on that recently launched in the UK. However, its use is, as in the UK, entirely voluntary.

The EPA does have 'Comprehensive Procurement Guidelines' designed to promote the use of recycled materials in product manufacturing. These are run alongside 'Tools for Local Government Recycling Programmes', 'Pay as You Throw' schemes. and recycling management. Again, however, all of these and other related initiatives are primarily advisory.

The EPA is required by Congress to produce a set of rules implementing the Law; these are published under Title 40 of the Code of Federal Regulations.[32] Handling of non-hazardous waste is covered under Parts 239 to 259. This is a diverse set of regulations, mostly concerned with the operation of waste recovery facilities, everything from the development of waste management plans to the thermal processing of solid waste.

In the specific case of recycled materials being used in food packaging, the US Food and Drug Administration (FDA) does have requirements governing their use. These can be found on the FDA web site under *Food Guidance, Compliance and Regulatory Information.*[35] Each recycling process requires FDA clearance before it can be put into operation. These requirements are in many ways parallel to those applied to virgin materials used to package food.

Parts 239 to 259 are then divided into further detailed guidelines, which are in themselves diverse. Part 246, for example, deals with recovery under 'high grade paper', 'residential materials' (lumping together glass, cans and mixed paper – but ignoring plastics) and 'corrugated containers'. Under 'residential materials', it covers topics as diverse as separation technology and public education.

In practice therefore, although Congress refers to 'regulations', the Act overall is best viewed as a code of practice for the 'Regulated Communities', that is primarily the individual states. Part 239 does set out 'requirements for state permit program determination of adequacy', but again this is concerned with operating procedures.

Although state legislation on packaging continues to form a complex patchwork of diverse and occasionally conflicting actions, there have been some actions to create more harmony. The Coalition of Northeastern Governors (CONEG)[36] is the best known. Dating from 1976, CONEG is a:

> . . . non-partisan association of the Governors of the seven Northeastern states. Members include the Governors of Connecticut, Maine, Massachusetts, New Hampshire, New York, Rhode Island and Vermont. CONEG encourages intergovernmental cooperation in the Northeast on issues relating to the economic, environmental and social well-being of the Northeast states.

CONEG activities go much wider than packaging but nonetheless they have pioneered an approach to packaging that is similar in many ways to that familiar in Europe. For example, their Toxics in Packaging Clearing House (TPCH) was set up in 1992 'to promote the Model Toxics in Packaging Legislation in an effort to reduce the amount of heavy metals in packaging and packaging components that are sold or distributed throughout the United States'. Its aim is to phase out the same heavy metals that have been highlighted in the EU Packaging and Packaging Waste Directive. The legislation has been adopted by California, Connecticut, Florida, Georgia, Illinois, Iowa, Maryland, Maine, Minnesota, Missouri, New Hampshire, New Jersey, New York, Pennsylvania, Rhode Island, Vermont, Virginia, Washington and Wisconsin, so far beyond the core CONEG membership but still only representing one-third of states and not national.

A number of states are proposing or have enacted bans on the use of plastic bags and non-reusable shopping bags. Few have been enacted to date after meeting strong opposition in state legislatures. In some states, such as Colorado and Texas, individual townships or counties have enacted bag bans. Much the same applies to attempts to ban certain types or materials of packaging, mostly on a local level.

There are two national campaigns seeking to create state bottle bills and bag bans, but at present they have had only limited impact.

The Society of the Plastics Industry has a resin identification system to assist plastics recycling. Thirty-nine states have formally adopted laws requiring rigid plastic containers to include the SPI code or a similar code.

There is an initiative by the Carton Council[37] (an industry body) to promote beverage carton recycling. This carton initiative has spread to 40 states, but nowhere is it yet backed by legislation. Similar initiatives have been started by PET bottle users such as Coca-Cola and PepsiCo. The extent of these industry-backed initiatives is such that it has prompted many to see them as a more effective alternative to state and national legislative action.

Beyond this, we move into the realm of specific state regulations. Some important examples of these are listed in the non-exhaustive list below, but it is always essential to check for legislation in any state where a product is likely to be sold. Note, this list does not repeat the heavy metals legislation mentioned above.

California

Rigid Plastic Packaging Container (RPPC) Law[38] Beverages Subject to California Refund Value (CRV).[39]

AB 3025 prohibits the sale of expanded polystyrene, aka Styrofoam, loose fill packaging, unless it is comprised of at least 60% post-consumer recycled material

Rigid plastic and glass containers must be either recyclable or reusable, or contain a specified percentage of recycled content

Extended Producer Responsibility laws are under consideration and these could eventually affect packaging if passed.

Connecticut

The Beverage Container Deposit and Redemption Law[40] covers beer, soda, and bottled water in metal, glass and plastic, excluding HDPE containers and non-carbonated beverage.

Delaware

Delaware previously had a bill allowing customers to return empty bottles to shops but it was repealed in 2010, being replaced with a temporary tax. At the time of writing, this repeal was being challenged.

Florida

Florida has legislation seeking to increase recycling to a 2020 target of 75% (now seen as highly unlikely given 2011 figures of 28%); however, this is not specific to packaging.

Hawaii

The Beverage Container Deposit Program places a 5 cent deposit on bottles under 64 ounces for soft drinks, beer, water, mixed spirits and wine, coffee and tea in metal, glass, PET or HDPE.[51]

Iowa
The Beverage Container Deposit Law IAC 4/17/02, 9/4/02 covers beer, soda, mineral water, wine and wine coolers and liquor in metal glass or plastic. The deposit is 5 cents.

Maine
The Returnable Beverage Container Law[42] covers all beverages (exempting dairy and cider) in metal, glass or plastic of less than 4 litres. The deposit is 5 cents.

A Product Stewardship Framework Law HP1159 has been passed, but it is unclear at the time of writing whether and if so in what way this would impact on packaging.

Massachusetts
The Beverage Container Recovery Law, General Law Section 323[43] covers beer, soda and mineral water in metal, plastic or glass (excluding biodegradable) containers. The deposit is 5 cents.

Massachusetts has a number of legislative proposals regarding the imposition of reusable or recyclable packaging, but none at the time of writing have become law.

Michigan
The Beverage Container Deposit Law – Michigan[44] Compiled Laws (MCL) Sections 445.571 through 445.576, covers beer, soda, carbonated and mineral water, wine coolers and canned cocktails in metal, plastic paper and glass of less than one gallon. The deposit is 10 cents.

Minnesota
Minnesota has a proposal for an Extended Producer Responsibility bill that would include packaging if passed.

New York
The Returnable Container Act[45] requires every beverage container sold in the state to carry a deposit of 5 cents. This law applies to all beverage containers made from glass, metal, aluminium, steel or plastic, in bottle, can or jar form, with volumes of one gallon or less.

North Carolina
North Carolina bans recyclable plastics containers from landfill.

Oregon
The Beverage Container Act[46] imposes a deposit of 5 cents on beer, soft drinks, water and flavoured water. Rigid plastic and glass containers must be recyclable or reusable or contain a specified percentage of recycled content.

Vermont
The Beverage Container Law[47] covers beer, soda, liquor and mixed wine drinks in metal, glass, plastic or paper (excluding biodegradable containers). There is a

refundable deposit of 5 cents. Manufacturers must pick up the beverage containers that were redeemed, or pay the redemption centres or retailers for handling the containers (a form of Extended Producer Responsibility).

8.4 The environmental and sustainability regulatory and legislative framework in the Asia-Pacific region

8.4.1 Australia

Australia has focused on the 'covenant' approach to packaging, whereby users of packaging (brand owners) are encouraged to take responsibility voluntarily for the packaging they place on the market and take action to mitigate its impacts. Uniquely in Australia, the government has also put in place the National Environment Protection Measure (NEPM)[48] to catch those seeking to avoid their responsibilities under the Covenant and encourage those that have not signed to do so.

The Australian Packaging Covenant[49] (APC), subtitled 'A commitment by governments and industry to the sustainable design, use and recovery of packaging', at the time of writing was in its third version, approved in July 2010. It is based on managing consumer packaging and paper products through all stages of their lifecycles, embodying the waste hierarchy approach of prioritising minimisation. However, it does acknowledge the benefits that packaging creates through protecting and preserving products. It covers:

- Performance indicators and targets (KPIs).
- Sustainable packaging guidelines.
- Roles and undertakings.
- Governance.
- Compliance, accountability and transparency.
- Funding.
- Monitoring and evaluation.

Much of the detail within the APC is set out in a series of six Schedules, appended to the document. The APC has three major aims within a concept of product stewardship, not dissimilar to the concept of extended producer responsibility:

- To foster design of packaging that is more resource efficient and more recyclable.
- To increase the recovery and recycling of used packaging from households and away-from-home sources.
- To take action to reduce the incidence and impacts of litter.

The KPIs and other requirements, such as details required within an action plan, need to be read in detail in the APC; they are partially hard targets such as recycling rates, partially subjective such as design criteria and partially aspirational such as working with others to achieve goals. However, the APC does require the preparation of an action plan, annual reporting of progress against that plan and that audits be allowed as required. The results must be prominently published on a readily accessible web site.

Also companies are required to:

- Make annual financial contributions to the Covenant Fund.
- Implement design and procurement processes that drive sustainable design of packaging, consistent with the Sustainable Packaging Guidelines (Schedule 2).
- Implement policies or procedures to buy products made from recycled materials.
- Establish collection and recycling programmes for used packaging materials generated on-site.
- Take action, where appropriate, to reduce litter.
- Assist the Covenant Council to respond to complaints from the public about the design and use of packaging materials.

The National Environmental Protection (Used Packaging Materials) Measure (NEPM) is enforced by each of the states and territories separately. Its primary purpose is to underpin the APC and combat 'free riders' defined within the Measure as:

> . . . a brand owner or organisation that is a participant in the packaging chain and is not a signatory to the Covenant, and is not producing equivalent outcomes to those achieved through the Covenant.

Also it aims to 'ensure that industry signatories do not suffer any competitive disadvantage as a result of fulfilling their commitments under the Covenant'.

The fundamental requirements of the Measure are that participating jurisdictions should oblige brand owners to:

- Undertake or assure the systematic recovery of consumer packaging in which the brand owner's products are sold.
- Undertake or assure the re-use, recycling or energy recovery of consumer packaging in which the brand owner's products are sold.
- Demonstrate that all materials that have been recovered by them or on their behalf have been utilised through (in order of preference):
  - re-use in the packaging of the brand owner's own products (if applicable), or
  - use within Australia as a secondary resource, or
  - export as a secondary resource, and
  - demonstrate that reasonable steps have been taken to ensure that consumers are adequately advised as to how the packaging is to be recovered.

8.4.2 China

There are two sets of relevant packaging regulations:

1. General Principles on Disposal and Utilization of Packaging Wastes (GB/16716-1996).

2. Tentative Administration Measures on Recycling and Reuse of Packaging Resources.

These regulations specify which packaging materials must be recycled and compel participation in the recycling process.

8.4.3 Japan

The Containers and Packaging Recycling Act[50] is administered by the Japan Packaging and Containers Recycling Association, a body specifically set up for the purpose. The Act was first established in 1995 with the latest revision published in June 2006. The Association is tasked with organising recycling for its members on a fee basis. The organisation deals with glass bottles, PET bottles, paper and plastics in four divisions. Steel and aluminium cans, drinks cartons and cardboard are excluded from the Act, as they are 'traded on the market'.

This Act works in a manner not dissimilar to the 'Green Dot' organisations of Europe. Each manufacturer, retailer, packaging manufacturer or importer is under obligation to recycle the containers that they use. An elaborate data-gathering operation determines the amounts of packaging to be recycled; the cost for that operation is determined and charged out in the form of fees that also cover other activities such as information dissemination, record keeping and education.

The Association then contracts with local authorities to carry out that recycling on its behalf. Consumers are obliged under the Act to sort containers to facilitate this recycling.

Material identification marks are mandatory on many forms of consumer packaging:

- Paper containers, except corrugated cardboard and paper drinks packs with no aluminium coating.
- Aluminium laminated paper beverage containers.
- Plastics containers and packaging, except for PET bottles for beverages and soy sauce.
- PET bottles.
- Steel cans.
- Aluminium cans.

There are exceptions for small packages, but not when those small packages form part of a larger packaging system such as ready meals. In Japan, such meals often consist of multiple dried and liquid components separately packed and then placed within the largest container, often made in the shape of a bowl. The marks are applied to this largest container.

In addition to the above, there are so-called Eco-marks that designate, for example, glass containers utilising more than 70% recycled materials and reusable containers. However, their use is not mandated by law.

8.4.4 New Zealand

For the five years up until 2009, New Zealand had a Packaging Accord that aimed to:

- Foster improved partnerships, policies and processes.
- Commit brand owners and retailers to take primary responsibility for product stewardship throughout the packaging lifecycle.
- Promote greater material and energy efficiency in the production, use and recovery of packaging materials.
- Create supply chain initiatives that foster markets for sustainably-made packaging, including more use of recovered packaging materials.

This was succeeded from July 2010 by the Packaging Product Stewardship Scheme, still a voluntary scheme but following the requirements of the Waste Minimisation Act of 2008. Apart from instituting a levy on landfill waste to fund waste reduction initiatives, the Act seeks to encourage producers, brand owners, importers, retailers, consumers and other parties to take responsibility for the environmental effects of their products through product stewardship schemes. Whilst the Act sets out primarily to encourage voluntary action, it also contains powers to force action where necessary.

The Stewardship Scheme is membership-based and seeks to follow the voluntary agreement route. Specifically it sets out to:

- Improve packaging design to reduce packaging waste.
- Improve systems to reduce packaging waste.
- Increase reuse of packaging.
- Increase recycled content in packaging to replace virgin material.
- Increase consumer awareness and understanding of sustainable packaging.

8.4.5 South Korea

Korea treats packaging under its extended Producer Responsibility Law No. 106 of 2005. According to 2006 Rules on the Standards of Product Packaging Materials and Methods, there are limits on the amount of void space in packaging according to the type of product, as well as limits on the number of layers of packaging that may be used. For example, processed foods 15%; beverages 10%; cosmetics 10%; OTC drugs 20%. The law also bans the use of PVC shrink wraps and laminates.

The South Korea Recycling Law Ministry of Environment Notification No. 2002–195 mandates a separate disposal label for packaging for foods and beverages, agricultural produce, dairy and fishery products, detergents, cosmetics (excluding glass containers), pet products, drugs, butane gas and pesticides. This applies only if the pack is composed of paper composites and laminates, metal, glass or plastic.

8.4.6 Taiwan

In Taiwan, a designated symbol is required on all containers that are subject to recycling requirements under the Waste Act. In addition to the four-arrow symbol, plastic containers should also carry a material identification mark such as the US SPI symbol.

The Waste Act states that:

For articles and the packaging and containers thereof that, after consumption or use, are sufficient to produce general waste possessing one of the following characteristics and cause concern of serious pollution to the environment, the manufacturer or importer of the articles and the packaging and containers thereof at issue or the manufacturer or importer of the raw materials shall bear responsibility for recycling, clearance and disposal and the vendor shall bear responsibility for recycling, clearance work:

1. Difficult to clear or dispose of.
2. Contains a component that does not readily decompose over a long period.
3. Contains a component that is a hazardous substance.
4. Is valuable for recycling and reuse.

The central competent authority shall officially announce the scopes for the articles and the packaging and containers thereof and the enterprises responsible for recycling, clearance and disposal in the foregoing paragraph.

8.4.7 Thailand

Thailand has rules regarding waste separation at source.

8.5 Future trends and conclusions

One important emerging trend is for corporations individually or collectively to start initiatives aimed at improving recycling and sustainability. For example, as we have seen in the US, a number of large and influential manufacturers and retailers have started up significant programmes to address their own recovery priorities, including creating their own infrastructure (with varying degrees of success). Some companies there have voiced support for increased regulation of packaging materials. In the same vein of voluntary action, there are also significant non-regulatory agreements such as the Global Packaging Project, which may have an impact on legislative thinking.

In a number of countries such as The Netherlands, Australia and New Zealand, coordinated action by industry is being applied in the place of legislated action. However, in some of those cases, the persistence of 'freeloading' by a proportion of companies has led to the creation of legislation to prevent the practice. The fact that in all cases these 'Covenants' have been renewed after expiry indicates that, freeloading apart, they are viewed as successful.

Other countries such as Japan have taken a half-way house approach modelled on the approach of many EU countries, whereby targets are mandated but the means of compliance are left open to industrial cooperation, subject to those targets being met.

Although the EU can reasonably be attributed the credit for the growing international action on packaging waste, there are questions regarding the universal applicability of its approach. For example, in East Asia, a Thematic Working Group on Solid and Hazardous Waste has come together under the auspices of the UNEP. It describes the need to redefine the task according to the needs of a region where cultures and practices can be different from those of Europe. A status report on this work was issued in 2010.[52]

South America has a different approach again. Led by Brazil, many countries are examining ways of intermeshing waste management practices with job and social benefit creation for the poor and underprivileged. In that case, the approach is based on a triangular relationship between local government, social enterprises and business. However, there are also highly successful examples where large producers and users of packaging have sponsored their own initiatives as part of local collaborations.

However, what this does is emphasise the degree to which international environmental and sustainability regulatory and legislative frameworks have emphasised recovery and recycling of packaging together with source reduction. In other words, they have primarily built the environmental pillar of sustainability. In doing so, they have largely ignored the issue of overall waste reduction and the prevention of product waste through packaging.

This battle to shift the focus of the debate to overall waste prevention is without doubt an uphill one. Even in Europe, where there is a pronounced shift towards addressing loss and waste of product, much legislative attention still rests with packaging reduction and recovery. Some legislators will freely admit this is the case because such action is popular and easy to enact. Unfortunately, scientific analysis and argument is too often a casualty in this debate.

We still therefore talk of 'sustainable packaging', even when from a rational point of view it is impossible to determine the contribution to sustainability without knowledge of and inclusion of its contents. Without such an approach, the inclusion of the Social and Economic pillars, already difficult, becomes next to impossible. Wasting product has economic, social and environmental impact. Where they can be measured, the impacts of the packaging needed to prevent such waste are much lower. For example, the environmental impact of a package is on average only one-tenth that of its contents. We see some movement on the sustainable sourcing of materials that constitute packaging, but very little on the packaged product.

However, two growing trends may change this. The first of these is the growing emphasis on preventing product waste in transport, in storage and in the home. Better physical protection of products, improved shelf life and better in-home preservation and dispensing would all help to significantly reduce these losses. Whether we will see these opportunities being acted on is still open to question but the second trend may help that recognition.

That noticeable second trend has been towards the inclusion of Key Performance Indicator models in company plans for packaging and sustainability. Whilst such models do still tend to emphasise the impact of packaging, they do at least include elements of the prevention of packaging damage. If sufficient attention is given to these KPIs in company assessments, then we could see a move towards talking about sustainability through packaging.

However, nowhere is legislation to mandate product sustainability through packaging even in contemplation. However, it is perfectly feasible and highly desirable and hence a positive note for the future on which to close this chapter.

8.6 References

1. Keep Britain Tidy (2012) Available from: http://www.keepbritaintidy.org [Accessed October 2012].
2. INCPEN (2012) 'The Industry Council for Packaging and the Environment'. Available from: http://www.incpen.org/ [Accessed October 2012].
3. The Commission to the European Parliament, The Council, The European Parliament, The Council, The European Economic and Social Committee and The Committee of the Regions. Roadmap to a Resource Efficient Europe. Brussels, The European Commission, 2011. Vol. {SEC(2011) 1067 final}; {SEC(2011) 1068 final}.
4. Potočnik, Janez (2011) 'Combating food waste in EU', conference. *Packaging News*. Brussels, Metropolis Business Media, 9 November.
5. C OAG Standing Council on Environment and Water (incorporating the National Environment Protection Council). National Environment Protection (Used Packaging Materials) Measure (2011) Canberra, Australian Government. F2011L02093.
6. CEMPRE (2012) 'Compromisso Empresarial Para Reciclagem'. Available from: http://www.cempre.org.br/ [Accessed October 2012].
7. European Commission (1994) 'Directive 94/62/EC on Packaging and Packaging Waste, as amended by Directive 2004/12/EC'. *EUROLEX*. Brussels, European Commission, 1994. 94/62/EC.
8. Directive 2008/1/EC of the European Parliament and of the Council of 15 January 2008, concerning integrated pollution prevention and control. *EUROLEX*. Brussels, European Commission. 2008/1/EC.
9. Regulation (EU) No 995/2010 of the European Parliament and of the Council of 20 October 2010, laying down the obligations of operators who place timber and timber products on the market. *EUROLEX*. Brussels, European Commission. Regulation (EU) No. 995/2010.
10. EU DIRECTORATE GENERAL ENVIRONMENT (2008) FLEGT Voluntary Partnership Agreements. *DG Environment Web Site*. Brussels, European Commission.
11. EUROPEAN COMMISSION (2010) Directive 2010/75/EU of the European Parliament and of the Council of 24 November 2010 on industrial emissions (integrated pollution prevention and control). *EUROLEX*. Brussels, European Commission. Directive 2010/75/EU.
12. Directive 1999/31/EC of 26 April 1999 on the landfill of waste. *EUROLEX*. Brussels, European Commission. Council Directive 99/31/EC.
13. Regulation (EC) No 1013/2006 of the European Parliament and of the Council of 14 June 2006 on the shipemnt of waste. *EUROLEX*. Brussels, European Commission. Regulation No. 1013/2006.
14. Communication from the Commission to the European Parliament, the Council, the European Economic and Social Committee and the Committee of the Regions on the Sustainable Consumption and Production and Sustainable Industrial Policy Action Plan. *EUROLEX*. Brussels, European Commission, 2008 (SEC(2008) 2110) (SEC(2008) 2111).
15. Regulation (EC) No 1221/2009 of the European Parliament and of the Council of 25 November 2009 on the voluntary participation by organisations in a Community eco-management and audit scheme (EMAS). *EUROLEX*. Brussels, European Commission, 2009. Regulation (EC) No. 1221/2009.

16. Council Directive 1999/13/EC of 11 March 1999 on the limitation of emissions of volatile organic compounds due to the use of organic solvents in certain activities and installations. *EUROLEX*. Brussels, European Commission. Council Directive 1999/13/EC.
17. Regulation (EC) No 1005/2009 of the European Parliament and of the Council of 16 September 2009 on substances that deplete the ozone layer. *EUROLEX*. Brussels, European Commission. Regulation (EC) No. 1005/2009.
18. Communication from the Commission to the Council, the European Parliament, the European Economic and Social Committee and the Committee of the Regions (2005) Thematic Strategy on the Sustainable Use of Natural Resource. *EUROLEX*. Brussels, European Commission. (SEC(2005) 1683) (SEC(2005) 1684).
19. Communication from the Commission to the Council, The European Parliament, The European Economic and Social Committee and the Committee of the Regions Taking sustainable use of resources forward (2005) 'A Thematic Strategy on the Prevention and Recycling of Waste'. *EUROLEX*. Brussels, European Commission. (SEC(2005) 1681) (SEC(2005) 1682).
20. Regulation (EC) No 66/2010 of The European Parliament and of The Council of 25 November 2009 on the EU Ecolabel. *EUROLEX*. Brussels, European Commission, 2010. Regulation (EC) No. 66/2010.
21. Communication from the Commission to the European Parliament, the Council and the European Economic and Social Committee (2006) 'Implementing the partnership for growth and jobs: making Europe a pole of excellence on corporate social responsibility'. *EUROLEX*. Brussels, European Commission. COM(2005) 136.
22. Directive 2005/29/EC of the European Parliament and of the Council of 11 May 2005 concerning unfair business-to-consumer commercial practices in the internal market and amending Council Directive 84/450/EEC, Directives 97/7/EC, 98/27/EC and 2002/65/EC. *EUROLEX*. Brussels, European Commission. Directive 2005/29/EC.
23. Directive 2004/35/EC of the European Parliament and of the Council of 21 April 2004 on environmental liability with regard to the prevention and remedying of environmental damage. *EUROLEX*. Brussels, European Commission. Directive 2004/35/EC.
24. Directive 2008/98/EC of the European Parliament and of the Council of 19 November 2008 on waste and repealing certain Directives. *EUROLEX*. Brussels, European Commission. Directive 2008/98/EC.
25. Duales System Deutschland. (2012) 'Der Grüne Punkt'. Available from: http://www.gruener-punkt.de/?L=1 [Accessed October 2012].
26. Pro Europe s.p.r.l. Packaging Recovery Organisation (2012) 'Pro Europe'. Available from: http://www.pro-e.org/ [Accessed October 2012].
27. European Commission. 97/129/EC: Commission Decision of 28 January 1997 establishing the identification system for packaging materials pursuant to European Parliament and Council Directive 94/62/EC on packaging and packaging waste. *EUROLEX*. Brussels: European Commission. Commission Decision 97/129/EC.
28. Swiss Federal Office for the Environment FOEN (2012) Available from: http://www.bafu.admin.ch/index.html?lang=en [Accessed October 2012].
29. Federal Office for the Environment (2000) 'Ordinance on Beverage Containers'. Bern, Federal Office for the Environment. 814.621.
30. Environment Canada (2011) Available from: http://www.ec.gc.ca [Accessed October 2012].
31. Canadian Council of Ministers of the Environment (2012) Available from: http://www.ccme.ca [Accessed October 2012].
32. United States Environmental Protection Agency (1976) Resource Conservation and Recovery Law 42 USC 6901. *EPA Wastes Laws and Regulations*. Washington DC, United States Environmental Protection Agency. 42 USC 6901.
33. Federal Trade Commission (2012) Available from: http://www.ftc.gov [Accessed October 2012].

34. Sustainable Packaging Coalition (2012) Available from: http://www.sustainablepackaging.org/ [Accessed October 2012].
35. US Food and Drug Administration (2012) 'Food Guidance, Compliance and Regulatory Information'. Available from: http://www.fda.gov/Food/GuidanceCompliance RegulatoryInformation/default.htm [Accessed October 2012].
36. CONEG (2012) 'Coalition of North Eastern Governors'. Available from: http://www.coneg.org/ [Accessed October 2012].
37. Carton Council USA (2012) 'We Recycle Cartons'. Available from: http://www.recyclecartons.com/ [Accessed October 2012].
38. Government of California (1991). 'Rigid Plastic Packaging Container Law 1991'. Public Resources Code (PRC) Sections 42300–42345). Available from: http://www.calrecycle.ca.gov/Plastics/rppc/ [Accessed October 2012].
39. Beverages Subject to California Refund Value. Sacramento: s.n. Available from: http://www.calrecycle.ca.gov/BevContainer/ProgramInfo/CRVList.htm [Accessed October 2012].
40. Government of Connecticut (1988) 'Beverage Container Deposit and Redemption Law'. Hartford, Department of Energy and Environmental Protection. Section 22a – 245. Available from: http://www.cga.ct.gov [Accessed October 2012].
41. Government of Iowa (1978) 'The Beverage Container Deposit Law'. Des Moins, Iowa, Department of Natural Resources. Code of Iowa Chapter 455C. Available from: http://www.legis.state.ia.us/Rules (does not work)
42. Government of Maine. Returnable Beverage Container Law. Augusta, Maine, Department of Agriculture, Food and Rural Resources. 32 MRSA Section 1865.
43. Government of Massachusetts. Beverage Container Recovery Law. s.l.: General Court of the Commonwealth of Massachusetts. General Law Section 323. Available from: http://www.nalegislature.govLaws (does not work)
44. Government of Michigan (1976) The Beverage Container Deposit Law. Lansing, Department of Licensing and Regulatory Affairs. Michigan Compiled Laws Sections 445.571 through 445.576.
45. Government of New York (1996) 'The Returnable Container Act'. Albany, Department of Environmental Conservation. Section 27–1007 (1)(b) of the Environmental Conservation Act.
46. Government of Oregon (1971) 'The Beverage Container Act'. s.l.: Department of Environmental Quality. House Bill 3145.
47. Government of Vermont (1972) 'Beverage Container Law'. Waterbury, Vermont Agency of Natural Resources.
48. Australian Government. National Environment Protection (Used Packaging Materials) Measure (2011) Canberra, National Environment Protection Council. F2011L02093.
49. Australian Packaging Covenant.
50. Govenment of Japan (1995) 'The Container and Packaging Recycling Law'. Tokyo, Ministry of Economy, Trade and Industry. Act No. 102 of 16 June 1995.
51. The Government of Hawaii. 'Beverage Container Deposit Program'. Available from: http://www.capitol.hawaii.gov/hrscurrent [Accessed October 2012].
52. AIT/UNEP Regional Resource Center for Asia and the Pacific (2010) 'Municipal Waste Management Report: Status-quo and Issues in Southeast and East Asia Countries'. Available from: http://www.environment-health.asia/ [Accessed October 2012].

9

Nanotechnology and the packaging of food and other fast-moving consumer goods

B. Park, GBP Consulting Ltd, UK

DOI: 10.1533/9780857098979.241

Abstract: Nanotechnology is the technology of the very small, with nanomaterials having at least one dimension in the range 1 to 100 nm where 1 nm is 1×10^{-9} m. Nanotechnology is making a large difference to products across industry, and nano-enabled products are now being used extensively in packaging with the Fast-Moving Consumer Goods sector driving developments of new products. These include nano-enabled packaging to deliver enhanced shelf life of food and pharma products and low-cost security tags. The list will continue to grow as companies realise the added value they can deliver with nano-enabled packaging.

Key words: packaging, nanotechnology, nanomaterial, manufactured, nanocomposite, food, pharma, consumer, regulatory status, active, smart.

9.1 Introduction

Fast-Moving Consumer Goods (FMCG) include a wide range of frequently purchased consumer products such as cosmetics, toiletries and household cleaning products, as well as other non-durables including glassware, light bulbs and batteries. FMCG also include certain pharmaceuticals and consumer electronics, but the largest sector comprises food and drink products that have a short shelf life, either as a result of high consumer demand or because the product deteriorates rapidly. Some of these food and drink products may be highly perishable.

Packaging for such products is therefore critical to prevent wastage and also damage to the product following manufacture, through transport to stores and then on to the consumer. It is also important that packaging is designed and produced from materials that can be disposed of safely and in an environmentally friendly way. In the world of FMCG, often the packaging contains as much complex technology as the goods themselves. Packaging producers seek to enhance material properties including mechanical strength, moisture retention, the barrier to gas and light, clarity and thermal properties.

The FMCG packaging market has been the subject of a market review in 2011 by Visiongain and they have indicated that the global market for FMCG packaging will have reached \$456 bn in 2011 (Visiongain, 2011).

Many different materials have been used in the packaging of FMCG, featuring glass, cardboard, paper and plastics of many varieties. Many technologies have fed into the production of these different types of packaging, but there is one more technology that is available today that has become a major focus for new developments. Nanotechnology is an enabling technology that is offering novel ways of producing packaging products for a range of applications and of ensuring an extended shelf life for those products.

9.1.1 Limitations of existing packaging

There is at least a preference, but more especially a need for an extended shelf life of the FMCG and safe disposal of the goods and the packaging without damage to the environment. This applies to all FMCG, but food safety is paramount and so the main focus for the developments and use of nanotechnology has been in packaging for food. There is a need to ensure that the quality and safety of food are maintained throughout manufacture, transportation and storage. In the pharma area, there is a growing demand for improved packaging materials that provide protection from the environment, as new medicines become more complex and less stable. Nanotechnology is already a major and growing component in food and beverage and also in pharma packaging. Where food and pharma are going, others will follow.

9.1.2 Nanotechnology

The term nanotechnology was coined in 1974 by the Japanese researcher Norio Taniguchi, to describe precision engineering with tolerances of a micron or less (Taniguchi, 1974). In the mid-1980s, Eric Drexler (1986) brought nanotechnology into the public domain with his book *Engines of Creation*. The prefix nano-derives form the Greek word for dwarf and one nanometre is equal to one billionth of a metre, that is, 1×10^{-9} metre.

Nanotechnology has been defined by the British Standards Institution as ‘the design, characterisation, production and application of structures, devices and systems by controlling shape and size in the nanoscale, which covers the size range from approximately 1 to 100 nm’ (BSI, 2009).

On 18 October 2011, the European Commission (EC) adopted the following ‘Recommendation on the definition of a nanomaterial’ (EC, 2011). A nanomaterial means:

> A natural, incidental or manufactured material containing particles, in an unbound state or as an aggregate or as an agglomerate and where, for 50% or more of the particles in the number size distribution, one or more external dimensions is in the size range 1 to 100 nm.
>
> In specific cases and where warranted by concerns for the environment, health, safety or competitiveness the number size distribution threshold of 50% may be replaced by a threshold between 1 and 50%.
>
> By derogation from the above, fullerenes, graphene flakes and single wall carbon nanotubes with one or more external dimensions below 1 nm should be considered as nanomaterials.

It is not clear how this definition will be incorporated into regulations and so it is likely that there will be many debates regarding its use in different market areas. It is intended that there will be a review in 2013 of how and where this recommended definition has been used. It will also be interesting to understand how this or future definitions for nanomaterials will influence such things as labelling, since it may be extremely difficult or even impossible to determine the exact structure of a material that may satisfy the definition prior to inclusion in a formulation or composite structure.

Nanotechnology is therefore the technology of the very small. To put this size into human context, the diameter of a red blood cell is ca. 7000 nm, the average thickness of a human hair is of the order of 80 000 to 100 000 nm and human nails grow at a rate of 1 to 10 nm per second.

The EC Recommended definition above referred to natural, incidental or manufactured nanomaterials. These can be considered based on the following.

- *Natural nanomaterials*: are created independently of man, and include a wide range of materials such as sea salt resulting from the evaporation of water from sea spray, soil dust and volcanic dust.
- *Incidental (or anthropogenic) nanomaterials*: are created as a result of action by man, with the main example being soot produced from the combustion of fossil fuels. Other incidental nanomaterials include welding fumes.
- *Manufactured (or engineered) nanomaterials*: have been designed and manufactured specifically by man. These have been synthesised for a specific purpose and can be found in several different shapes including platelet-like, wire-like and spheroidal-like, i.e. they are of nano size in one, two or three dimensions, respectively.

It is this last category of manufactured nanomaterials that will largely be the focus for this chapter.

Standards covering various aspects of nanomaterials and their use continue to be developed by bodies such as the International Organization for Standardization (ISO) and they help to ensure the open, safe and responsible development of nanotechnology. The UK has been playing a key role in leading the development of nanotechnology standards through its national committee, NTI/1 'Nanotechnologies', and has held both the chair and secretariat of the committee ISO TC 229 'Nanotechnologies'. Through these committees, the UK has been able to support this emerging discipline and use standardisation to help ensure its safe and successful global development and growth.

9.2 Regulatory status of nanotechnology

9.2.1 Europe

In Europe, Registration, Evaluation, Authorization and Restriction of Chemicals (REACH) provides an over-arching legislation applicable to the manufacture, placing on the market and use of substances on their own, in preparations or in articles. Nanomaterials are covered by the definition of a 'substance' in REACH, even though there is no explicit reference to nanomaterials. The general obligations in REACH, such as registration of substances manufactured at one tonne or more and providing information in the supply chain apply as for any other substance.

In 2009 the EC began a Registration, Evaluation, Authorization and Restriction of Chemicals Implementation Project on Nanomaterials (RIPoN), which was intended to provide advice on key aspects of the implementation of REACH with regard to nanomaterials. The EC posted final reports in October 2011 concerning nanomaterials and information requirements (RIPoN 2) and chemical safety assessment (RIPoN 3) (EC, 2011).

The EC stated that, 'based on the scientific and technical state-of-the-art with regard to nanomaterials, the reports were developed so that advice on specific issues related to nanomaterials could be integrated into the existing REACH guidance documents'. While the reports were stated not to represent European Chemicals Agency (ECHA) guidance, companies were invited to consult the two reports and take the recommendations into account as appropriate when preparing or updating REACH registration dossiers and evaluating hazard information for potential classification under Classification Labelling Packaging (CLP) regulations.

Under a separate process, a third report from the RIPoN activity relating to Substance Identity (RIPoN 1) has also been published. According to the Commission, it was not possible to reach consensus on the recommendations amongst the experts working on this project and so further work is required before recommendations can be forwarded to ECHA.

Many of the related provisions, including safety data sheets and classification and labelling, already apply today, independently of the tonnage in which the substances are manufactured or imported.

9.2.2 US

The Food and Drug Administration (FDA) regulates, *inter alia*, food and food packaging to help ensure food safety. In 2006 the FDA formed the Nanotechnology Task Force to identify and address ways to evaluate the potential effects on health from FDA-regulated nanotechnology products (FDA, 2006).

A year later, the task force recommended that FDA issue guidelines to industry and take steps to address the potential risks and benefits of drugs, medical devices, cosmetics and other FDA-regulated products that incorporate nanotechnology (FDA, 2007). The proposed guidelines were the first step toward developing policies that might guide regulation of products using nanotechnology.

A report from the Woodrow Wilson International Center for Scholars Project on Emerging Nanotechnologies, called *Assuring the Safety of Nanomaterials in Food Packaging: The Regulatory Process and Key Issues*, was published in 2008. The author commented that the food-packaging industry, food companies and consumers share an interest in assuring that any safety questions are identified and, if present, are carefully evaluated and resolved before a new packaging material is marketed (Taylor, 2008). The FDA and Environmental Protection Agency (EPA) regulatory processes and data requirements in relation to food packaging were summarised in this report.

FDA has subsequently been working with the White House, the National Nanotechnology Initiative, other US government agencies and international regulators to focus on generating data and coordinating policy approaches to ensure the safety and effectiveness of products using nanomaterials.

In June 2011, the FDA released draft guidance to provide regulated industries with greater certainty about the use of nanotechnology (FDA, 2011). This did not announce a regulatory definition of nanotechnology, but is the first step to narrowing the discussion and working with industry to determine how to proceed.

It was suggested that nano-stakeholders, when considering whether an FDA-regulated product contains nanomaterials or otherwise involves the application of nanotechnology, would be asked by the FDA:

- whether an engineered material or end product has at least one dimension in the nanoscale range (~1–100 nm), or
- whether an engineered material or end product exhibits properties or phenomena, including physical or chemical properties or biological effects, which are attributable to its dimension(s), even if these dimensions fall outside the nanoscale range, up to one micrometer.

These considerations apply not only to new products, but also may apply when manufacturing changes alter the dimensions, properties or effects of an FDA-regulated product or any of its components. In addition, they are subject to change in the future as new information becomes available, and to refinement in future product-specific guidance documents.

Environmental regulation covering chemicals in commercial use is controlled by the Toxic Substances Control Act (TSCA). Under this Act, the EPA maintains a list of all commercial chemical substances that are manufactured in the US or are imported to or exported out of the US. Given that manipulation of known chemicals on the nanoscale does not create new chemical substances, nanoparticles and nanoscale manipulation of existing chemicals would not create a new chemical substance and therefore would not trigger the new substance notification requirement established by TSCA. The nanoscale substances and manipulation of materials will seldom create new chemicals, but will often likely be considered for significant new uses, and thus subject to TSCA notification and testing requirements.

9.2.3 Australia

A new regulatory process, enacted under the National Industrial Chemicals Notification and Assessment Scheme (NICNAS), took effect on 1 January 2011 (NICNAS, 2010) and applies to any new material/chemical that falls under the following working definition of an 'industrial nanomaterial', i.e.:

> . . . industrial materials intentionally produced, manufactured or engineered to have unique properties or specific composition at the nanoscale, that is a size range typically between 1 and 100 nm, and is either a nano-object (i.e. that is confined in one, two or three dimensions at the nanoscale) or is nanostructured (i.e. having an internal or surface structure at the nanoscale).

The following are notes to this working definition:

- 'Intentionally produced, manufactured or engineered' materials are distinct from accidentally produced materials.
- 'Unique properties' refers to chemical and/or physical properties that are different, because of their nanoscale features as compared to the same material without nanoscale features, and result in unique phenomena (e.g. increased strength, chemical reactivity or conductivity) that enable novel applications.
- Aggregates and agglomerates are considered to be nanostructured substances.
- Where size distribution shows 10% or more of a substance (based on number of particles) is at the nanoscale, NICNAS will consider this substance to be a nanomaterial for risk assessment purposes.

Any substances meeting the definition will require a NICNAS permit or certificate for product introductions after this date.

These administrative amendments to the new chemicals notification and assessment framework are components of the NICNAS strategy for regulating industrial nanomaterials. According to the Australian government, these are necessary to 'address the uncertainty surrounding the risks posed by industrial nanomaterials and the appropriateness of current risk assessment protocols and practices'. These regulations are also needed 'to maintain or enhance existing levels of public health, worker safety and environmental protection in relation to industrial nanomaterials' (NICNAS, 2009).

The new regulations have been developed in consultation with the NICNAS Nanotechnology Advisory Group, comprising industry, community, government and scientific experts.

9.2.4 China

The regulatory regime in China for the management of nanotechnology chemicals will likely manage risks in a comparable way to those identified under the EU's REACH regulation. Day-to-day enforcement of chemical production and food safety regulations falls directly into local governments, through national agencies such as the State Food and Drug Administration (Richmond *et al.*, 2010).

9.2.5 Japan

In November 2011, The Ministry of Economy, Trade and Industry (METI) announced that a 'Committee on Safety Management for Nanomaterials' will be established on the basis of the results obtained from the project 'Research and Development of Nanoparticle Characterization Methods' commissioned by the New Energy and Industrial Technology Development Organization (NEDO) and the latest scientific knowledge (METI, 2011). Specifically, the committee will focus on risks (especially in view of exposure to nanomaterials in products) caused by nanomaterials and study appropriate management procedures for nanomaterials considering the actual usage and lifecycles. This will be a generic approach to determining safety management procedures for nanomaterials, but clearly nanomaterials for packaging will fall within its scope.

9.2.6 Nanocodes

It is clear that there is no consistency across the world in terms of regulations concerning nanomaterials, whether generally or specifically (Matsuura, 2006). It is also worth noting that different countries will move at different speeds in the future in seeking to develop acceptable and consistent regulations for nanomaterials, which will meet the needs of all stakeholders. In the absence of such consistent regulations, nanocodes have been developed to provide a set of voluntary principles describing best practice that an organisation should follow. One example is the Responsible NanoCode, which was developed in the UK in 2008 and includes seven Principles providing strategic guidance on governance of nanotechnology (Responsible NanoCode, 2008).

These principles comprise the following:

- Board Accountability.
- Stakeholder Involvement.
- Worker Health and Safety.
- Public Health, Safety & Environmental Risks.
- Wider Social, Ethical Environmental & Health Impacts.
- Engaging with Business Partners.
- Transparency and Disclosure.

Each organisation should be open and transparent about its involvement with and management of nanotechnologies and report regularly and clearly on how it implements the Principles of the Responsible NanoCode, even if it does not actively publicise its use of the Code. This most likely will be done as part of the Company Social Responsibility (CSR) Report.

9.3 Issues and concerns around nanotechnology

9.3.1 House of Lords report

In January 2010, the UK's House of Lords Science and Technology Committee published a report entitled *Nanotechnologies and Food*, which considered whether the use of nanotechnologies may play a valuable role in the food sector, whether effective systems are in place to ensure that consumers are aware of and protected against any potential risks and to understand and address some of the concerns that the public may have about nanotechnologies (Nanotechnologies and Food, 2010).

As part of their evidence collecting, they did receive evidence about how and where nanotechnologies might be used in food packaging. However, their conclusions mainly centred on the use of nanotechnologies in food and these formed the basis of their recommendations covering Health and Safety, Regulatory Coverage and Enforcement, and Effective Communication. There were no specific recommendations regarding nanotechnologies and food packaging, but the UK's Food Standards Agency (FSA) was recommended to create and maintain a list of publicly available food and food packaging products containing nanomaterials that have been approved by the European Food Safety Authority.

The UK Government welcomed the Report and set out actions that would be taken to address the points raised by the report. The FSA agreed to engage with stakeholders and publish the outcome of these engagements, the major focus of which was a series of citizens' forums set out to establish a dialogue with the public on food standards (FSA, 2011). While concerns were raised in these forums about the application of nanotechnology directly on foods, there was greater acceptance of nanotechnology-based packaging, although this was predicated on there being minimal impact of the nanotechnology-based packaging on the environment or on product cost.

Research was conducted in 2008 at ETH Zurich Institute for Environmental Decisions in Consumer Behaviour (Siegrist *et al.*, 2008). A survey of 350 people showed that, although people were critical of the use of nanotechnology in food, there was a relatively high acceptance of nanotechnology in food packaging. Surprisingly, the argument that packaging may transfer nanoparticles to the product was not raised as a threat, despite this typically being the main criticism used in the argument against nanotechnology by critics. An earlier study (Siegrist *et al.*, 2007) involving 153 people found that the benefits of packaging using nanotechnology were perceived as being greater than those for foodstuffs containing nanomaterials. Results further suggested that social trust in the food industry is an important factor directly influencing the effect evoked by these new products.

Ensuring the public understands more about how and where nanotechnology is being included in FMCG will mean greater engagement from the retailers who are the link between products and the public. To date, the retailers would argue that although they may have policies related to nanotechnolology, including requesting suppliers to disclose information to them in advance of product supply, they have had limited information on which they may base a public awareness campaign. Packaging may be the exception, but has almost certainly not had the same focus as the goods themselves.

9.3.2 Non-Governmental Organisations (NGO) views

Various Non-Governmental Organisations (NGOs) have expressed views on nanotechnology and they have consistently requested that there should be a compulsory reporting scheme for all nano-enabled products.

The UK consumer organisation Which? has expressed its views on what it wants to see in relation to controls on the use of nanotechnology. It recognises that nanotechnology has the potential to change lives for the better, but advises that until there is greater transparency and better understanding of risk, consumers will still be in the dark about what they see as the hidden aspects of nanotechnology (*Which?*, 2011).

Alongside more research into emerging nanotechnologies, they have called for:

- Mandatory registration of all products that contain nanomaterials.
- A public database containing all products that have been registered.
- A definition of nanotechnology that will enable everyone to ascertain what products should and should not be included on this database.

Specific NGOs, including Friends of the Earth (FOE) and the Soil Association, have gone further. In March 2008, FOE called for:

> . . . a moratorium on the further commercial release of food products, food packaging, food contact materials and agrochemicals that contain manufactured nanomaterials until nanotechnology-specific regulation is introduced to protect the public, workers and the environment from their risks, and until the public is involved in decision-making.

A US-based organisation called 'As you Sow' published a document in December 2011, called *Sourcing Framework for Food and Food Packaging Products containing Nanomaterials*, where they suggested that food packaging companies need to assess the potential risks associated with exposure to nanoparticles used in the manufacture of food packaging materials. They claim that current regulatory controls are inadequate to assess or ensure safety and that there is a lack of knowledge regarding how nanomaterials interact at the molecular or physiological levels and their impact on health and the environment. As You Sow noted that the guidelines were created with food companies including Kraft, McDonald's, Whole Foods, Yum! Brands and Pepsi (As You Sow, 2011).

There have been no actions from governments or industry in response to this call. Industry manages its business on the basis of risk. Risk is based on a combination of hazard and exposure and so industry, including packaging producers, seeks to minimise risk by using materials such as nanomaterials of intrinsically low hazard and by limiting exposure.

9.4 Nano-enabled products of the future

Nano-enabled products of the future will include nanocomposites, barrier layers, active components and smart features. Active packaging usually means having functions beyond the inert passive containment and protection of the product. Intelligent or smart packaging usually involve the ability to sense or measure an attribute of the product, the inner atmosphere of the package or the shipping environment (Soroka, 2008).

9.4.1 Nanocomposites

Nanocomposites incorporating nanoparticles have been developed for:

- Barrier properties.
- Mechanical strength.
- Temperature and moisture stability.
- Durability.

Available in natural and synthetic varieties, nanoclay is gaining popularity in plastic packaging. Nanoclays based on montmorillonite account for the largest use of nanomaterials in packaging. Nanoclays are in the form of platelets, ca. 1 nm thick with a surface of ca. 1000 nm by ca. 1000 nm. The structure of packaging containing nanoclay platelets leads to gases following a tortuous path, slowing their transmission and thereby significantly extending the shelf life of stored food and beverages. Difficult for moisture and gas molecules to navigate, the tortuous path also provides challenges for flavour, odour and colour molecules to pass through.

Increased gas barrier properties are not only of interest in the food market, but also in the beverage market. More specifically, nanocomposites could capture a large portion of the carbonated beverage, such as beer, soft drinks and carbonated water markets. Because nanoclays do not differentiate between oxygen and carbon dioxide in the way that scavengers do, they are more universally applicable (Brody, 2007a,b).

Thus, for a carbonated beverage such as beer, a barrier to ingress of oxygen and loss of carbon dioxide is achieved. Mechanical properties are also enhanced, even at lower weight. Nanoclays are used at levels of 2–5% in polyethylene (PE), polypropylene (PP), polyamide (PA), polystyrene (PS), polyurethane (PU) and polyethylene terephthalate (PET). It should be remembered that nanocomposite technology can be applied to flexible or rigid packaging. However, key to achieving ultimate performance is realising even dispersion of particles and novel processes are being commercialised to ensure optimal dispersion for maximum performance.

The size of these nanoclays does not allow them to scatter light. Because the light does not scatter, the material does not appear cloudy. Hence, the small particles do not affect clarity, allowing packages to be designed with improved barrier properties and still have the ability to show off the product inside (Brody, 2006).

9.4.2 Barrier layers

A 40–50 nm thick layer of aluminium nanoparticles applied directly to plastic film can replace the need for foil laminates. The nanoparticles are commonly applied by vapour deposition. They provide a barrier, especially to oxygen and permit downgrading of the film thickness, limit disposal impact and save cost (Fisher, 2008).

In a development from Fraunhofer Institute in association with Vacuum Technology Dresden and Applied Materials Inc., 10 nm alumina transparent coatings have been applied to a range of different polymers including PET, PP, biaxially-oriented polypropylene (BOPP) and polylactic acid (PLA). The coatings can be applied on a vacuum-based roll-to-roll process using plasma deposition of the alumina (Food Production Daily, 2010). The process produces much thinner layers compared to traditional coating processes, creating a barrier to humidity and oxygen, leading to extended freshness. This technology is claimed to provide one of the best and most effective barriers against moisture for transparent foils in the packaging sector to date (Hirvikorpi *et al.*, 2010).

In 2001 Tetra Pak announced it had received acceptance from the US FDA for Glaskin™ packaging barrier coating as a safe food contact substance. Glaskin's silicon dioxide coating for PET bottles created a clear glass-like interior coating, enabling liquid food manufacturers to package oxygen sensitive beverages in PET containers without sacrificing product quality. Glaskin was noted as being 'safe for consumers, while maintaining the taste and quality of the products, cutting costs and was environmentally sound' (Tetra Pak, 2001).

Glaskin was used in packaging to extend the shelf life and maintain the taste and nutritional value of beverages such as beer, carbonated soft drinks, fruit juices and other oxygen sensitive beverages.

When aluminum cans and tubes are used, additional interior and possibly even exterior coating is required, especially in the food industry, because direct contact between the product and the packaging can lead to aluminum corrosion and the food could spoil. Processing steps like pasteurization increase corrosive impact on packaging as well. With the PlasmaPlus® coating technique, plasma jets deposit micro-fine glass-like nano-coats. These nano-coats form a highly effective protective film on the packaging with minimum material usage. Plasma coats also provide a perfect primer for quality imprinting (Plasmatreat, 2011).

9.4.3 Nanocellulose

Nanocellulose is a material derived from wood fibres. It comprises uniform, redispersible natural nanoparticles obtained from the crystalline regions of cellulose fibres. It has exceptional strength characteristics and is completely renewable. Nanocellulose can form transparent films with excellent barrier properties, allowing it to be competitive with petroleum-based plastics in food and other packaging. Due to its reinforcing properties, nanocellulose can also be used in bio-composites and other matrix materials. Nanocellulose is now being produced on a large scale and so is available for industrial use (Food Manufacture, 2010; EMPA, 2011; Science Daily, 2011).

9.4.4 Decorative applications

Pearl effect pigments in nature are the result of nanotechnology. The colours in butterfly wings and the pearl effects of shells are due to the light being bounced off nanoscale layers in the structure of the wing or shell. Synthetic versions of pearlescent or nacreous coatings or pigments possess optical effects that not only serve decorative purposes but also can provide important functional roles, such as in security labels. Applications have included perfume bottles.

The pearlescent effect is produced by the specular reflection of light from the many surfaces of platelets, such as mica with parallel orientation at various depths within the coating. Light striking the platelets is partially reflected and partially transmitted through the platelets. A pearly lustre effect is produced by the dependence of reflection on viewing angle, and the sense of depth is created by reflection from many layers.

9.4.5 Active materials

Active packaging usually means having functions beyond the inert passive containment and protection of the product. Intelligent and smart packaging usually involve the ability to sense or measure an attribute of the product, the inner atmosphere of the package or the shipping environment (Soroka, 2008).

Antimicrobials

Silver is one of the oldest known antimicrobials and has been used in the treatment of wounds for centuries. Nanoparticulate silver has gained wide acceptance as an antimicrobial agent and has been incorporated in food packaging and also immobilised onto the surface of packaging leading to claims of extended shelf life of the food by inhibiting the growth of microorganisms. Product examples containing nano silver include food packaging and food storage containers. Nanoparticulate silver has also been applied to paper for food packaging (Gottesman *et al.*, 2011).

The antimicrobial properties of nanoparticulate zinc oxide and magnesium oxide have also been investigated. Compared to nanoparticulate silver, the nanoparticles of zinc oxide and magnesium oxide are expected to provide more affordable and safe food packaging solutions in the future (Food Production Daily, 2005).

UV absorbers

UV protection is important in plastics as excessive exposure to ultraviolet (UV) light can damage the plastic leading to loss of mechanical properties, change in colour, generation of 'off tastes and odours' or print fading. UV radiation can also adversely affect substances contained within the plastic, leading to colour fading of contents including food, accelerated oxidation of oils and fats and loss of vitamin content in, for example, fruit juice and milk. Nanoparticulate titanium dioxide and zinc oxide deliver superior UV protection and significantly improved transparency compared to traditional larger particle metal oxide powders (Croda, 2011).

Primarily aimed at clear film and bottle packaging applications, these nanoparticulate UV absorbers protect both the packaging and the contents from the harmful effects of UV light, giving consumer products greater shelf appeal and longer shelf life.

Oxygen scavenging materials

The removal of oxygen from packaging ensures the retention of the natural colour and preserves nutritive value of food. Spain-based NanoBioMatters have recently introduced O_2 Block – a new family of advanced organoclay oxygen scavengers, which the company claims simplifies and reduces the cost of scavenging technology for applications in food and pharmaceutical packaging (European Plastics News, 2010).

The new system is based on modified clay that is functionalised with an active iron to create a highly efficient oxygen scavenger. The company claims that the key to the effectiveness of the new system is its use of a purified and layered clay-based carrier for the iron, making it possible to disperse the active ingredient directly into the polymer during production. O_2 Block is suitable for use with packaging materials based on PE, PP and PET, as well as PLA, at loading levels of 1–10%.

Controlled released packaging

LaCoste *et al.* have reported on 'smart blending' of two or more polymers to manipulate packaging film morphologies to provide a wide range of properties. This approach provides packaging releasing active compounds such as antimicrobials, antioxidants and flavour compounds to extend the shelf life of food (LaCoste *et al.*, 2005).

9.4.6 Intelligent or smart packaging

Intelligent or smart packaging incorporates nanosensors to monitor the condition of the food or other contents. The key factor common to general use of 'smart' packaging is reducing the cost of sensors. Price can be low if the sensor is printed directly onto the inside of the package using the same inkjet technology that is used to print the label on the outside of the package.

Scientists at Strathclyde University have developed oxygen sensitive ink that can be used for food packaging. The ink is usually a bright colour, but can be bleached colourless when exposed to UV light. It will then remain colourless until exposed to oxygen. Consequently, the inks could be used as a tamper-evident or food package integrity indicator. The key feature of the ink is the utilisation of nanoparticulate titanium dioxide (Lee *et al.*, 2005; Mills, 2005; Mills *et al.*, 2010).

Recently the team was reported to have developed an ink that has a delay in reacting to oxygen. One application of such a feature is that it will show how long a package has been open and provide some additional reassurance over the best-before date that the food is still safe to eat. They call this delayed indicator a 'consume within' indicator and think it may help enormously with the high level of household waste, which includes a significant proportion of perfectly safe-to-eat food (*Packaging News*, 2010).

Nanotechnology can provide added functionality to barcodes by the inclusion of nanoparticles within the barcode ink that can provide additional information on interrogation to authenticate products and provide tracking and traceability information. RFID tags consist of a silicon microchip providing memory for data and a coupling element that acts as an antenna to access this information (Rice University, 2010).

Nanotechnology can bring additional functionality to the antenna. Key to their future success will be reducing the unit cost to compete with barcodes. Ink jet printing of antenna can provide a low-cost process to deliver conductive inks. These conductive inks can be based on silver, but copper is lower cost and Intrinsiq have developed a nanocopper-based ink, Intrinsiq CI, which can be applied in RFID antenna (Intrinsiq, 2010). Rice University has developed RFID tags with carbon nanotube infused ink and Vorbeck Materials is planning to begin producing the first graphene-based conductive inks to be used for imprinting RFID antennas on substrates (Rice University, 2010; Vorbeck, 2011). More information on RFID tags can be found in Chapter 3.

Nanotechnologies will play a key role in the future use of printing technologies, especially in printed electronics. High-throughput printing processes will dramatically reduce the cost of fabricating electronic devices. Providing the necessary components for smart packaging could contribute to end-users not only providing improved products, but also enhanced consumer experiences and brand recognition.

Ntera's NanoChromics™ technology combines nanocrystalline and colour change materials in novel printable devices. A NanoChromics Display (NCD) can be printed with industry-standard printing equipment on virtually any kind of printable surface and utilises the natural colours of electrochromic inks. The product is claimed to be the world's thinnest display technology at less than 30 microns thick with low power and low voltage, directly compatible with 1.5 V power systems, which can be integrated with the structure. Organic solar technology can easily provide enough voltage, even in low light conditions, to power the NCD display. The technology is based on something called electrochromism, where materials change colour when a charge is applied (Ntera, 2010).

Ntera refers to the new technology as an example of 'printegration', where a display, a sensor and a charge-storing layer are integrated by way of successive layered printing steps. NTERA's cost-effective printed NanoChromics™ display technology is claimed to have the potential to revolutionise security solutions (brand protection) and product marketing (consumer interaction) within the field of smart packaging.

9.4.7 Nanoparticle migration

Questions have been asked regarding the potential for nanoparticles to migrate out of packaging into the food or other product within the packaging. Studies conducted at the Food and Environment Research Agency (Simon *et al.*, 2008) in the UK found no detectable migration of nanoclay from PET and no detectable silver migration from PP. Lack of migration of TiN from PET was also noted. On the basis of modelling, it could be predicted that any detectable migration of nanoparticles from packaging into food might take place only in the case of very small nanoparticles – of the order of 1 nm – that are not bound in the polymer matrices. Otherwise, appreciable migration would be highly unlikely.

9.5 Future trends

9.5.1 Market forecasts and views from market analysts

There have been various market reports over the years that have tried to estimate the scale of the future value of products that are 'nano-enabled'. A report from Lux Research in 2006 entitled *The Nanotech Report 4th Edition* noted that nanotechnology was incorporated into more than $30 billion in manufactured goods in 2005 (Lux, 2006). The projection was that in 2014, £2.6 trillion in manufactured goods would incorporate nanotechnology. In 2009 Lux went on to report that they believed that as the economic downturn drained demand in several nano-enabled market segments, total revenues from products incorporating nanotechnology should still reach £2.5 trillion in 2015. This was claimed to reflect a 21% drop from earlier projections (Lux, 2009).

According to a report, *Nanotechnology Market Forecast to 2013*, published in January 2011 from RNCOS, the global nanotechnology market was projected to grow at a compound annual growth rate (CAGR) of around 19% during 2011 to 2013 (RNCOS, 2011). This report expected that the global market for nanotechnology-based manufactured goods would be worth $1.6 trillion in 2013, which is not inconsistent with the updated report from Lux Research.

Another report, this time from the Helmut Kaiser Consultancy, has concluded that global growth rates for nano-enabled products will amount to 8–21% per year in the next 15 years (Helmut Kaiser, 2005a).

The FMCG sector is one with continued pressures to deliver 'new and improved' products on short timelines and packaging can be considered as one way to contribute to the 'new and improved' appearance of these products. Therefore, packaging is a major factor in supporting growth of the whole FMCG sector. The global market for FMCG packaging has been forecast in May 2011 to reach $456 billion in 2011 and nanotechnology is seen as a major driver for growth for the next decade.

Visiongain has also reported in May 2011 the global value of the food and beverage packaging market will reach $316.7 billion in 2011 (Visiongain, 2011).

9.5.2 Food and beverage packaging

According to a market study published by Helmut Kaiser Consultancy in 2005, nanotechnology has been significantly increasing its impact on the food and beverage packaging industry during the previous three years (Helmut Kaiser, 2005b). The sales of the nano-related packaging products have risen from $150 million in 2002 to $860 million in 2004 worldwide. Nonetheless, compared with the over $100 billion food and beverage packaging industry, the growth potential of the nano-enabled packaging is still enormous. It was predicted that nanotechnology would change 25% of the food packaging business over the next decade.

In a July 2009 study from iRAP, Inc., *Nano-enabled Packaging for the Food and Beverage Industry – a Global Technology, Industry and Market Analysis*, the total nano-enabled food and beverage packaging market in the year 2008 was $4.13 billion, and was expected to grow in 2009 to $4.21 billion and forecasted to grow to $7.30 billion by 2014, at a CAGR of 11.65% (iRAP, 2009). Active technology was believed to represent the largest share of the market, and would continue to do so in 2014, with $4.35 billion in sales, and the intelligent segment was forecast to grow to $2.47 billion in sales.

The Asia/Pacific region, in particular Japan, was the market leader in active nano-enabled packaging, with 45% of the current market, valued at $1.86 billion in 2008 and projected to grow to $3.43 billion by 2014, at a CAGR of 12.63%.

According to a report by Priority Metrics Group, 'Nanotechnology in Packaging', in 2004, the nano-related food and beverage packaging market represented $860 million in sales from more than 250 nano-packaging products (Priority Metrics Group, 2009). The total nano-enabled food and beverage packaging market has grown to over $4 billion in 2009 and is forecast to grow to over $7 billion by 2014. Active technology currently represents the largest share of the market, and will likely continue to do so by 2014.

9.5.3 Pharma packaging

Packaging in the pharmaceutical industry has gone through major changes in the past decade. The advent of new drug delivery systems and the development of new biochemical compounds have resulted in a need not only for enhanced protection against factors such as moisture, light, oxygen and mechanical forces, but also for packaging forms to play a more integral role in the drug delivery process. Packaging design today provides stability and shelf life to the drug and the delivery system, which becomes fundamental to the safety, convenience and compliance of drug use.

Basic categories of nanotechnology applications and functionalities appear in development of pharmaceutical (or pharma) packaging in terms of enhancement of plastic materials' barriers, incorporation of active components that can deliver functional attributes beyond those of conventional active packaging, and sensing and signalling of relevant information.

According to a March 2010 report from iRAP, Inc., *Nano-enabled Packaging for the Pharmaceutical Industry – a Global Technology, Industry and Market Analysis*, the total market for nano-enabled packaging for pharma was \$3.8 billion in 2009, and this was expected to grow to \$8.1 billion by 2014, at a CAGR of 16.5% from 2009 to 2014 (iRAP, 2010).

China's growth opportunities were considered to be among the strongest, based on rapidly expanding pharmaceutical manufacturing capabilities there and the phasing-in of a government programme designed to upgrade the quality and integrity of nationally produced medicines. The market for nano-enabled packaging for pharma will grow from \$360 million to \$2.03 billion by 2014.

9.6 Conclusions

It is clear from these forecasts of growth for nano-enabled products across all market areas and the specific forecasts for growth in nano-enabled packaging for food and beverage and also pharma product packaging that these two areas are leading in terms of having products already in the market place and also in terms of expectations of significant growth. However, it must be realised that where food and beverage and also pharma packaging go today, other sectors will follow.

Mundal *et al.* (2009) have noted that nanotechnology is expected to change the entire packaging industry with self-assembly reducing fabrication costs and infrastructure, and more flexible packaging methods providing consumers with fresher and more customised products. There are many new developments that are delivering new and attractive packaging options to companies to entice consumers to buy their products and this is clearly the case for packaging for FMCG. Nanotechnology will be an enabler to deliver smart, novel packaging that can benefit not only the product producer, but also the consumer by providing extended shelf life with additional product information and enhanced security at a cost that is acceptable both to the producer and the consumer.

Given the enormous amount of FMCG packaging that is used, there is obviously concern about the huge amount of waste. Alongside this, concerns have been expressed about the inclusion of nanomaterials and the potential for free nanomaterials getting into the environment, but when bound within packaging there are no free nanomaterials that can escape into the environment. In addition, because nanotechnology leads to enhanced performance for lower total packaging weights, there is likely to be less waste for disposal.

Nanotechnology is exciting and is already being used extensively in packaging, and the future nano-enabled packaging developments for FMCG are expected to include:

- Nanocomposites containing nanoclays to improve the strength and flexibility of packing materials, while providing effective barrier properties.
- Nanosensors embedded in packaging to detect pathogens and contaminants, while microbial coatings will protect contents from bacteria.

- Packages with tracking functions based on nanocomponents to help producers monitor products as they move through the supply chain.
- Smart and active packaging systems to allow producers to trigger flavour (nano)releases at certain times, or under certain conditions.
- Light, translucent and highly durable nanocellulose-based packaging to provide an alternative to conventional plastics packaging.

9.7 References

AS YOU SOW (2011) 'Sourcing Framework for Food Packaging and Products Containing Nanomaterials'. Available from: http://www.foodsafetynews.com/nano%20in%20food%20framework.pdf [Accessed October 2012].

BRODY A. L. (2006) 'Nano and food packaging technologies converge', *Food Technology*. 60(3), 92–4.

BRODY A. L. (2007a) 'Case studies on nanotechnologies for food packaging', *Food Technology*, 61(7), 102–7.

BRODY A. L (2007b) 'Nanocomposite technology in food packaging', *Food Technology*, 61(10), 80–4.

BSI (2009) 'DD CEN ISO/TS 27687:2009'. Available from: http://shop.bsigroup.com/ProductDetail/?pid=000000000030199918 [Accessed October 2012].

CRODA (2011) 'UV Protection'. Available from: www.croda.com/home.aspx?s=136&r=208&p=1525 [Accessed October 2012].

DREXLER K. E. (1986) *Engines of Creation: The Coming Era of Nanotechnology*, New York, Anchor Press/Doubleday.

EMPA (2011) 'Using Nanocellulose to Create Novel Composite Materials'. Available from: www.empa.ch/plugin/template/empa/3/113491/---/l=2 [Accessed October 2012].

EUROPEAN COMMISSION (2011) 'Nanomaterials'. Available from: http://ec.europa.eu/environment/chemicals/nanotech/index.htm [Accessed October 2012].

EUROPEAN PLASTICS NEWS (2010) 'NanoBioMatters Unveils New Oxygen Scavenger'. Available from: www.europeanplasticsnews.com/subscriber/headlines2.html?id . . .

FDA (2006) FDA 'Opens Dialogue on "Nano" Regulations, 2006'. Available from: www.fda.gov/ForConsumers/ConsumerUpdates/ucm258462.htm [Accessed October 2012].

FDA (2007) 'Nanotechnology Task Force Report, 2007'. Available from: www.fda.gov/. . ./Nanotechnology/NanotechnologyTaskForceReport2007/default.htm

FDA (2011) 'Considering Whether an FDA-Regulated Product Involves the Application of Nanotechnology – Guidance for Industry'. Available from: www.fda.gov/RegulatoryInformation/Guidances/ucm257698.htm [Accessed October 2012].

FISHER, D. (2008) 'The Next Innovative Wave: Nanotechnology in Packaging'. Available from: www.iopp.org/files/public/FisherRITNanotechology

FOOD MANUFACTURE (2010) 'Swedish Firm Seeks Partner to take Nanocellulose to Market'. Available from: www.foodmanufacture.co.uk/content/view/print/323084 [Accessed October 2012].

FOOD PRODUCTION DAILY (2005) 'Nanotech Discovery Promises Safer Food Packaging'. Available from: www.foodproductiondaily.com/Packaging/Nanotech-discovery-promises-safer-food-packaging [Accessed October 2012].

FOOD PRODUCTION DAILY (2010) 'Fraunhofer Develops Nano Thin Coating to Enhanced Shelf Life'. www.foodproductiondaily.com/. . ./Fraunhofer-develops-nano-thin-coating-to-enhance-shelf-life

FSA (2011) 'FSA Citizens Forums: Nanotechnology and Food', TNS-BMRB Report.

FRIENDS OF THE EARTH AUSTRALIA, EUROPE AND USA (2008) 'Out of the Laboratory and onto our Plates: Nanotechnology in Food and Agriculture'. Available from: http://nano.foe.org.au/node/219 [Accessed October 2012].

GOTTESMAN R., SHUKLA S., PERKAS N., SOLOVYOV L. A., NITZAN Y. and GEDANKEN A. (2011) 'Coating of paper by microbiocidal silver nanoparticles', *Langmuir*, 27(2), 720–6.

HELMUT KAISER (2005a) 'Summary about the State of Nanotechnology Industry Worldwide 2006–2015'. Available from: www.hkc22.com/nanomarkets.html [Accessed October 2012].

HELMUT KAISER (2005b) 'Nanotechnology in Food and Beverage Packaging Industry Worldwide 2008–2010–2015'. Available from: www.hkc22.com/nanopackaging.html [Accessed October 2012].

HIRVIKORPI T., VAHA-NISSI M., MUSTONEN T., IISKOLA E. and KARPINEN M. (2010) 'Atomic layer deposited aluminium oxide barrier coatings for packaging materials', *Thin Solid Films*, 518, 2654–68.

INTRINSIQ (2010) 'New product launch – "Intrinsiq CI" Copper Nanoparticles-based Ink for Printed Electronics Applications'. Available from: www.intrinsiqmaterials.com/. . ./Intrinsiq%20CI%20copper%20inks%20launch.pdf

IRAP (2009) 'Nano-enabled Packaging for the Food and Beverage Industry – a Global Technology, Industry and Market Analysis'. Available from: www.innoresearch.net/Press_Release.aspx?id=18

IRAP (2010) 'Nano-enabled Packaging for the Pharmaceutical Industry – a Global Technology, Industry and Market Analysis'. Available from: www.innoresearch.net/Press_Release.aspx?id=22 [Accessed October 2012].

LACOSTE A., SCHALCH K. M., ZUMBRUNNEN D. and YAM K. L. (2005) 'Advanced controlled release packaging through smart blending', *Packaging Technology and Science*, 18, 77–87.

LEE S. K., SHERIDAN M. and MILLS A. (2005) 'Novel UV-activated colorimetric oxygen indicator', *Chemical Materials*, 17, 2744–51.

LUX RESEARCH (2006) 'The Nanotech Report, 4th Edition'. Available from: https://portal.luxresearchinc.com/research/document_excerpt/2778 [Accessed October 2012].

LUX RESEARCH (2009) 'Nanomaterials State of the Market Q1'. Available from: https://portal.luxresearchinc.com/research/document_excerpt/4248 [Accessed October 2012].

MATSUURA J. (2006) *Nanotechnology Regulation and Policy Worldwide*, Boston, MA, Artech House.

METI (2011) 'Establishing the Committee on Safety Management for Nanomaterials'. Available from: www.meti.go.jp/english/press/2011/1130_02.html [Accessed October 2012].

MILLS A. (2005) 'Oxygen indicators and intelligent inks for packaging food', *Chemistry Society Review*, 34, 1003–11.

MILLS A., HAZAFY D. and LAWRIE K. (2010) 'Novel photocatalyst-based colorimetric indicator for oxygen', *Catalysis Today*, 161(1), 59–63.

MUNDAL P. K., CHOI K., MIN S. G. and LEE C. H. (2009) 'Application of nanotechnology in food industry: an overview', *Korean Journal for Food Science of Animal Resources*, 29(4), 403–8.

NANOTECHNOLOGIES AND FOOD (2010) *House of Lords Science and Technology Committee 1st Report of Session 2009/2010*, January, London, The Stationery Office Limited.

NICNAS (2009) 'Nanotechnology – Discussion Paper: Proposal for Regulatory Reform of Industrial Nanomaterials', 4. Available from: http://www.nicnas.gov.au/Publications/Chemical_Gazette/pdf/2010oct_whole.pdf#page=14 [Accessed October 2012]. www.foodsafetynews.com/nano%20in%20food%20framework.pdf [Accessed October 2012].

NICNAS (2010) 'Adjustments to NICNAS New Chemicals Processes for Industrial Nanomaterials'. Available from: www.nicnas.gov.au/. . ./Nanotechnology/FAQs_Nano_Adjustments_for_New_Chemicals_Processes_Dec_2010.pdf

NTERA (2010) 'NTERA and plastic electronic GmbH enter an agreement to develop advanced printed electronics products and announce the availability of an Evaluation Kit based on NanoChromics™ display technologies'. Available from: www.ntera.com/press/pepressrelease.php

PACKAGING NEWS (2010) 'Nanotechnology: Micro Science Thinks Big for Future'. Available from: www.packagingnews.co.uk/. . ./nanotechnology-micro-science-thinks-big-for-future/

PLASMATREAT (2011) 'Food-safe Nano-coating Appropriate for Food with for Optimal Corrosion Protection and Quality Imprinting on Tubes and Cans'. Available from: www.plasmatreat.com/. . ./nano-coating-of-beverage-cans.html

PRIORITY METRICS GROUP (2009) 'Nanotechnology in Packaging'. Available from: http://www.pmgco.com/Publications/NanoTechnologyinPackaging/tabid/156/Default.aspx [Accessed October 2012].

RESPONSIBLE NANOCODE (2008) 'The Responsible NanoCode – Update May 2008'. Available from: http://www.responsiblenanocode.org/documents/TheResponsibleNanoCodeUpdateAnnoucement.pdf

RICE UNIVERSITY (2010) 'Nano-based RFID Tag'. Available from: http://www.media.rice.edu/media/NewsBot.asp?MODE=VIEW&ID=13899 [Accessed October 2012].

RICHMOND N., PURRA M. and JARVIS D. (2010) 'Mapping Emerging Nanotechnology Policies and Regulations: The Case of the People's Republic of China'. Available from: http://www.risk-and-regulation.com/wp-content/uploads/2010/07/Case-study-of-China.pdf [Accessed October 2012].

RNCOS (2011) 'Nanotechnology Market Forecast to 2013'. Available from: www.rncos.com/Report/IM185.htm [Accessed October 2012].

SCIENCE DAILY (2011) 26 October. 'New Process for Manufacturing Nanocellulose: Using Nanocellulose to Create Novel Composite Materials'. Available from: www.sciencedaily.com/releases/2011/10/111026091004.htm [Accessed October 2012].

SIEGRIST M., STAMPFLI A., KASTENHOLZ H. and KELLER C. (2008) 'Perceived risks and perceived benefits of different nanotechnology foods and nanotechnology food packaging', *Appetite*, 51(2), 283–90.

SIEGRIST M., COUSIN M. E., KASTENHOLZ H. and WIEK A. (2007) 'Public acceptance of nanotechnology foods and food packaging: the influence of affect and trust', *Appetite*, 49, 459–66.

SIMON P., CHAUDHRY Q. and BAKOS D. (2008) 'Migration of engineered nanoparticles from polymer packaging to food – a physiochemical view', *Journal of Food and Nutrition Research*, 47(3) 105–13.

SOROKA W. (2008) *Illustrated Glossary of Packaging Terms*, Illinois, US, Institute of Packaging Professionals.

TANIGUCHI N. (1974) *Proceedings of the International Conference on Prod. Eng. Tokyo*, Part II, Japan Society of Precision Engineering.

TAYLOR M. R. (2008) *Assuring the Safety of Nanomaterials in Food Packaging: The Regulatory Process and Key Issues*, Woodrow Wilson International Center for Scholars Project on Emerging Nanotechnologies, July.

TETRA PAK (2001) 'Tetra Pak's Glaskin™ Seals FDA Acceptance'. Available from: www.thefreelibrary.com/Tetra+Pak's+Glaskin(TM)+Seals+FDA+Acceptance,+Introduces+new+Class+of. . .-a080561014

VISIONGAIN (2011) 'The Fast-Moving Consumer Goods (FMCG) Packaging Market 2011–2021'. Available from: www.visiongain.com/. . ./The-Fast-Moving-Consumer-Goods-(FMCG)-Packaging-Market-2011-2021

VORBECK (2011) 'Bringing Graphene to Market. A Start-up's Conductive Graphene Inks can be used to Print RFID Antennas'. Available from: www.vorbeck.com/news/graphene_market.html [Accessed October 2012].

WHICH? (2011) 'Nanotechnology – What Which? Wants'. Available from: http://www.which.co.uk/campaigns/technology/what-you-need-to-know-about-nanotechnology/what-which-wants/ [Accessed October 2012].

10

Smart and interactive packaging developments for enhanced communication at the packaging/ user interface

P. Butler, Packaging Materials & Technologies, USA

DOI: 10.1533/9780857098979.261

Abstract: Conventional text-based and graphic packaging communication is reaching its natural limit, suggesting a future trend towards more effective visual and sensory forms of communicating, particularly at the packaging/user interface. A number of smart and interactive packaging developments are described that offer enhanced communication to consumers and brand owners. Of particular note are recent developments in interactive packaging and augmented reality aimed at cementing brand experiences and spreading those via social networking.

Key words: smart packaging, interactive packaging, lenticular labels, colour change labels, electroluminescent labels, consumer/packaging interface, augmented reality; smart phones, wireless power.

10.1 Introduction

For years packaging has fulfilled the important needs of the supply chain in protecting the product by retaining its integrity, bringing it safely to market without damage and providing communication to the consumer regarding the product and its attributes. Without packaging, many food and other perishable products would be unavailable for much of the year and others would be subject to massive amounts of spoilage and waste. A further important role for packaging is as a key brand ambassador for the product.

This chapter is concerned with packaging for the consumer and brand owner and focuses on improvements and innovation at the packaging/user interface, principally to enhance communication. The central thesis is that overall packaging needs to change and become smarter and more interactive. It needs to communicate more effectively both on shelf and at home, to support brands in a more innovative manner and be designed more with the consumer in mind.

10.1.1 The state of the current packaging/user interface

Increasingly, for consumers, the existing functions of bringing the product safely to the shelves and then being part of the buying decision are taken for granted. Consumers buy the product, not the packaging, so although the packaging can sometimes form part of the reason to buy, once home the packaging can be an item of frustration, with lack of openability topping the list and having the dubious honour of its own descriptive term, 'wrap rage'.

From a consumer perspective, far too many packaging formats are poorly designed, so that they are:

- Difficult to open, e.g. rigid 'clamshell' packaging, vacuum screw top jars.
- Difficult to reseal, e.g. food packaging where the food is not consumed at one sitting.
- Difficult to read, e.g. those with instructions or useful data printed in minute font sizes.
- Difficult to completely empty, so that some of the product is wasted, e.g. toothpaste tubes, high-on-shoulder glass jars containing viscous food products.

Finally there is the problem of disposal. Once the packaging has been opened and the product has been used, the packaging becomes just another nuisance to be disposed of. Although consumers are getting more educated and want to do the right thing and recycle packaging, many packaging items are difficult or confusing to dispose of properly.

Improvements are needed in all these areas but the need for clearer, legible and more innovative communication channels is the focus of this chapter.

10.1.2 The need for clearer communication

Providing clear, effective communication is fundamental to good packaging design, both to assist consumers and to promote a brand. But with consumers searching for increased amounts of information about products and legislation requiring labels to carry more data on nutrition, health and recycling, sometimes in multiple languages, the net result is a printed packaging format close to overload. To make matters worse, there is a trend towards smaller pack sizes, driven by source reduction strategies as part of packaging sustainability initiatives, and more single serve convenience sizes. The net result is less area of packaging on which to place graphics and text. Exacerbating the problem further are the future demographic changes to the consumer population, which will dramatically increase the percentage of older consumers with reduced visual acuity, incapable of reading fine print on packaging.

10.1.3 Brands feeling the pinch – the need for differentiation

In these recessionary times, the squeeze on brands has never being greater. Traditionally, there were always consumers who were of the opinion that brands were not worth their extra cost, and this trend appears to be growing as financial pressures force others to trade-down to own and private label products. Increasingly, own-label products are looking to aspire to more highly priced brands and using their packaging to imitate this. In addition, brands have to be constantly on the look out for copycat packaging and, via their packaging at point of sale, do all they can to maintain their presence and stand out on the shelves.

Differentiation and emotional engagement with the consumer are essential features of brands and need to be strongly supported by their packaging. In general, consumers making non-planned purchases are not attracted to packaging that is a little better but to packaging that is different, and catches the eye as a result. The use of colour and shape can be powerful agents in this form of differentiation, and shaped metal packaging provides a good example of the sort of impact it can make on supermarket shelves. As already discussed, conventional text and graphics will struggle to differentiate a product in the future via its packaging, suggesting the need for new channels of communication and differentiation.

10.1.4 Smart or interactive packaging?

Smart packaging has been defined as packaging that does more than the important though traditional functions of storing, protecting and providing information about the product (Kerry and Butler, 2008). This kind of packaging has been classified in many other different ways – the terms 'active', 'controlled', 'intelligent', 'diagnostic', 'functional', 'communicative' and 'enhanced' have all been used. The all-embracing term 'smart packaging' is preferred here, encompassing aspects of packaging design, the use of smart materials, and the incorporation of mechanical, chemical, electrical and electronic forces, or a combination of these, within or on the package.

Technologies, such as colour-change chemistry, lenticular graphics and two-dimensional (2D) bar codes are simple examples of smart packaging to improve communication. They offer a conduit to improved story-telling about a brand product's attributes, provenance, how the product can be used, or provide useful product information in an easily understandable communication format for the consumer. Electronic smart packaging adds the possibility of moving displays, lighting and sound, and can have serious applications in the healthcare sector to improve the human/packaging interface as well as providing novelty or adding new functionality for fast-moving consumer goods (FMCGs).

Interactive packaging, as defined in this chapter, differs from smart packaging in that the 'smart' elements are largely external to the packaging. With this approach, the packaging remains essentially unchanged but the external smart device can only operate in conjunction with the packaging. An image recognition coding system, such as a quick response (QR) code printed on the packaging that enables a smart phone to interact with the packaging and connect to the internet, would be one such example.

The next sections will suggest ways in which it is possible with smart and interactive packaging to provide clearer, more effective communication at the consumer/packaging interface, by adding to, and not replacing, existing graphics and printed text. The approach adopted will be to show and discuss current packaging examples, document innovative trials that did not fully commercialise, and highlight future concepts, particularly those that are using interactive packaging and augmented reality (AR) to cement brand experiences and spread those via social networking.

10.2 Smart packaging – enhanced communication directly via packaging

10.2.1 Animated graphics using lenticular labels

A lenticular label is a simple low-cost optical way of providing striking three-dimensional (3D) optical effects, image flipping, zooming and morphing, to enhance visual communication between a consumer and the packaging. A series of small strips is taken from each image and interlaced into one congruent image, on top of which is placed a plastic lens consisting of an array of optical elements (lenticules) that project a slightly different image to each eye. When viewed from different directions, different images are visible, so the effects can be arresting when a consumer walks past the packaging. The technology is not new and has been applied over the years to CD, movie and magazine covers, toys, greetings cards, book covers, postage stamps, bottle labels and other novelty items. Much of these applications have been little more than visual tricks in the marketers tool kit, designed to differentiate products in a crowded shelf space. But with recent improvements in lenticular technology to provide better resolution, brand marketers and package designers are re-evaluating the potential of lenticular effects (and also holographic images) on packaging as a more sophisticated way to communicate product attributes and provenance to a consumer (Swientek, 2002). As an example, the eye make-up kit from Stila Cosmetics has a lenticular flip image on the outer packaging, showing the eye before and as it should look like after make-up application (Fig. 10.1).

10.2.2 Animated graphics using optical Moiré effects

A Moiré pattern is an interference effect as one line or grid pattern is rotated or moved over another of closely similar dimensions, giving the illusion of movement or revealing secondary images. Initially developed as a security printing method to provide anti-counterfeiting features on packaging, the technology has potential but remains a curiosity, particularly for FMCGs.

However, for certain niche products, the effects can be arresting. Widex, a manufacturer of high definition hearing aids, has used the Moiré effect on

(a)

(b)

Fig. 10.1 A lenticular label on an eye make-up kit has a flip image that shows how the eye should look before (a), and after (b), the application of various cosmetic colours.

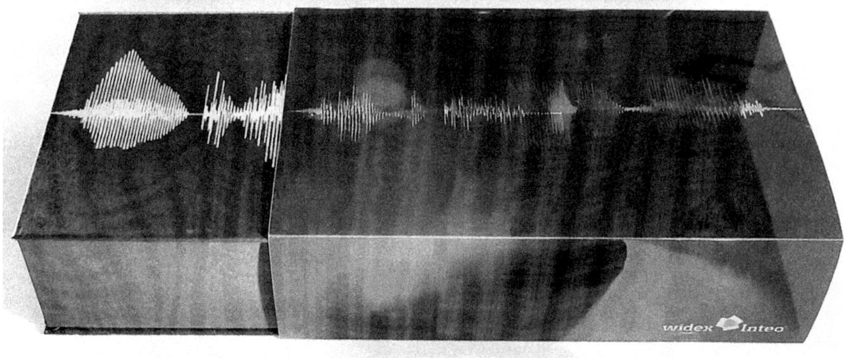

Fig. 10.2 Sliding the outer semi-translucent sleeve of this box containing a hearing aid to open it creates the optical illusion of an animated sound wave.

packaging where the consumer has to slide an outer sleeve over a fixed inner carton as part of opening the packaging to take out the hearing aid (Fig. 10.2). Designed by the Danish firm Goodmorning Technology, the packaging visually shows a moving soundwave as the user pulls out the outer transparent sleeve, with the graphics for this raster effect being the soundwave equivalent of Widex's slogan, 'High Definition Hearing'.

10.2.3 Colour-change labels for freshness and ripeness alerts

The role of packaging in positively preventing food waste is unfortunately largely unrecognized (Butler, 2011). What is clear from the post-consumer food waste data in the UK (WRAP, 2009) is that a significant component of the total amount wasted arises because consumers are in a state of confusion over date coding and frequently throw perfectly good food away that has been carefully stored at the correct low temperature but has reached its 'use by' or 'best before' date. In addition, fresh fruit and vegetables are commonly wasted because either quantities are overbought or consumers are not capable of evaluating their degree of ripeness. The extent of both of these waste streams can be diminished by clearer and more accurate smart colour-change label communication to indicate accurate shelf life or indicate ripeness (Butler, 2006).

The current date coding system used in the UK has two major flaws: one, it is largely incomprehensible to consumers; and two, it is inaccurate as a measurement of shelf life. Between 2001 and 2007, the UK Food Standards Agency conducted annual surveys of consumer attitudes to food standards, including asking consumers how well use by and best before dates were understood. The results every year showed there was sustained and widespread confusion surrounding the dates. It has been estimated that date code confusion is directly responsible for about 5% of UK household avoidable food waste (Bloom, 2010).

Date code inaccuracy arises from the fact that food degradation processes generally obey Arrhenius kinetics, which are exponentially sensitive to temperature and only linearly sensitive to time. To accurately evaluate shelf life, both temperature and time need to be considered, with temperature being far more important than time. Date coding is only accurate when food is consistently stored at microbiological dormant temperatures between 0 and 4°C.

(a)

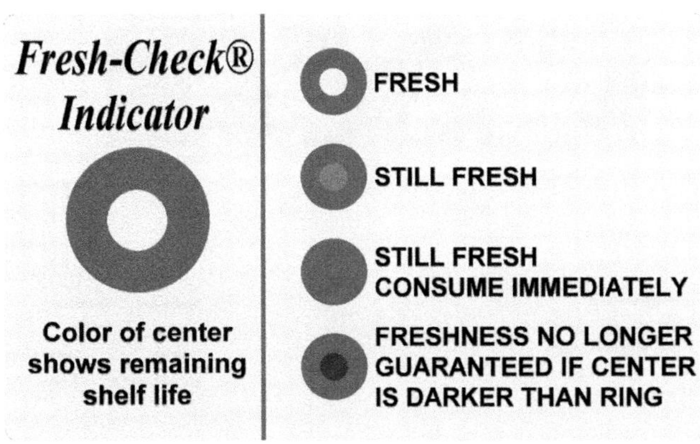

(b)

Fig. 10.3 (a) The French supermarket group Monoprix has had considerable success in the use of the Fresh-Check® time-temperature indicator labels, shown in close up in (b), on perishable foods in their stores.

Time-temperature indicators (TTIs) use chemical reactions to exhibit an irreversible change of colour in response to the combined effects of time and temperature, so offer the opportunity to measure cumulative abuse of products along the cold chain. Applied as labels to the outside of the pack, they provide manufacturers, retailers and consumers with a simple, yet effective method of controlling the safety of perishable food product inventories. The consumer can then look for a simple colour change rather than attempting to decipher esoteric date codes. The most successful commercial use of TTIs has been on perishable food products sold in the French grocery chain Monoprix (Fig. 10.3) and polio vaccine vials distributed throughout the developing world by the WHO. There is sustained and continually development of different TTI variants of increasing accuracy (Taoukis, 2008), but widespread commercial application has been elusive largely due to the resistance of large food retailers, most of whom do not want consumers to have this sort of information.

Freshness indicators, by contrast, sit inside the pack and monitor quality via the presence of degradation volatiles in the headspace, such as amines given off from fish and meat, or hydrogen sulphide from poultry, and change colour according to their concentration (Smolander, 2008). Ripeness indicators for fruit work in a similar way, detecting the ethylene gas generated by ripening fruit. This sensor is not necessary for fruits like bananas whose skin is their own ripeness colour change indicator, but has been developed for fruits such as avocados, melons and pears (Fig. 10.4).

Fig. 10.4 The front of packaging for Concorde pears with an active ripeness indicator. The label inside the pack changes colour from dark red through orange to yellow as the pears ripen.

10.2.4 Electronic animated lighting effects on packaging

Moving up in terms of cost and level of complexity, are packaging communication possibilities using today's miniature electronics, circuitry and power sources. The first of these to be considered are powered visual labels using the phenomenon of electroluminescence (EL), whereby a material emits light in response to the passage of an electric current. EL devices can be produced in thin flexible formats that produce a light of different colours when stimulated by AC current, whilst at the same time producing no heat. For label purposes, a colour image can be sandwiched on top of the EL device so it becomes back-lit when the device is activated, creating a 2D illuminated display.

This effect is particularly effective in branding and engagement with young night club goers who are receptive and always looking for new ideas. Night club environments are ideal for colour and flashing light associations with a particular beverage. Ballantine's 'Listen to Your Beat' campaign for Scotch whisky included an EL label with a blue graphic equalizer display on a striking dark blue bottle. Powered by batteries, or mains power, the bottle display reacts to musical sounds to create drama and increase brand presence in darkened bars (Fig. 10.5). Similar effects can be achieved by LED lighting (Cognifex, 2011).

Other packaged products, apart from alcoholic beverages, have trialled illumination effects as a 'wow' factor to assist marketing, many as promotions having just a short lifetime on the shelves. NXT shaving products for men is an example (Fig. 10.6). The specially designed triangular package has an LED in the bottom that lights up once every 15 seconds, illuminating air bubbles suspended in the clear gel. The plastic is tinted blue, and when the batteries in the base activate the LED, the packaging glows like a miniature lava lamp. In the crowded category of men's grooming products, the shelf impact is novel and arresting and clearly aimed squarely at the 18- to 24-year-old male audience.

One disadvantage of this type of packaging is the quantity of waste generated at disposal. The base contains two AAA batteries and associated electronics housed in a relatively heavy plastic casing (Fig. 10.6(b)), which cannot be considered a good example of sustainable packaging, which more and more consumers are looking for. This factor is a basic problem with the use of conventional electronics in disposable packaging, leading to significant consumer resistance to purchase, which could partially explain why this type of packaging is so transient.

10.2.5 Electronic animated graphics and sound on packaging

Combining light and sound effects is also possible using conventional electronics but the complexity is likely to add at least several dollars to the cost of a package, putting it out of bounds for all but high value products.

One of the most successful is the packaging for the Barbadian rum 'Coyopa', made in small batches in the R L Seale Distillery in Christ Church, Barbados,

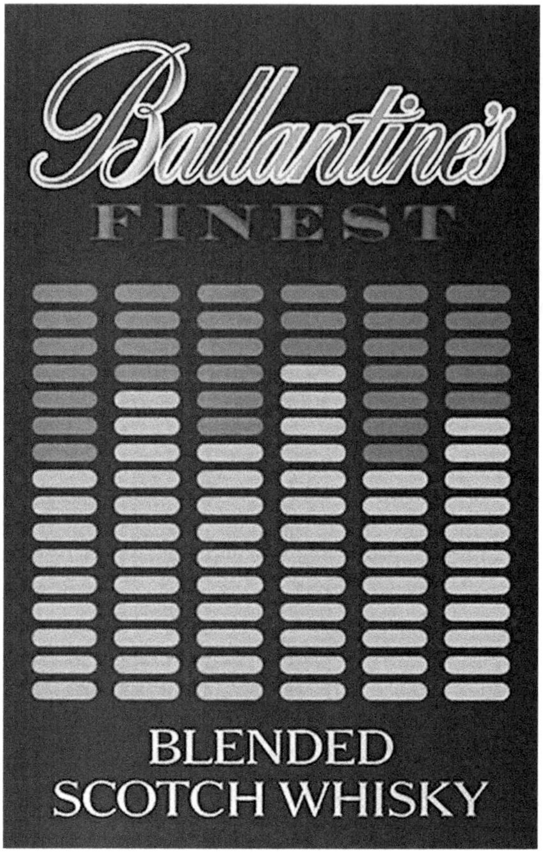

Fig. 10.5 Electroluminescent label on a whisky bottle that sequences in response to music, creates an additional 'wow' factor in bars and night-club environments.

which retails for around $50. The rum is aged for ten years in oak casks to gain character and depth and create a premium rum that embodies the spirit of Barbados. To add to the provenance and brand appeal, the packaging was designed to create a 'dancing bottle' that would appeal to all the senses – sight, hearing, as well as taste. From the base assembly of the bottle, containing the electronics and sound chip, thick film printed circuitry leads up the paper-thin composite label consisting of four EL backlit picture quadrants showing Caribbean dancers in action. Each time the bottle is lifted, the song and illuminated label dance routine plays for around 15 seconds and then shuts off automatically.

Using conventional silicon device circuitry with LED or EL illumination and sound chip technology, all driven by coin cell battery power on packaging, enables speech and sound on packaging. It is just a question of time before electronic smart packaging becomes more widely used and useful to society, moving away

(a)

(b)

Fig. 10.6 NXT shaving products for men come in a triangular plastic container with intermittent LED lighting from the base unit, creating eye-catching shelf appeal (a). Less impressive is the packaging waste components when the container becomes empty and has to be disposed of (b).

from novelty towards more serious applications. There are many potentially life-saving applications in the healthcare sector, where smart packaging could significantly assist patients to adhere to medication regimes by visual communication and spoken electronic reminders.

For FMCGs, the question of lack of sustainability and cost remain, which could lessen with the advent of printed electronics. In all forms of electronic smart packaging however, a significant technological problem to overcome is the power drain from miniature or even printed batteries that can severely limit the lifetime and scope of the sound and vision effects. One solution to this problem for disposable packaged items is to separate the power source from the packaging and couple these together via the use of smart shelves or other surfaces that incorporate wireless power.

10.2.6 Electronic animation with integrated wireless power

In a transformer, coil-to-coil inductive coupling permits the transfer via magnetic fields of electric current from one coil to another. This principle can be applied to provide power to packaging to drive displays and sound when the package sits on an activated surface below which there is a powered primary coil. This interacts with a printed copper receiver coil inside the packaging. Back-lit EL display labels would be the technology of choice for this arrangement, since they produce no heat when illuminated and are voltage compatible with 110 V AC.

At the Consumer Electronics Show in Las Vegas in 2011, Fulton Innovation (2011) showed what might be possible in the future in terms of real-time communication from the package to the consumer to differentiate a brand in a unique eye-catching way (Fig. 10.7).

However, the greatest commercial application of this technology probably lies with the retailer, as the pack could also be configured to communicate via the store shelf back to the store's inventory management system. Product quantities could be identified and tracked, expiration dates monitored, and new stock automatically ordered when supplies are low, all to help reduce lost sales. Consumer benefits might accrue in the smart home of the future (Section 10.4.3).

10.3 Interactive packaging – enhanced communication via internet connectivity

10.3.1 Augmented reality (AR)

Augmented reality (AR) refers to the technology of superimposing computer generated content, such as sound, video, graphics or GPS data, over a real-world view of the environment (How Stuff Works, 2011), familiar to sports TV fans worldwide as arrows indicating the movement of players or yardage lines marked out that do not actually exist on the playing surface. From a packaging perspective, the augmented content can be superimposed on real-world images, such as the

Fig. 10.7 Back-lit electroluminescent parts of the cereal graphic light up in sequence when the inductively coupled item of packaging is connected to a wirelessly powered smart surface.

packaging itself. The fused real and virtual content can be viewed on display screens, which can be a video, TV, PC, smart phone or tablet screen, either in the home or in-store.

As we will see in the next section, when brand-developed applications are available and can be downloaded for free to a smart phone or smart tablet, together with logo-recognition software, the stage is set for some exciting developments in interactive packaging.

10.3.2 Webcam-based augmented reality

The most successful use of webcam-based AR on packaging to date has been initiated in France on Nestlé children's cereal packets, where software from Dassault Systèmes has allowed true customer interaction in the form of a range of on-line interactive games and interactions. In these promotions, the cereal package is purchased and taken home where AR can be experienced via a personal PC and webcam. This currently comes in two variants as far as the packaging is concerned.

In the first, the packaging itself is the platform on which the AR content is projected. Consumers hold the package in front of the webcam and the computer generated game is superimposed on it. Playing the game involves watching the webcam image and moving and tilting the packaging, so the packaging becomes an AR game console. The AR campaign started in 2009 on packets of Chocapic cereal and featured characters from an animated 3D film *Arthur and the Revenge of Maltazard* (Fig. 10.8). The promotion, which involved more than two million cereal packages in France, generated intense interest with consumers and increased the market share of Nestlé France by 1.6%. Players reported feeling more immersed in the AR game than with traditional video games, on account of the merging of their own environment with the virtual world of the game. The branded box of packaged cereal sits at the centre of this experience.

In a similar move, in 2010 Lego installed terminals in key stores, with image recognition software from Metaio, allowing customers to hold up special Lego 'digital boxes' – containing typically a construction project – and watch a 3D animation of the kit superimposed on the packaging. This animates the sequence of completion of the build as the package is moved in front of the terminal screen, fusing with the live video of the actual product packaging. Customers get to see

Fig. 10.8 A screen shot of AR using the packaging of a Nestlé cereal product, where the packaging is held and manipulated to play the virtual game as it is overlaid on the real image of the pack.

exactly what the completed kit looks like in 3D before they buy the product, and via this engagement are much more likely to purchase.

In a second Nestlé/Dassault AR iteration, the cereal package is not used to control aspects of the game; instead unique coded tags are cut out from the back of the packaging and affixed to either the hand or head of the player. These codes are recognised by the software program via the webcam and automatically launch the player into the game or interactive experience.

Over 26 million packages in 53 countries (excluding the US), with children's cereal brands, such as Nesquik and Chocapic, carried the promotion throughout 2011 for the new computer-animated film *Rio*, featuring Blu the parrot, from 20th Century Fox and Blue Sky studios. Each cereal pack (Fig. 10.9) had a coded hand card printed on the back, which could be cut out, placed on the hand and held up to a webcam with the PC set to the URL, www.nestle-rio.com, to start the interaction. Once into the game, by manipulating the hand card, it is possible to interact with the 3D bird and even feed it by tipping the tag to 'fill' a bowl on the screen. Players can also discover another Rio character called Jewel, and make the characters come to life in their own personal environment.

Other promotional games of increasing sophistication have followed, as owners of megabrands such as Nestlé step up their efforts in digital marketing. In a third Nestlé/Dassault iteration, some 4.2 million cereal packs were produced offering the chance to play an AR video racing game, the first to be developed in 3D, to coincide with the launch of the film *Arthur 3 – The War of Two Worlds*.

In this game, users drive a virtual motor-cycle in three different racing environments, dependent on the type of cereal bought. A pair of 3D glasses, together with a coded square, is first cut from the package. The game works in 3D on any type of screen, using the anaglyph (red and blue) glasses cut out from packages; however, for owners of 3D TVs who want a more immersive experience without loss of colour, there is also a 3D TV mode. In either mode, the coded square is fixed to the 3D glasses and game control is via movement of the head to tilt, movement forward to accelerate and movement backward to decelerate, eliminating the need for keyboard, mouse or joystick control. The different components of the set-up are shown in Fig. 10.10.

There is no doubt that these AR developments are attracting a lot of attention and, being interactive, they have the advantage of engaging the consumer for a longer period than some of the other light and sound packaging effects discussed earlier in this chapter. AR strongly differentiates brands from the competition and has the potential to bridge the communication and emotional gap between a brand and a consumer. The digital marketing of brands is developing rapidly and for a game the PC/webcam/TV environment can provide the best experience, particularly if large screen and 3D effects become more widespread. However, the smart phone and tablet platform has the advantage of mobility and has been quick to respond to AR, as will be discussed in the next section.

Fig. 10.9 AR cereal package (a) having a unique printed code on the back (b), which is cut-out and fixed to the hand to control the game via the webcam image.

Fig. 10.10 The components of 3D AR are the cereal package having a unique printed code on the back, which is cut-out together with the 3D glasses (bottom of picture) to control the virtual motor cycle racing game.

10.3.3 Smart phone/tablet-based augmented reality

All that is needed to enter the mobile marketing world of AR is an internet-capable smart phone or tablet with a camera and a previously downloaded software app. When the object is scanned with the camera, the device immediately displays the additional AR content.

Ben & Jerry's was an early adopter of this type of AR with a free iPhone app containing a programme called *Moo Vision*, entertaining consumers by giving them product information, facts and AR content, by pointing the iPhone camera at the carton lid of one of the flavours (Fig. 10.11). For example, imaging the lid of the New York Super Fudge Chunk flavour presents a virtual tour of New York, with the Statue of Liberty holding an ice cream cone. By collecting pop-ups, such as 'cage free eggs' and 'fair trade cocoa', the user can unlock various iPhone wallpapers as prizes. The app also allows users to share adventures and new flavours with friends via Facebook and Twitter and find retail Scoop Shops in their area.

This software development marked an important future trend – that of eliminating the need for markers such as QR codes, or coded squares like those

Fig. 10.11 Moo Vision, a free iPhone app that recognises the brand logo of Ben & Jerry's, allows smart phone users to access augmented content superimposed on the carton lid of various ice cream flavours.

used on the Nestlé cereal packages, to trigger an AR response. Since this markerless technology does not rely on computer generated symbols for identification, relying instead on software to search for existing visual patterns on the product packaging, the existing packaging does not need to be specially designed and printed for the user to access the AR content.

Software from UK-based technology firm Blippar – a free iPhone and Android image-recognition app – builds on this concept by recognising existing brand logos on packaging and converting them into virtual experiences, some say in effect spelling the end for QR codes. Cadbury has used the app to create an AR game that is triggered by imaging one of several Cadbury chocolate bar products, such as Dairy Milk, Crunchie or Twirl. An introductory instruction screen of the game *Qwak Smack* appears on the screen as if stuck to the chocolate bar. The game lasts for 30 seconds, and challenges players to tap quacking cartoon ducks as they randomly appear. The user can then submit their score to go into a draw to win an array of prizes.

Other major brand owners, apart from Cadbury (Kraft), have been quick to appreciate the power to engage consumers provided by this new innovative technology. For example, Unilever, owner of the 'Love it or Hate it' Marmite brand, has developed an interactive campaign using the Blippar app featuring the 'genie in the Marmite bottle', whereby the instruction 'Rub to Reveal' produces a range of favourite Marmite-containing recipes (Fig. 10.12).

10.4 Future technology and societal trends affecting brand differentiation and consumer communication

10.4.1 The promise of printed electronics

Conventional silicon-based electronics and lithium battery power sources are today's basic ingredients for electronic smart packaging. A few high-end consumer brands have flirted with products assembled in this way, most detailed in this chapter, but there has been no application that has lasted or achieved significant volumes. This is believed to be due in part to the increased on-cost and also increasingly to the view that disposable conventional electronics on FMCGs have no place in a sustainable world. Printed electronics promises to change that.

Already e-ink and electrochromic displays, thin-film flexible batteries and printed sensors are in the early stages of commercialization and further advances, such as the ability to directly print batteries and circuitry on packaging, are predicted. More sophisticated indicator labels could be developed, combining chemical sensing with simple electronic processing and low-cost displays in a flexible, disposable format. Power will come from printed polymer photovoltaics, printed thin-film batteries or energy-harvesting devices. The technology of printed electronics is advancing rapidly and many observers believe that printed batteries, displays, circuitry and electronic components will eventually bring down costs to acceptable levels, allowing applications for mid-range food and beverage products. As an example, Innovia Films and imprinted logic circuits firm Pragmatic Printing have developed interactive printed electronic labels that activate a sequence of flashing lights when a beverage bottle is held, using printed organic light emitting diodes (OLEDs) and a small printed battery to create a thin 2D label on a BOPP substrate.

Full commercial realization of colour-change sensors on packaging to indicate accurate shelf life, or food quality and freshness, may be difficult to achieve. European-based research (SustainPack, 2005) suggests that retailers are afraid that the dynamic nature of the data could confuse consumers and create discord between the consumer and the brand owner or retailer. Some retailers feel that in extreme cases cunning consumers would find ways to cheat the system.

The key technology breakthrough necessary for printed electronics, in order to make electronic smart packaging more commonplace, is the development of low-cost, high-speed processes enabling roll-to-roll printing, either of an adhesive label or directly onto the packaging, of thin-film transistor circuits, complete with printed power source, sensors, sound capabilities and high-brightness visual displays. Although this is not an insurmountable challenge, and there is much activity in this technological area, it is likely that the first applications will be

Fig. 10.12 AR on an iPad. The Blippar app recognises the Marmite logo and creates a number of animated brand interactions, including favourite Marmite recipe suggestions.

niche on either high-end or promotional products and it might take five to ten years or more before it is realizable for high-volume commodity packaging.

10.4.2 Will augmented reality be a future reality?

By contrast, the technology for AR is already here. Explosive growth is predicted with apps on every smart phone by 2014, according to Metaio (2011), which also believes that the tablet will emerge as the device of choice for many of the applications. Certainly major brands are taking AR seriously as a new addition to their packaging and overall digital marketing strategies. Table 10.1 captures the state of development as far as the author is able to establish at the time of writing (October 2011), and includes links to internet videos where the AR effects can be more readily appreciated.

AR will overcome many of the existing limitations of today's packaging (Section 10.1), relieving some of the frustrations of consumers, providing the differentiation a brand manager is looking for and linking seamlessly into the arena of social networking. But will it be a fad or will it last?

The concept of using AR to engage with consumers through product packaging will more likely be successful and long-lasting if it is well executed and represents true value in the mind of the consumer. As an example, in certain cases it could represent significant assistance to consumers simply by showing how to use or assemble a product, thereby acting as a visual instruction manual. Videos, interactive features, and for food products, nutritional information and recipes, once hidden away on a company's web site, can be brought to life on the packaging itself using technology the consumer already has. And with image recognition software, a package does not need be specially designed and printed; only the free downloadable app is necessary. For major sporting, promotional or even local events, the AR content could be changed to reflect the marketing presence or support to that event, rather than printing special edition promotional packaging.

Just as QR codes appear to be getting established and used by consumers with smart phones, the technological shift to AR could threaten to make QR codes yesterday's news. If there is a fly in the ointment for AR future growth however, it would appear to be the current lack of standardisation in software, forcing consumers to download specific apps for specific products for specific smart phones. In addition, some education will be necessary to raise awareness of AR compliant print or packaging. Once that is accomplished, pointing a phone or tablet at a publication, poster or piece of packaging should become more widespread.

10.4.3 Packaging and smart devices – smart phones, smart homes and smart shopping

Smart phones are predicted to play an increasingly important role in bringing together smarter packaging and consumers and to be an integral part of the mobile marketing of packaged products in the future, especially for brands. The

Table 10.1 Status of augmented reality and interactive packaging, as of October 2011

Company	Brand	Platform	Year/region launched	AR experience	Software partner	View video
Nestlé	Nesquik, Chocapic, etc.	PC/webcam	2009/France	Player tilts cereal box to play and complete a game	Dassault Systèmes	http://www.youtube.com/watch?v=p3QgigeSE1s
Nestlé	Nesquik, Chocapic, etc.	PC/webcam	2011/France, worldwide	Rio game experience with game control by hand movements of player	Dassault Systèmes	http://www.youtube.com/watch?v=DYaDONOktLM
Nestlé	Nesquik, Chocapic, etc.	PC/webcam/3D TV	2010/France	Arthur 3 racing game in 3D with game control by head movements of player	Dassault Systèmes	http://www.youtube.com/watch?NR=1&v=jQ5dR_SHPHU
Lego	Selected Lego construction kits	In-store digital kiosk	2010/USA	Use Lego box to view completed construction kit before purchase	Metaio	http://www.youtube.com/watch?v=mUuVvY4c4-A
Edeka Hessenring	Perquell own-brand Cool Cola-Orange soft drink	iPhone	2011/Germany	3D cartoon animated animal character released from can and poses in photos taken with camera	Metaio	not available
Pepsico	Walkers crisps	smart phone/tablet	2011/UK	Local weather forecast for tomorrow plus suggestions to eat in or out	Blippar	http://www.youtube.com/watch?v=PQwdxuOeqKA&feature=related
Unilever	Marmite	smart phone/tablet	2011/UK	recipes	Blippar	http://www.youtube.com/user/blippar1?blend=6&ob=5#p/a/u/2/FpczQ7tJ-L8
Kraft	Cadburys	smart phone (iPhone & Android)	2011/UK	'Qwak smack' game on selected chocolate bars	Blippar	http://www.youtube.com/user/blippar1?blend=6&ob=5#p/a/f/2/TtnNgRSu4gE

(*Continued overleaf*)

Table 10.1 Continued

Company	Brand	Platform	Year/region launched	AR experience	Software partner	View video
P&G	Pringles	PC/webcam	2010/China	Football game controlled by Pringles tube	Total Immersion/ Axis3D	http://www.youtube.com/watch?v=deUDtw_GHew
Unilever	Ben & Jerry's	iPhone	2010/USA	Carton lid of ice cream reveals various flavour-related augmented realities	Edelman digital/ Circ.us/ Metaio	http://www.youtube.com/watch?v=VLKLg2AnvGE
Pepsico	Doritos Sweet Chili	not known	2009?/Brazil	Release of a Doritos Lover, a 3D toy-art-inspired-monster	Cubocc	http://vimeo.com/10290321
Red Bull	Red Bull	iPhone	2011/USA	Racing game with own designed track	Circ.us	http://circ.us/clients/red-bull-augmented-racing/

phenomenal growth of the smart phone shows no sign of abating. In the US, estimates suggested (Nielsen, 2011) that by the end of 2011 smart phones would overtake regular cell phones, with one out of every two Americans possessing a smart phone by Christmas of that year, compared to just one in ten in the summer of 2008. In the UK, 27% of British adults and 47% of teenagers now own a smart phone, with 59% having acquired one in the last 12 months (OFCOM, 2011), suggesting possession is becoming something of an obsession.

Portable media tablets, such as the iPad, will swell these numbers and add to the mobile marketing opportunities by providing a greater screen size and computing power than smart phones, allowing more textural information to be displayed legibly and sophisticated graphics to be displayed, and permitting the brand owner to move away from gimmickry and simple gameplay towards more sustained and useful communication that could still include a sense of fun.

Integration in the home between smart packaging and smart appliances such as refrigerators and microwaves has been a simmering thought in the minds of the futurist for some time, and some attempts have been made to bring the concept to reality. In 2008 Marks & Spencer launched a range of microwaveable products printed with a ‘smart code’ that, when scanned by a smart microwave made by Samsung, could be automatically heated and browned through their optimal cooking sequence without intervention from the consumer (Butler, 2009).

Removing the power needs from the packaging, as is possible using wireless power technology, would open up the possibility of smart cupboards, devices and work surfaces in kitchens and bathrooms sharing information and data through various surfaces, including glass, wood and plastic. In the kitchen for example, smart pressure sensors would allow the consumer to check how full milk or cereal cartons were, reminding them to replenish products via their shopping lists. However, there are major uncertainties over the adoption of wireless power for disposable consumer goods, and few would wish to fit out a kitchen with inductively coupled surfaces until the technology becomes more established.

A less trivial and potentially life-saving application, more worthy of the technology perhaps, is that of electronic compliance packaging for pills and tablets taken in the home as part of a medication regime. Here we might envision a smart medication cabinet in a bathroom connected to the mains power and the internet, fitted with smart shelves with wireless power and sensing/communication capabilities. Electronic compliance packaging would be scanned into the cabinet and the appropriate medical regime recorded and displayed as a reminder on the packaging, including when to discard out-of-date drugs. Other reminder messages could be displayed (Fig. 10.13). More compliant application of medication for the aged, unwell or infirm consumer could result in a significant improvement in the outcome of medical treatments, leading to a reduction in waste and inefficiency and overall costs for healthcare providers.

On-line grocery shopping, currently confined to home PCs via the store’s web site, is another area where smart phones and packaging may have a significant future together. In an experiment that could easily have come from the futurist film *Minority Report*, set in 2054, consumers in South Korea are testing out

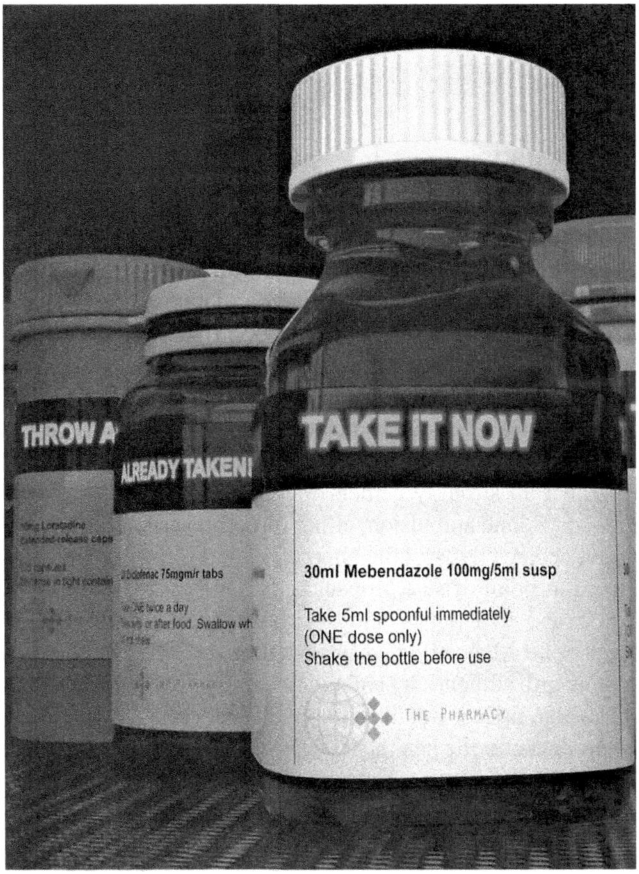

Fig. 10.13 Electronic smart packaging is likely to have the greatest societal benefits in the healthcare sector, as a positive and potentially therapeutic assistance to patients undergoing complex and long-term medication.

shopping in a virtual supermarket in one of Seoul's subway stations. Brightly lit billboards line the platform with over 500 pictures showing all the common items of food and drink a shopper might buy. Under each picture is a price and QR code, inviting the consumer, while waiting for the train, to order groceries on-line via their smart phone for delivery that evening at home.

Using unproductive time this way in an already busy day appears to be the key advantage of this scheme and has been taken up enthusiastically by younger Koreans, who are increasingly using smart phones to take care of many of their daily tasks and improve their productivity, in this case converting waiting time to shopping time. Bringing the store to the consumer in this way may not work in other countries or cultures, because the essential social interaction component of shopping is absent, so for the moment virtual stores, in subways or in other public spaces, are a niche rather than the next wave of global retailing. If they do become mainstream however, the implications for branding and packaging design could be profound.

10.5 Conclusions

Effective text-based packaging communication is reaching its natural limit and will need to be augmented by visual forms of imagery, such as changing colour, lighting and moving displays on packaging to remain effective. Beyond these eye-catching innovations lie the other as yet largely untapped senses for shelf appeal and differentiation – the sense of touch, smell and sound. Some of these are already commercial, such as tactile surfaces that improve grip on packaging and enhance a brand feeling of comfort with a consumer, and embossing and raised ink printing. The sense of smell can be engaged with encapsulated fragrance inks and scratch and sniff surfaces. The act of handling or opening a package can generate fragrance to strongly cement the emotional attraction of a brand and what it stands for. Sound and vision, either directly on the packaging or indirectly through wireless connectivity of the internet, as in the case of AR, can be expected to be combined at point of sale on packaging, and in the home, to heighten the total sensory brand experience and support the essential brand promise and provenance. It will allow effective storytelling in high-end products that can infuse a product with authenticity and uniqueness, appealing to the discriminating consumer for whom price is not the number one criterion, as well as adding a little bit of fun to products. In the healthcare sector, with a growing ageing population, electronic smart packaging could have major benefits to patients undergoing complex and long-term medication. The era of ubiquitous smart and interactive consumer packaging is rapidly approaching.

10.6 References

BLOOM J. (2010) *American Wasteland*, Philadelphia, Da Capo Press, 166.

BUTLER P. (2006) 'Developing a new dating game', *Materials World*, May 2006, 24–5.

BUTLER P. (2009) 'Smart packaging – providing improved convenience and new functional benefits for consumers', in *The Wiley Encyclopedia of Packaging Technology*, 3rd Edition, New York, Wiley, 1124–34.

BUTLER P. (2011) 'Smart packaging solutions to food waste reduction', *Food Science & Technology*, 25(3), 36–8.

COGNIFEX (2011) Available from: www.cognifex.com [Accessed October 2012].

FULTON INNOVATION (2011) Available from: http://ecoupled.com/ces/packing-serious-technology

HOW STUFF WORKS (2011) 'How Augmented Reality Works'. Available from: http://www.howstuffworks.com/augmented-reality.htm [Accessed October 2012].

KERRY J. and BUTLER P. (2008) *Smart Packaging Technologies for Fast Moving Consumer Goods*, Chichester, Wiley.

METAIO (2011) 'By 2014: Augmented Reality will be on every Smart Phone', Press Release 28 September 2011. Available from: http://www.metaio.com/press/press-release/2011/by-2014-augmented-reality-will-be-on-every-smart phone

NIELSEN (2011) 'Smart Phones to Overtake Feature Phones in US by 2011'. Available from: http://blog.nielsen.com/nielsenwire/consumer/smartphones-to-overtake-feature-phones-in-u-s-by-2011.

OFCOM (2011) 'A Nation Addicted to Smart Phones'. Available from: http://consumers.ofcom.org.uk/2011/08/a-nation-addicted-to-smart phones/

SMOLANDER M. (2008) 'Freshness indicators for food packaging', in Kerry J. and Butler P., *Smart Packaging Technologies for Fast Moving Consumer Goods*, Chichester, Wiley, 111–27.

SUSTAINPACK (2005) 'Innovation and Sustainable Development in the Fibre-based Packaging Value Chain Instrument', IP D 6.23 Report on Retail Survey Results. Funded by the Nanotechnology RTD programme of the European Union. Available from: http://www.sustainpack.com/project%20reports/A&F_D6.23.pdf

SWIENTEK B. (2002) 'Lenticular Labels Add Motion to Static Products on Shelves', Brand Packaging 22–24 March 2002. Available from: http://www.apap-inc.com/A_R_PU/brand_article.pdf [Accessed October 2012].

TAOUKIS P. S. (2008) 'Application of time-temperature integrators for monitoring and management of perishable product quality in the cold chain', in Kerry J. and Butler P., *Smart Packaging Technologies for Fast Moving Consumer Goods*, Chichester, Wiley, 61–74.

WRAP (2009) 'Household Food and Drink Waste in the UK', final report, Banbury, Waste & Resources Action Programme. Available from: http://www.wrap.org.uk/retail_supply_chain/research_tools/research/report_household.html [Accessed October 2012].

11

The future: global trends and analysis for the international packaging market in relation to the speed of impact of packaging innovation and likely material changes

N. Farmer, Consultant, UK

DOI: 10.1533/9780857098979.288

Abstract: This chapter will consider the likely impact within the international fast-moving consumer goods (FMCG) market of innovations, technologies and materials. It will also contain statistics information and analysis on trends in global packaging markets and highlight important companies who are driving technological developments forward and are considered to be some of the important players in the FMCG sector. The chapter will also contain an overview of where packaging material usage could be by 2020. Innovations in today's packaging market include developments in sustainability and the environment, recycling of materials, material reduction and waste disposal. The innovative use of bioplastic materials, biodegradability and compostability of materials are also considered in terms of market trends and analysis. It is intended that the evaluation of markets and technologies will be beneficial for companies, academic institutions, research organisations and professional associations who are seeking to understand where key areas of the international FMCG packaging market could be in ten years time and assisting in their information, research and decision making processes. It is believed that the chapter will round-off the book by assessing the future markets for some of the innovations identified in earlier chapters.

Key words: bioplastics, flexible packaging, biodegradable materials, compostable materials, active and intelligent packaging, recycled PET (RPET), PET, rigid plastics, glass packaging, glass containers, metal packaging, metal cans, cartonboard and cartons, paper and paperboard packaging, globalisation of packaging markets, light-weighting of packaging; material reductions, recycling and waste initiatives.

11.1 Introduction

The packaging market to the year 2020 is likely to be greatly influenced by techno-economic trends. The pace of change in material usage and the likely speed of development of innovative solutions are both likely to be subject to much external influence.

What is without doubt is that the current emphasis on material reduction, light-weighting, recycling, recycled content and cost savings will still be major issues by 2020. Increased shelf life, by virtue of further developments in active and intelligent packaging, improved barrier performance with enhanced use of new coatings, additives and barrier materials, will also impact greatly on our lives. In many cases, the greater use of bioplastics will help ensure that the markets for these materials will become more developed. Greater clarity in relation to regulatory issues and consumer safety will help the development of the market for nanotechnology and, to a lesser extent, GM Foods over the forecast period.

11.1.1 Nanotechnology

Nanotechnology has great potential in fast-moving consumer goods (FMCG) applications. However, until a substantial launch by a major FMCG producer publicly declaring that a nanomaterial is used in a food and drink packaging application, it is likely that the known multi-million investment in R&D by leading raw material and consumer goods groups will stay below the radar.

11.1.2 GM foods

GM Foods are also poised for growth on the international stage. Whilst there is still wide public opposition in many countries in Europe to a product once dubbed 'Frankenstein's Food', with tighter regulation and testing procedures in place allied to a commitment by companies in the supply chain to investigate new technologies, GM Foods will have an impact, resulting in the need, amongst other things, for greater labelling. Indeed reports in 2010 indicated that some products on supermarket shelves in the UK could contain indirect GM ingredients and not be labelled as such, as there is currently no legal obligation for this when the GM link is further back in the supply chain.

11.1.3 Consumer technologies

In consumer technologies, there will be a continuing move to purchase more on-line over the next ten years. Factors driving this development forward will include improved ease of access to on-line stores and enhanced customer service, all influenced by the time constraint factor, causing more and more consumers to make their weekly shopping purchases in this way.

As regards social media such as Facebook, LinkedIn, Twitter and Youtube, the benefits for the packaging industry of these consumer phenomena are hard to accurately quantify. It probably comes down to understanding the business needs of each individual organisation and its client base. It is clear that some of the biggest brands have a strong presence on Twitter and are getting their messages across to a wider audience as a result. It is also a great way to monitor what is being tweeted about the company, effectively enabling consumer research to be undertaken and analysis to be conducted. Of course, the internet is the driving force in all consumer technology developments. With more money spent on on-line advertising than on television advertising in the UK in the first half of 2011 (according to the Internet Advertising Bureau (2011) and PricewaterhouseCoopers), there is still more to come in the global market, as FMCG companies increasingly turn to the internet to boost sales.

11.1.4 Packaging design and digital technology

The whole landscape of packaging design has radically changed in recent years. Digital technology has been the catalyst for change. The need for greater product differentiation at the point of purchase in the FMCG market has meant that new designs and innovations have to be produced more quickly and efficiently than ever before. This important trend will continue at a rapid pace over the forecast period. As a result, three-dimensional (3D) design and imaging, prototyping and sampling will all show further growth. Virtual proofing, prototyping and the creation of new 3D shapes are all part of the shortening of the design process. This in turn is inextricably linked to the need for faster speed to market and optimisation of the packaging supply chain. A company, such as FFEI of Hemel Hempstead, UK, with their RealPro 3D-Packager for design and communication of print and packaging concepts and RealVue 3D-Packager for creation of accurate visual prototypes, including elements such as embossing, substrate textures, cut-outs and metallic inks, is a good example of an international player leading the field in these types of innovations.

11.1.5 Packaging and digital print

In the future, the FMCG market will be more about speed of response, shorter production runs and special brand promotions, brand extensions and more pack varieties, all linked to maintaining and enhancing brand values. High-quality digital print will play an increasing part in all of this, being ideal for these types of applications.

In their 2011 report, *The Future of Digital Printing for Packaging to 2016*, Smithers Pira said that globally, digital printed packaging would have a compound annual growth rate (CAGR) of 20.6% between 2011 and 2016, to reach a market value of $12.26 billion in 2016, underscoring the above market analysis.

11.1.6 Research and development

Over the forecast period, packaging companies who are seeking to innovate will require their R&D to be synchronised with the market specific needs of customers, to allow greater targeted development to take place. A good illustration of how this could work was the opening in March 2012 by MeadWestvaco Corporation, a leading multinational integrated packaging solutions group, of a new $13 million state-of-the-art R&D plant in Richmond, Virginia. The company decided to relocate its Centre for Packaging Innovation (CPI) and its pilot plant from North Carolina to Richmond to be closer to the group's global headquarters, to boost cooperation between the company, its customers and its suppliers. The focus will be on exploring the use of innovative food packaging materials and the study of polymers and coatings. One specific project mentioned was a new type of packaging for snack goods.

Over the forecast period, the need to get innovative products rapidly to market will prompt the creation of more of these types of R&D facilities in international packaging businesses who supply the FMCG sector. The emphasis will be on speed of response, partnerships with customers and the investigation of new materials and technologies, as the above example demonstrates.

11.1.7 Augmented reality (AR) technology and quick response (QR) codes

What is beyond doubt is that major brand owners will continue to embrace consumer technologies in a multiplicity of ways, such as the utilisation of apps for smart phones. By these types of methods, FMCG producers aim to connect more with their customers. The massive rise in the use of smart phones has assisted developments such as QR codes, an important trend for the future. In late 2011 and early 2012, QR codes were to be found in newspapers, magazines and on different items of packaging. These black and white two-dimensional (2D) pixelated squares, when scanned by a smart phone camera, enable the user to go to a web link, which then provides many different interactive possibilities. It is important to note that already some packaging and branding experts believe that the future could lie in AR technology, a rival that could replace QR codes, as Chapter 10 discusses in detail.

As advances in consumer technologies continue at a rapid pace, the potential for AR in the international packaging market, as illustrated by the usage by major groups such as Nestlé and Cadbury's (Kraft Foods), shows how this type of innovation, driven by major FMCG companies and their suppliers, could have global implications for the industry over the next ten years.

11.2 The consumer packaging market and the challenge of glass containers versus plastic containers

Improved shelf life, light-weighting and recyclability of plastics will expedite the move away from glass, with further market share gains likely for plastics. This is illustrated by figures in *Packaging Unwrapped*, the Rexam Consumer Packaging Report (2011/12), using statistics from Smithers Pira (2012). These show that of the \$395 billion global consumer packaging market in 2010, rigid and flexible plastics accounted for 37% of all global packaging sales, with glass at 11%. Rigid plastics is forecast to be the fastest growing packaging material during the 2010 to 2015 period, with an estimated annual growth rate of 4.1%. Light-weighting, safety and convenience are seen as important factors in this expansion. Flexible plastics annual average growth is put at 3.8%. Glass is forecast to grow by 2.7% per annum over the same period, slightly less than the annual average growth for total global packaging sales, which is put at 2.9%. Total consumer packaging sales are forecast to reach $456 billion in 2015.

While glass will continue to lose market share in some sectors, it will still have success in segments of these markets where aesthetic appeal and quality image are important. Glass is perceived as a natural material with intrinsic quality. Alcoholic drinks are one of the main sectors for glass demand, as a result of longstanding usage over many product categories. The glass manufacturing industry therefore continues to believe that the material has a buoyant future. For example, a survey undertaken by Belgian market research agency In Sites in December 2010 for the European Glass Container Federation, FEVE (2010), concluded that glass was the preferred material for food and drink for more than 75% of the 8600 participants in the survey, which was carried out across 17 countries. FEVE said that consumers preferred glass because of three characteristics: its taste preservation, its health benefits due to it being an inert material and its environmental friendliness. The same body also reported that European glass packaging production volumes grew by 4.7% in the first half of 2011.

Nevertheless, the use of rigid plastic containers in markets such as soft drinks, juices, wines, sauces, jams and other similar sectors will advance at a considerable pace over the next decade. In supermarkets particularly, PET containers will continue to grow in demand. Generally the convenience, light-weighting, shatter-resistance and extended shelf-life features of new plastic packs will be a major advantage.

For example, in October 2011, Sainsbury's in the UK said that it had replaced glass peanut butter jars with plastic, cutting the packaging weight for the product by 83%. This move is part of the company's objective to cut its total packaging weight by a third by 2015. Sainsbury's, like many others, are committed to not only reducing packaging volumes, but also to ensuring that functional pack performance is maintained.

Amcor Rigid Plastics announced in October 2011 a new 240 oz round, wide-mouth PET jar for the food industry. The stock hot fill container is seen by the company as a lightweight replacement for glass for sauces, salsas, jams, jellies and other food products filled at temperatures up to 205°F. These are precisely the markets that glass is most at risk from substitution into plastics, for the reasons already identified.

There is, of course, considerable debate as to the relative merits of the two competing materials and glass does have benefits in terms of relatively stable pricing and excellent sustainability and environmental features in terms of high recycling rates. However, for end-uses such as wine in small size packs for airlines, concerts, outdoor events, picnics and leisure consumption, plus the retail food markets identified above, plastics will triumph.

There are other markets where glass is being challenged. The beer market is still heavily dominated by glass containers. Indeed, a multi-client study, published by Smithers Pira in 2011, said that PET beer bottle usage totalled 6 billion units in 2010, just 2% of the 293 billion world beer-packaging market volumes. However, the report forecast that global PET beer bottle consumption would grow at a CAGR of 5.3%, to reach 7.7 billion containers by 2015. Clearly there is growth in this sector for plastics, particularly in territories such as Central and Eastern Europe where PET bottles are popular. However, generally in markets such as Western Europe and the USA, there is a long way to go to achieve volumes anywhere approaching that of glass.

11.3 The flexible packaging market

11.3.1 Overview

In flexible packaging, the convenience of packs, including improved ease of opening, reclosure and reuse features, will drive the market forward in a big way. The continued rise of pouch packs will play a large part in this development, as will improved performance of materials.

According to Smithers Pira in their August 2011 report, *The Future of Flexible Packaging to 2016*, the global consumer flexible packaging market will grow at a CAGR of 4.1%, much in line with other forecasts, to reach US$ 71.3 billion in value. This is a higher value figure than other survey estimates, as it includes sectors such as tobacco, pharmaceuticals and medical products. Smithers Pira say the growth will be fuelled by flexible packaging replacing many traditional types of pack, including metal cans, glass and plastic bottles and liquid cartons, in a diversity of end-use sectors. The report illustrates the considerable advantages of flexible packaging, in providing innovative new packs, which offer consumer convenience, safety and enhanced barrier performance, resulting in extended shelf life. The tonnage for the global consumer flexible packaging market is put at 18.1 million tonnes in 2011 and is forecast to reach 22.5 million tonnes by 2016.

An interesting point made by the report is that whilst flexible packaging may have reached maturity in much of North America and Western Europe, where future growth will as a result be more modest, stronger growth will be achieved in developing countries. In this respect, Asia, which was the largest regional market with 29.1% of global market volume in 2011 (followed by Western Europe and North America), is also the fastest growing market, with a forecast CAGR of 7.9% for 2011 to 2016. The region is forecast to represent an impressive 55% of total world flexible packaging consumption growth during the period 2011 to 2016, according to Smithers Pira. India and China are identified as the fastest growing national markets for consumer flexible packaging, accounting for 44% of world flexible packaging consumption growth during this period.

In their report, *World Converted Flexible Packaging*, September 2009, The Freedonia Group of Cleveland, Ohio, named Asia, Latin America, Eastern Europe and the Africa/Mideast regions as the areas in the world that would outpace the global average growth figures.

When looking at the USA in their *Converted Flexible Packaging* report, October 2009, Freedonia Group identified pouch packs as the flexible packaging products most likely to experience the fastest growth. Demand was expected to grow by 4.7% per year to $7.9 billion in 2013 (Freedonia Group, 2009b). Whilst this is a healthy growth rate, particularly in a recessionary climate, it was a reduction in the rate of demand growth compared to the 2003–2008 period reflecting, amongst other things, the established presence of pouch packs in some markets. Thus, when looking ahead towards 2020 and the end of the period under review, it is likely that the pouch pack will further establish its presence in new markets, away from its base of food and drink, achieving further volume growth. As was noted in Chapter 1, its success in household products and pet food in Europe demonstrates that the pack still has many new markets applications. Pouch packs have many benefits, including light weight, flexible shape, print and decoration options. Consumers find them easy to open and convenient to handle. However, they are less easy to recycle than, for example, traditional metal cans, although this factor does not appear to have affected their popularity in the international market to any significant degree.

11.3.2 Material developments

As regards materials, we have noted the likely growth of higher-barrier plastics over the coming decade. One of these materials is biaxially-oriented polyester film (BOPET). Of the companies in this sector, the developments by Indian flexible packaging manufacturer Uflex are worthy of note. In 2011 the company said that its Global Films Business Division would invest $90 million each in two phases of a new greenfield facility in Elizabethtown, Kentucky. The first phase will be a state-of-the-art BOPET line, followed by a plasma-enhanced high barrier metalliser, to be commissioned by the end of 2012. The company is one of the world's largest film producers and already has plants in India, Mexico, Egypt and Dubai.

Similarly, developments such as the announcement by Octal of Oman of expansion at its Salalah, Oman production plant, are of significance. As a result of the investment at the site, the company, already one of the largest global producers of PET sheet, will also become one of the world's biggest manufacturers of PET resin on a single site. Octal is forecasting a sales growth to $1.5 billion (1.2 billion Euros) in 2012. The company is already a significant exporter on an international basis. It has targeted new packaging applications for growth, including yoghurt, meat and cheese.

11.4 Active and intelligent packaging developments

As was noted in Chapter 1, the active and intelligent packaging market is poised for growth over the next decade. Some of the innovations that have been under development are now coming to fruition, with major retailers leading the way.

Such an example is the development by 'It's Fresh', part of the Food Freshness Technology (FFT) group. This patented ethylene remover technology has been undergoing trials for several years. Ethylene is the ripening hormone that causes fruit to ripen and then turn mouldy. In late 2011/early 2012, it is being used by both Marks and Spencer and Tesco in the UK for fresh fruit and berries. For Marks and Spencer it is being used for strawberry punnets, the first application in the packaging of consumer products.

The technology guarantees that by placing a small 'It's Fresh' strip, for example in a punnet of Marks and Spencer strawberries, they will stay fresh for longer. This is possible because the technology reduces the level of ethylene being produced, reducing the deterioration of the fruit, extending the shelf life and the storage period. In the case of the Marks and Spencer punnets, the 'It's Fresh' strip measures 8 cm × 4.5 cm and does not affect the recyclability of the pack. The active ingredient in the strip is a patented mixture of clay and other minerals providing, it is claimed, to have over a 100 times greater ethylene absorption capacity compared with any other known materials. In the case of Tesco, the technology was initially used for transit packs of Argentine avocados, the advantages being extended shelf life in store of up to 6 extra days and 30 days increased cold storage. Both retailers plan to use it for more fruit and vegetable consumer packs in 2012.

This innovation is a great example of how active packaging can play a major part in a cost-effective way in extending shelf life and helping to reduce food waste, two of the most important areas of technological development over the next decade.

11.5 An overview of the bioplastics market

The big debate will be the speed of growth of the bioplastics market which, whilst currently forecast to be in double digits, is still starting from a low base. There are several scenarios to consider when assessing likely growth over the period. The current commodity price in boom shows little sign of abating, driven by the strong demand for materials in China, the Middle East and the Far East. Petrochemical prices are likely to remain at high levels over the whole of the next decade, which will provide those bioplastics that come from sugar cane and corn starch with a golden opportunity to gain market share. However, the converse of this argument is that forecasts for material shortages and higher prices for these staple materials is also likely to escalate the costs of production and material availability. It will then all come down to a matter of material performance and, as volumes increase, economies of scale for bio-materials. However, notwithstanding these factors, greater use of bioplastics with enhanced performance features and greater cost competitiveness should accrue over the period. This is confirmed by figures from the trade association European Bioplastics, which claims that bioplastics production, in all its forms, exceeded one million tonnes in 2011; overall production will more than double between 2010 and 2015. The organisation said that growth will be attributed to expansion in many sectors, including packaging. The survey, which was conducted in conjunction with the University of Applied Sciences and Arts of Hanover, Germany, also revealed that production rates of bioplastics had reached a run-rate of 900 000 tonnes in the first half of 2011, compared with 700 000 tonnes in 2010. It also said that production of bioplastics is expected to reach 1.7 million tonnes by 2015 (the University of Applied Science and Arts, 2011). The survey also found that in 2010, biodegradable materials accounted for 400 000 tonnes of total bioplastic production, with bioplastic commodity plastics making up the remaining 300 000 tonnes. The association believe that this ratio is expected to reverse, with bio-based commodity plastics making up the majority of production in 2015, although biodegradable materials will still grow considerably.

However, a report in 2011 by BCC Research entitled *Biodegradable Polymers* said that whilst the biopolymers market is set to grow, and the packaging sector is expected to reach almost £1.7 billion in 2016, there were problems to be overcome in terms of relatively high prices and difficulties concerning infrastructure for effective composition (BCC, 2011). This illustrates the difficulty that bioplastics is likely to face over the coming decade to achieve its growth targets.

11.6 Biodegradable and compostable materials

For the reasons mentioned in the paragraph above and for other factors (not least of all questions over returns on investment and profitability of businesses), biodegradability and compostability of materials will continue to be an area of great debate over the forecast period, with claims and counter claims being made about materials and their performance. Nevertheless, it is forecast that further significant developments will be made in the bioplastics field over the next 10 years.

11.6.1 First biodegradable cling film

As an illustration, in May 2011, the first biodegradable cling film went on sale in the UK, manufactured by Baco. The company claims its Baco Biowrap is completely biodegradable and can be broken down within two years, if put on the compost heap or sent to landfill. Normal plastic cling film can take decades to fully decompose. The new product is more expensive than standard cling film and is made from controlled life, oxo-biodegradable polyethylene. Nevertheless it has much to commend it, particularly as large amounts of cling film are used by households each year in the Western world.

11.6.2 Compostable confectionery wrapper

A compostable confectionery wrapper, believed to be the first in this sector, was introduced in the summer of 2011 in a joint project by Innovia Films, BASF SE, DFTA and Flint Group. BASF SE supplied the newly developed and fully compostable lamination adhesive Epotal ECO, which is certified by DIN CERTCO. The Nature-Flex lamination films for the confectionery bags and wrappers are supplied by Innovia Films and are fully compostable. Innovia is a good example of a company that seeks to innovate in sustainable flexible packaging. Nature-Flex is a compostable film with great potential, as has been observed in Chapter 1, although starting from a relatively low base point in terms of market volumes.

11.6.3 Polyhydroxyalkanoate (PHA) materials

Polyhydroxyalkanoate (PHA) is a class of plastics that are biobased and compostable and could be poised for growth in packaging applications over the next 10 years. PHAs are linear polyesters naturally produced by bacterial fermentation of sugars or lipids and provide sustainability benefits compared to conventional plastics. In the US, a business called Telles was created in 2006 as a joint venture between Metabolix, a renewable chemical company and agri-business Archer Daniels Midland. The Telles plant in Clinton, Iowa, was claimed to be the world's largest PHA operation, designed to produce 50 000 tonnes of the material per year. The brand name for the material is Mirel; it is commercially available in many grades including thermoforming, injection moulding, cast and blown film and sheet. It was intended that the plant would be running to full capacity by mid-2013. However, in January 2012, Archer Daniels Midland dissolved the joint venture, saying capital costs were more than anticipated, commercialisation of the product had taken too long and financial returns were uncertain. As a result, Metabolix plans to scale down its PHA operations. Metabolix has subsequently signed a letter of intent with Antibioticos of Spain to manufacture Mirel at its plant in Leon, Spain.

This is an example of the difficulties experienced by bioplastics start-up businesses in uncertain economic times. The market for PHA is small, at around 1.5% of the total global bioplastic packaging market. Investment in new technology is never inexpensive and market development and competitive pricing are always going to be issues in these types of situations. The bioplastics market is growing at a rapid rate, as we have seen elsewhere in this chapter, but materials such as PHA still have a long way to go to achieve commercial success on a volume basis (Chapter 5).

11.6.4 Polylactic acid (PLA)

Use of polylactic acid (PLA) (Chapter 1), a biodegradable material derived from renewable sources such as corn-starch or sugar cane, has up to 2011 been restricted by several factors, including low melting temperature of the material. However, developments in bioplastic materials are ongoing and in PLA films technological advances are being made (Chapter 5). European demand could support the production of over 650 000 tonnes of PLA by 2025, according to the National Non-Food Crops Centre (2011). This figure seems ambitious and further technological developments will need to be considerable to achieve these types of targets.

One of the largest challenges being faced by the industry is to improve the thermal stability of PLA films, as previously noted. Currently they are only stable up to temperatures of around 60°C. Furthermore, PLA films have a high stiffness, are limited in flexibility, are brittle, and because of poor tear resistance, difficult to process on some packaging machines. However, when used in food or non-food applications, PLA films have excellent sealing qualities when tight sealing of a pack is needed. Of course, like all bioplastics, PLA films will need to offer a good barrier to oxygen and be cost-effective, something which in the latter case has still to be proved. Their growth will therefore depend on the technological developments mentioned in Chapter 5, being more widely adopted in a profitable and fully cost-effective way over the forecast period.

11.7 Recycled PET materials – the market over the next ten years

In Chapter 1, reference was made to the growth of the recycling of PET bottles and the development of closed-loop recycling systems. These types of initiatives are likely to grow at a rapid rate over the next ten years, as the plans announced by Coca-Cola, with their joint venture recycling plant in Spartanburg, South Carolina, testify. However, words of caution should be applied when assuming that these types of developments will rapidly and smoothly succeed. For example, it was reported in April 2011 that Coca-Cola had curtailed operations at the Spartanburg plant, a joint venture with the United Resource Recovery Corporation, LLC. Reports at the time that the site had closed down were refuted by Coca-Cola, who said that the operations were being restructured and new equipment was being installed. Through an official statement at the time, the company also reaffirmed its commitment to 'recovering, recycling and reusing our packaging and to supporting the recycling industry'. The plant has the capacity to produce 100 million pounds of recycled plastic a year for reuse as bottles. Only time will tell as to whether the whole project will succeed. Indeed, in August 2011 it was reported that the plant was reopening, refuting earlier allegations in June that it might be sold. Nevertheless, it is vital that a company as large and successful as Coca-Cola does continue with its ambitious plans and that closed-loop recycling achieves its objectives internationally.

The current recessionary economic climate will undoubtedly be a factor in the short term and any new and innovative development is likely to find the going tough in its first few years of life. The Spartanburg plant, which opened in January 2009, is the world's largest PET bottle-to-bottle recycling plant; it is key to the closed-loop project that Coca-Cola and others are pioneering. It will be interesting to see the next stage of its development and indeed how the whole PET recycling industry proceeds over the period under review. Certainly the drive for greater sustainability in the FMCG market is dependent on these types of schemes and innovations succeeding.

It was announced in May 2011 that Coca-Cola Enterprises in the UK were following the initiative already established in the US, undertaking a joint venture with ECO Plastics to develop a new, purpose built, recycling facility in Lincolnshire. Some 35 000 tonnes of PET bottles were reprocessed in the UK in 2010 and, when fully operational, the new plant will increase this figure to over 75 000 tonnes. It is claimed that the new facility will also supply Coca-Cola Enterprises with enough UK-sourced quality rPET to achieve its target of including 25% of this material in all the company's plastic packaging in Britain by 2012. The longer-term objectives of Coca-Cola corporately include plans for all of its plastic bottles to be made from a combination of plant-based materials and rPET (recycled PET) by 2020. Indeed, in September 2011, Coca-Cola announced it was rolling out its PlantBottle in Great Britain. From September, all Coca-Cola, Coke Zero and Diet Coke in 500 ml bottles would be sold in PlantBottle packaging, i.e. PET plastic bottles made partially from plants. Coca-Cola said that more than 200 million PlantBottles would be on the shelves in the UK in 2011. As previously noted in Chapter 1, the PlantBottle will be on the market in 2011 in 20 different countries. Global volumes are put at more than five billion.

11.8 International PET developments

Further capacity for PET is coming on-stream throughout the world. In May 2011, leading converter and manufacturer of PET bottles and performs, APPE (the packaging division of La Seda of Spain), announced it was investing a further 10 million Euros at its PET recycling plant in Beaune, France, increasing its annual output from 25 000 to 35 000 tonnes. The company said the investment reflected the demand for recycled materials in the food and drink market.

APPE is now successfully developing multilayer wine bottles with an oxygen scavenger between the PET layers, to prevent oxygen passing through the bottle and to avoid oxygen in the headspace. The company won an innovation award for the development from judges at Emballage, November 2010, in Paris.

APPE's introduction of a PET beer keg for international brands Heineken and Amstel in August 2011 was also a significant development. Launched initially in The Netherlands, the 'Tapje' keg is a small draft beer system, which consumers can use at home. It features APPE's proprietary monoBLOXTM barrier technology with oxygen scavengers, which provides good barrier performance, keeping oxygen out and the carbonation in; it is also claimed that the beer will stay fresh for 30 days once opened. This pack is an important example of further advances being made in barrier performance in the PET market and how the beer sector is being targeted by leading plastic packaging producers.

This is further confirmed by the developments being undertaken by Petainer, a leading pan-European specialist plastics technology business, with operations in Sweden and the Czech Republic and with customers throughout Europe. The company exhibited its Petainer Keg, a lightweight container manufactured in recyclable PET, designed for one-way use, at the Brau Beviale exhibition in

Nuremberg, Germany in November 2011. Petainer stressed the advantages of the PET pack against the traditional metal keg, including its low environmental impact, low cost and low weight. In October 2011, Petainer supplied the new lightweight, recyclable PET beverage container to Pivovarinny Zavod Samko, an independent brewery in central Russia, which cited benefits of the pack that included lower transport costs, easy recycling and lower environmental impacts, again when compared with metal containers. To support its expansion plans in developing large PET containers for beverages and other applications, the company opened a new corporate head office in Peterborough, England in September 2011.

These market expansion plans are likely to continue over the forecast period, as plastics container producers target beer and other market sectors for further growth, to the detriment of metal and glass containers.

Notwithstanding some difficulties in the current recessionary climate and assuming the pace of economic recovery is on course, it is forecast that the recycled plastic market will experience substantial growth to 2020, as a result of these types of developments.

11.9 The growth of recycling and utilisation of waste resources

There is clearly a need for joined-up thinking in the whole area of utilisation of waste resources. This is irrevocably linked to the holistic approach that more leading consumer goods producers are adopting (Chapter 1). In terms of innovations, the recycling field has achieved much, with use of recycled content rising and performance characteristics of plastic materials increasing regularly. Generally the recycling rates for Europe and the US are upward. There is a genuine commitment to progress more rapidly and to take all these and other initiatives to the next stage.

11.9.1 UK Government/DEFRA recycling rates 2013–2017

This commitment is illustrated by the UK Government's announcement in December 2011 of proposals for increased recycling targets from 2013 to 2017. A consultation document published by DEFRA proposed three options, including the Government's preferred option of higher recycling targets for aluminium, plastic and steel (DEFRA, 2011). These would increase to 55% for aluminium, from 40% in 2012, 57% for plastics from 32% now and 76% for steel from the current 71%, all by 2017; new sub-targets for glass by end-use were proposed and the overall packaging recovery rates increased from 74% to 79% in 2017. These figures were all confirmed in the UK Government's budget of March 2012. Nevertheless, questions still persist from the plastics industry particularly as to how plastics recycling targets can be increased from the current figure of 32% to the target of 57% by 2017.

One of the concerns of the industry, as expressed by the British Plastics Federation (2011), the leading trade association, is that as a result of the higher targets the large proportion of plastic waste being exported will inevitably continue and in so doing impose high cost burdens on the UK supply chain with little benefit. The corollary of this was the feeling amongst those in the recycling industry that the government had done what it originally set out to do with the higher recycling rates and as a result environmental benefits would follow, underscoring investment in processing infrastructure. Thus, the debate continues and is destined to do so over the forecast period. However, the upward trend in recycling rates is to be applauded and a sign of further commitment to a more sustainable future. A word of caution is necessary here. In November 2012 China indicated that it could refuse to take low quality plastic material for recycling from the UK. It is believed that 67% of Britain's plastic waste is sent to the Far East. If implemented, this policy could jeopardize the UK's chances of meeting the government's new recycling targets.

The drive to reduce pack weight, material content and the whole recycling and sustainability debate has been discussed at length in Chapter 1. Particular reference was made to markets, such as Easter eggs, where considerable improvements have been made in these areas. A recent example of how this trend will continue over the forecast period was illustrated by the announcement at Easter 2012 that Nestlé UK and Ireland had became the first major confectionery producer to use 100% recyclable packaging across its entire Easter egg range. The company replaced 48 tonnes of rigid plastic used in its mug eggs with FSC approved cartonboard; as a result, its complete Easter range became fully recyclable in all waste streams. By also using compostable film for the windows, a reduction of 30% in packaging weight was achieved for the Nestlé mug egg range. This is an excellent example of the holistic approach to resource replenishment and sustainability by a leading FMCG producer.

11.9.2 Need for consensus across complete packaging supply chain

Consensus is still needed across all in the packaging supply chain and there are few signs that this is happening, as the plastics industry and others continue to argue their case and defend their corner in the debate.

As an example, in the spring of 2011, the European Council of Ministers were discussing the alleged waste problems caused by single-use carrier bags and the hazards discarded carriers cause to marine life (European Council of Ministers, 2011). A Europe-wide approach to the issue was also being considered. European Union officials are assessing options that include a ban on single-use carrier bags and other alternatives such as price and tax incentives. Clearly the debate about the benefits of plastics materials and otherwise will continue throughout the forecast period.

11.10 Metal packaging

11.10.1 Market overview

Metal packaging, which includes food and beverage cans, accounts for 15% of the consumer packaging market by value in 2010, according to statistics in the Rexam Consumer Packaging Report (2011/12) supplied by Smithers Pira. Of this figure, metal beverage cans account for 6%, with other metal packaging at 9%. The forecast annual growth rate over the 2010 to 2015 period is 1.6% for beverage cans and 1.9% for other metal packaging. Thus, whilst growth will be achieved in metals, it is likely to be at a slower rate than other competing materials and indeed the average figure for the whole market.

It likely that the long-established, low-margin metal container food market, such as canned vegetables and processed foods, will continue to lose out to newer plastic packaging technologies. These will benefit from greater consumer acceptance, advances in technical performance, including enhanced extended shelf-life features and other added value characteristics, such as ease of opening and aesthetic shelf appeal. However, cans will remain an important segment of the food container market, due to their long shelf life, efficient production methods, and as a result, often lower pricing than other competing options such as plastics. In a recessionary climate this is an important consideration.

Recycling rates for cans are excellent. Germany has already achieved 80% can recycling. Generally the European target is for 75% of cans to be recycled by 2015. On the wider front, statistics from the Metal Packaging Manufacturers Association (2011) in the UK and The Can Makers (2011), also in the UK, indicate a metal packaging industry targeted recycling rate for 2020 of 80%. The important point here is that major metal packaging producers, who serve international markets, have recognised their role in the global sustainability programme and are pledged to continue to increase recycling rates to even higher levels. As noted elsewhere in this chapter and in Chapter 1, the drive for improved sustainability is a key part of the trends and innovations in the worldwide consumer goods packaging market.

11.10.2 World market opportunities for metal packaging

Metal packaging should be presented with niche market opportunities for growth and innovation over the next ten years. Added value opportunities in terms of advanced decorative and shaping of containers, new coating technology and increased use of digital print will impact in a positive way. The opportunities presented by digital technology to create speedy brand creation and on-line graphic designs should ensure that metal cans are at the forefront of marketing campaigns by leading FMCG producers. Investment in can-making equipment will continue around the world, as emerging nations, such as the MERCOSUR countries (Argentina, Brazil, Uruguay and, in July 2012, Venezuela, Paraguay suspended in 2012), develop their economies and, as a consequence, consumer packaging requirements. An example is Brazil, which in 2011 overtook the UK as the world's sixth largest economy, although predicted to revert to seventh position in 2012, due to the weakness of its currency. Here it was announced in May 2011 that Rexam is to build a new can-making facility near Belem in the north of Brazil. It would be operational by mid-2012 and have an annual capacity of 1.2 billion cans. The plant will bring the company's total production capacity in the country to approximately 14 billion units. Cans are also expanding into the wine market, particularly for picnics and outdoor events. Socialising at home in Western countries, aided by tougher drink and drive laws, will assist the growth in can volumes as well over the forecast period.

Investment by leading packaging supply chain companies in China is also a significant innovative development for the future, as we shall see later in the chapter. However a word of caution. In February 2012 Crown Holdings announced plans to build a new two-piece beverage can plant in Nanning, Guangxi Province, China. At that time the company already had five beverage plants operating in the country. Subsequently in October 2012, citing sluggish economic conditions and unfavourable weather in its key markets, Crown indefinitely postponed plans to build the new site at Nanning and one at Xinxiang also. The company said this was consistent with its 'prudent approach to investing capital'.

11.10.3 Market for aluminium bottles

An important innovative development is the growth in popularity of aluminium bottles in the beverage market. Several leading players, such as Rexam, Boxal (acquired by the Ardagh Group in 2012) and CCL Container, have produced attractive aluminium drinks packs, which consumers like because of their convenience, re-sealability and attractive appearance. They are also 100% recycled and available in a variety of sizes. Their growth is expected to continue over the forecast period.

A good example of this is Rexam's aluminium FUSION bottle, which won the gold award at The Cans of the Year Awards (2011), sponsored by *The Canmaker* magazine, the third award FUSION had received in that year. It had previously won the Best Packaging Award at the Beverage Innovation Functional Drinks Awards 2011 held in Washington DC in March. The bottle essentially combines the functional and graphic design potential of a can with the look and feel traditionally associated with a bottle. It was designed for Ocoo Beauty Nectar, a light and fruity beauty enhancing drink, currently sold in 250 ml FUSION bottles in Germany.

11.10.4 New metal packaging concepts

New metal packaging concepts were dealt with from a technical development viewpoint in Chapter 6. In terms of market trends and analysis, an innovation worthy of note is the self-heating Hotcan pack, from Heat Food and Drink of the UK. This food pack concept was originally supplied to emergency services and used by them as contingency stock in emergencies in the UK. In early 2012, the company opened new production facilities at Burton-on-Trent, Staffordshire and is targeting retailers and the on-the-go convenience food sector. The pack depends on the can's endothermic heating mechanism, which relies on a reaction between granular limestone and water in the space between an inner and outer can. The pack, which does not need a microwave to heat up, is ideal for people on work sites or in an office. Whilst originally a niche market product, the new pack could create new opportunities in the on-the-go food packaging sector, which is growing rapidly as consumers become more and more time constrained. Although it is starting from a low volume base and extra cost compared with other packs which could be a prohibitive factor, the development is a good example of innovation in the metal packaging sector. The company has linked up with Sheffield Hallam University's in-house design consultancy Design Futures to develop the new pack (Sheffield Hallam University, 2011).

11.10.5 BPA replacement

The quest to find replacement materials for BPA in food cans (Chapter 1) will be a challenge for the whole metal packaging industry. From a market analysis viewpoint, once a suitable new material has been found, the safety aspects of this will be a major boon to the industry. Any viable solutions discovered will help end consumer safety concerns and aid growth within the metal can market in the period under review.

11.11 Paper and paperboard market

11.11.1 Market overview

Figures in *Packaging Unwrapped*, the Rexam Consumer Packaging Report 2011/12, based on statistics provided by Smithers Pira, indicate that paper and board accounted for 34% of all global packaging sales in 2010, putting the material slightly behind plastics (rigid and flexible), which accounted for 37% of global packaging sales. However, as previously noted, whilst plastics (both rigid and flexible) are forecast to be the fastest growing packaging materials during the 2010 to 2015 period, paper and board is forecast to grow at an annual rate of 2.4% over the same period. This compares to an annual average growth rate of 2.9% for all packaging materials between 2010 and 2015. Thus, whilst growth in paper and board sales will still be positive, they will be at a slower rate than competing plastic materials, but still faster than metals. However, cartons will come under increased pressure from plastic materials, notably flexible packaging, which can offer increased shelf life, aesthetic appeal, and ease of opening, pouring and re-closing, issues which carton producers have traditionally had difficulty in providing innovative solutions for. The continuing growth of eat-on-the-go consumption and food in smaller packs should benefit the paper and paperboard market, with convenience and ease of handling important factors.

In specific sectors, higher growth rates than those stated above will be achieved. For example, according to a report from Smithers Pira in September 2011, global demand for retail-ready packs (RRPs) will rise from 19 to 27 million tonnes over the next five years. This 42% increase will be generated by, amongst others, retailers seeking to reduce labour costs to help achieve demanding profitability targets. Consumer focused packs with enhanced branding opportunities will continue to transform the retail ready sector into a key part of the modern packaging innovations world.

11.11.2 Aseptic packaging developments

As previously noted, paper and paperboard packaging has greatly benefited from its ability to provide high recycling rates, ensuring greater sustainability advantages in use of the material. Improved technological developments in materials will continue to be important for the market's development over the forecast period.

In the aseptic packaging market, technology and innovation are moving forward rapidly. Flexible and high-speed applications are the technological driving forces in this sector, as are food safety, quality and hygiene issues (Chapter 1). These types of developments are also likely to be an important part of the innovations in paper and paperboard over the next ten years.

The announcement in the summer of 2011, that Tetra Pak had launched the world's first aseptic carton bottle for milk, is therefore of particular significance. The Tetra Evero Aseptic combines one-step opening and closing, easy handling and pouring – in essence, the functionality of a bottle with the benefits of a carton. There are three separate components to the pack – plastic cap, plastic closure and carton body. Innovative injection moulding techniques help to combine the closure, carton sleeve and capped neck into a package that is ready-to-fill. The Evero Aseptic offers many advantages, including environmental benefits. It is recyclable and made from FSC certified renewable paperboard. Most importantly, it will keep milk fresh in ambient conditions for up to six months. Whilst initially aimed at the ambient white milk market and non-oxygen sensitive milks, the pack is likely to be available later for other beverages, including flavoured milk, cream and oxygen sensitive milk products with active ingredients such as Omega-3, iron, zinc and vitamin C. For dairy companies, capital investment in a Tetra Pak A6 iLine is necessary to produce the Tetra Evero.

This is a great example of an innovation that combines the benefits of a carton with those of a bottle into one integrated pack. There is a capital cost, as there often is with new technology investments, but it is believed that production efficiencies and cost savings will accrue over the longer period.

11.11.3 Cartonboard safety concerns

There were scares in the spring of 2011 that mineral oils from printing inks had found their way into food sold in recycled cartonboard packs for cereals. The pan-European cartonboard producer MM Carton subsequently claimed in October 2011 that it had developed an innovative new recycled cartonboard that protects against migration of mineral oils into foodstuffs; the outcome of tests are awaited. Many food packaging companies and the European Food Standards Authority (EFSA) are investigating the situation and seeking solutions across all sectors of the food contact supply chain (European Food Standards Authority, 2011; Food Standards Authority, 2011). The EFSA published a scientific opinion on the risk from mineral oils in June 2012. It broadly said that the presence of mineral oil hydrocarbons (MOH) in dry foods may be partially attributable to the use of recycled paper/cardboard packaging. The Food Standards

Authority (FSA) of the UK said in December 2011 that a study they conducted during the year found that mineral oil in cartonboard food packaging posed no threat to the health of consumers. In June 2012, the FSA said it was not advising consumers to change their eating habits based on the EFSA opinion. Nevertheless, the industry needs to be increasingly aware of these types of threats over the forecast period and the potential risks to the integrity of the consumer packaging supply chain.

11.11.4 Demand growth at a slower rate

Notwithstanding these findings, food safety and security will continue to be a vital issue throughout the forecast period, particularly in direct food contact packaging applications.

Nonetheless, with good environmental credentials and improved functional and material performance, in some instances in conjunction with other materials (including barrier coated board for products such as hot, cold and greasy products) and in markets such as RRPs where corrugated board packs have great advantages (by virtue of their material reduction, via combined primary and secondary usage, improved print quality to help increase in-store sales and promotion of a quality brand image), paper and paperboard packaging products will experience increased demand over the forecast period, as the figures from *Packaging Unwrapped* indicate, albeit at a slower rate than the average for the market overall. Developments in the ambient market will also aid growth, as we have seen with Tetra Pak's innovation.

Leading users of cartonboard are creating demanding targets for sustainability up to 2020. Tetra Pak is setting its sights on providing sustainable packaging 'using only renewable materials, a minimal carbon footprint and creating zero waste'. Its 2020 targets include reducing emissions by 40%, recycling 100 billion used beverage cartons and developing packaging materials based on 100% renewable materials.

11.11.5 New markets for cartonboard

Despite the likely higher growth rates for both rigid and flexible plastics compared to paper and paperboard, there are developments in certain markets in the cartonboard field in relation to substitution from plastics. For example, it was announced in May 2011 that Green Bottle, a small Suffolk, England company had agreed a deal to supply Asda Stores in the UK with paper milk bottles. The bottles would consist of a paper shell and an inner plastic liner that holds the liquid. Whilst plastic bottles take considerable time to decompose and cannot be multi-recycled, a Green Bottle paper shell would decompose in approximately five weeks, it is claimed. The company believe the bottle could ease the growing landfill crisis, as it has a carbon footprint of around one-third smaller than a plastic bottle. Green Bottle say that more than 15 million plastic bottles are used every day in the UK and even if they are recycled the vast majority end up in land fill. The paper casing can also be recycled up to five times or can be composted domestically.

The initial launch was for a small number of Asda stores; the 100 000th bottle was produced towards the end of 2011, at Green Bottle's manufacturing unit in Cornwall, England. The company is starting to establish a niche market and plan to roll out the bottle nationally. A Green Bottle for the wine market is another innovation and time will tell how these developments will progress. The environmental credentials are considerable and an indication of the way the market could develop. However, there is a long way to go for substantial volume growth to be achieved in this particular case and it is starting from a low base point, as many new developments often do. Innovations in packaging often start from these types of beginnings, driven by entrepreneurs who see a niche opportunity that grows to something much bigger.

11.12 Globalisation of business opportunities, innovations and world markets

Innovations in materials markets, including metals, glass and plastics, could manifest themselves in many ways over the forecast period. One of these is by corporate mergers and acquisition activity. In so doing, markets that are mature and potentially under threat from competing materials could benefit by new investment and exposure enhanced technological advances through the injection of new capital and R&D. Equally, companies who have innovative products but lack capital investment and funding can benefit by being acquired by larger groups which have the financial resources to allow greater investment in new technology to take place.

11.12.1 Major international groups getting bigger

By consolidation of the industry, multinational groups are seeking ways of achieving that elusive objective: volume growth in a recession and times of market down-turn. During the latter months of 2010 through to early 2012, several corporate deals were concluded that could have significant implications for the pace of change of the global packaging industry over the next ten years, particularly in mature markets such as metal and glass.

The deal between Ardagh Glass and the metal packaging group Impress created a 3.2 billion Euros corporate business, with massive resources and a global span of operations and employees. The transaction will undoubtedly bring consolidation and rationalisation, but what it will also provide is the opportunity for businesses with huge resources to invest in new technologies, including greater R&D. Subsequent to this, in May 2011, Ardagh announced that it would be undertaking an Initial Public Offering (IPO), probably in the medium term. To demonstrate the truly global nature of the business and its plans for the future, Ardagh said that the floatation could take place on the New York stock exchange rather than in Europe. A greater pool of potential public companies as potential acquisition targets in the US was a major reason given at the time for the decision. However, it subsequently announced early in 2012 that it had agreed to acquire Boxal, a European supplier of aluminium containers, for 85 million Euros, showing that its international spending spree showed no signs of abating, despite some concerns about the group's debt levels. A global player with intent to put together a mega-group, with interests in metals, glass and other materials, is clearly a pointer to the future shape of the global packaging market.

In June 2011, US plastic container manufacturer Graham Packaging was purchased by Reynolds Group for approximately $1.68 billion. In March 2011, the Reynolds Group, which is owned by the Rank Group of New Zealand, announced sales of almost $10 billion, positioning it in the world's top 10 packaging companies. Graham employs over 8300 people in 15 countries with 97 manufacturing operations. As a result of the deal, one of the world's largest packaging manufacturing businesses just got bigger and in so doing is poised to create strategic opportunities and purchasing synergies for both companies. Companies such as these, with a track record for innovation and growth, should drive the pace of change in materials and technologies over the forecast period.

This trend is further illustrated by RPC Group, a leading pan-European plastic packaging producer, known for their success in innovations and packaging developments (Chapter 1). The company increased its revenues by almost 14% to £819 million in the year to December 2011. It intends to explore further acquisition targets, not just in Europe but also in growing economies outside Europe. Revenues of £1 billion in 2012 could well be on the horizon.

Similarly, the purchase of the Packaging Division of SCA of Sweden by DS Smith for £1.4 billion in January 2012, will mean that the company will become one of Europe's leading suppliers of recycled packaging for consumer goods and also make the company the second largest corrugated packaging group in Europe, with sales of over £4 billion and 25 000 employees. This acquisition is about fulfilling the needs of international consumer goods markets from a sustainable viewpoint with the supply of recycled materials – a major theme of the international packaging market.

Viewed from a different angle, private equity group Sun Capital Partners, based in Florida, now has a global packaging business with a turnover of $4 billion (£2.5 billion). Since 2005, the group has bought companies in many packaging material markets, including rigid plastics, paper and board and flexible packaging on a multinational basis. Sun Capital owns, amongst many others, Albea the previous beauty packaging business of Rio Tinto Alcan, the Britton Group and Paccor, a rigid plastics group, which includes businesses such as the rigid packaging operations of Huhtamaki. This is once again globalisation on a massive scale, achieved by astute acquisitions and targeting of innovative companies with added value products and services.

This type of private equity-owned business model is likely to play a significant part in the global packaging industry of the next ten years. Of course, demanding profit performance and returns on capital investment and planned exit strategies will be very much part of the business model.

11.12.2 Chinese market expansion

In the development of new innovative opportunities and the identification of growth markets, China represents a global packaging phenomenon. As reported in *Packaging News*, UK, in October 2011, a report in September of the same year by financial advisory company Catalyst Corporate Finance, entitled *Paper and Packaging Sector M&A Update*, said that China aims to become the world's largest packaging producer by the end of 2012. According to the report, China's output of paper and cardboard products now exceeds 80 million tonnes, making it the world's largest paper manufacturer. Smithers Pira say the fastest expansion for RRPs in the world will occur in China, where their market share will rise from 15% to 18% over the 2011 to 2016 period. Early in 2011, China reached $53.4 billion worth of annual plastic production, $12.8 billion of which is exported, while imports account for $7.1 billion of plastic products, also according to Catalyst Corporate Finance, as reported in *Packaging News*, UK, October 2011 (*Packaging News*, 2011).

As previously noted in this chapter, more international businesses are investing in China. In March 2012, Coca-Cola inaugurated the opening of its 42nd bottling plant in the country. The plant represents a $160 million investment by the company in China and is part of a three year $4 billion investment announced in 2011. Coca-Cola and its China-based bottling partners have identified China as one of the most important growth markets in the world. In March 2012, Tetra Pak announced the start-up of its second production line in China. The line in Hohhot will increase production capacity at the plant by 10 billion cartons per year to a total of 20 billion cartons.

These are impressive numbers, reflecting the drive to innovate with packaging that is technically superior to that which was used in the country before. With disposable income increasing there is a burgeoning consumer demand, which shows only marginal signs of slowing down, all aiding the innovative process.

From a different angle but still part of the theme, The International Hologram Manufacturers Association (IHMA) observed in the spring of 2012 that China was well placed to become a leading player in the world holography market, with a growing interest in exporting its technology, including UV embossing for packaging, to which the Western world is responding (IHMA, 2012).

Along with the other BRIC countries (Brazil, Russia and India), China is poised to expand in a big way in the packaging sector, as befits a country that in 2011 became the world's second largest economy in terms of GDP. Joint ventures and technological innovation are key to any market development in China. As this section of the chapter shows, great opportunities exist for innovations in the growing consumer market of the 'middle kingdom', as it is called by the Chinese, over the forecast period.

11.13 Conclusions: market trends and usage of major packaging materials to 2020

11.13.1 Flexible packaging

Significant growth in flexible packaging over the whole period attributed to continuing improvements in performance, cost and technical advances in materials, including improved barrier performance and barrier resins. However, rates of growth in volume terms will be slower than pre-recession times, due to economic factors, reductions in packaging usage and reduction in the numbers of flexible packaging layers used, allied to material reduction initiatives by leading international consumer goods companies.

11.13.2 Rigid plastic containers

Growth in rigid plastic containers is due to new markets being established, technical advances in material performance, providing extended shelf life and improved barrier features. There will be a substitution of rigid plastics for glass containers, due to light-weighting, safety, convenience and improved barrier performance. An accelerated rate of market development for bioplastics materials will be driven by initiatives from leading FMCG producers.

11.13.3 Paperboard packaging

Continued growth of retail ready packs (RRP), advances in aseptic packaging technology, recycling and sustainability benefits, improvements in material performance (sometimes in conjunction with other materials), new technological developments, improvements in convenience and ease of handling will all assist the market, albeit with growth at a slower rate than plastics materials.

11.13.4 Glass container packaging

Despite overall market growth, losses to competing materials will continue, particularly to rigid plastics in sectors where light-weighting, convenience and shatter-resistance are important. Glass will still be popular in markets where aesthetic appeal and quality are important, such as alcoholic drinks and spirits. Excellent sustainability and environmental features, including high recycling rates, will benefit the material. In markets such as soft drinks and juices, wines, sauces, jams, spreads and other similar sectors, plus airlines, outdoor events and entertainment venues, glass will continue to come under considerable threat, as more major brands switch to rigid plastics over the period.

11.13.5 Metal packaging

Growth in metal packaging will be achieved over the forecast period, but at a slower rate than the average for all packaging materials. Low-margin metal packaging sectors, such as canned vegetables and processed foods, will lose out to newer plastics technologies. However, cans, particularly beverage cans, will continue to be an important sector due to long shelf life and efficient production, with investment over the period in new can-making equipment in emerging markets. Niche added value opportunities will be presented with advanced decorative and shaping techniques, coating technology and increased use of digital print.

11.14 References

BCC RESEARCH (2011) 'Report on Biodegradable Polymers'. Available from: www.bccresearch.com/report/biodegrable-polymers-pls025c.html [Accessed October 2012].

BRITISH PLASTICS FEDERATION (2011) Available from: www.bpf.co.uk [Accessed October 2012].

CATALYST CORPORATE FINANCE (2011) 'Paper and Packaging Sector M&A Update'. Available from: www.catalystcf.co.uk>LatestNews [Accessed October 2012].

DEFRA (2011) 'Increased Recycling Targets from 2013–2017'. Available from: www.defra.gov.uk/consult/files/packaging-consult-doc.pdf [Accessed October 2012].

EUROPEAN COUNCIL OF MINISTERS (2011) 'Waste Problems caused by Single-use Carrier Bags'. Available from: www.envicura.com/. . ./EU+Officials+to+consider+European+plastic+. . .

EUROPEAN FOOD STANDARDS AUTHORITY (2011) 'Migration of Mineral Oils into Foodstuffs'. Available from: www.food.gov.uk/multimedia/faq/mineraloils/ [Accessed October 2012].

FEVE (EUROPEAN GLASS CONTAINER FEDERATION), (2010) 'Consumer Survey for Glass Industry'. Available from: www.feve.org/index.php?option=com_content&view. . .id. . .

FOOD STANDARDS AUTHORITY (2011) 'Printing Inks and Mineral Oil Survey Results'. Available from: www.food.gov.uk/newsarchive/2011/dec/mineraloils

FREEDONIA GROUP (2009a) 'World Converted Flexible Packaging', Converted Flexible Packaging Report, September 2009. Available from: www.freedoniagroup.com/brochure/25xx/2556smwe.pdf [Accessed October 2012].

FREEDONIA GROUP (2009b) 'USA Converted Flexible Packaging', Converted Flexible Packaging Report, October 2009. Available from: www.marketresearch.com/Freedonia-Group. . ./Converted-Flexible-Packaging

IHMA (INTERNATIONAL HOLOGRAM MANUFACTURERS ASSOCIATION) (2012) Available from: www.ihma.org/ [Accessed October 2012].

INTERNET ADVERTISING BUREAU (2011) 'Money Spent on Internet Advertising'. Available from: www.iabuk.net [Accessed October 2012].

METAL PACKAGING MANUFACTURERS ASSOCIATION, UK (2011). Available from: www.mpma.org.uk [Accessed October 2012].

NATIONAL NON-FOOD CROPS CENTRE (2011) Available from: www.nnfcc.co.uk [Accessed October 2012].

PACKAGING NEWS (2011) 'China Aims to Become World's Largest Packaging Producer'. Available from: www.packagingnews.co.uk/news/china-largest-producer/ [Accessed October 2012].

REXAM CONSUMER PACKAGING REPORT (2011/12) 'Packaging Unwrapped'. Available from: www.rexam.com/files/pdf/packaging_unwrapped_2011.pdf [Accessed October 2012].

SHEFFIELD HALLAM UNIVERSITY (2011) 'Design Futures' Packaging Design Consultancy. Available from: www.designfuturesgroup.com/packaging-design [Accessed October 2012].

SMITHERS PIRA (2011) 'The Future of Digital Printing for Packaging to 2016'. Available from: www.Smitherspira.com/future-of-digital-printing-for-packaging-to-2016.aspx [Accessed December 2012].

SMITHERS PIRA (2012) 'Growth in Retail Ready Packs'. Available from: www.smitherspira.com/growth-in-retail-ready-packaging

THE CAN MAKERS, UK (2011) Available from: www.canmakers.co.uk [Accessed October 2012].

THE CANS OF THE YEAR AWARDS (2011) Sponsored by *The Canmaker* magazine. Available from: www.canmaker.com [Accessed October 2012].

UNIVERSITY OF APPLIED SCIENCE AND ARTS (2011) 'Production Rates of Bioplastics', Hanover, Germany. Available from: www.european-bioplastics.org/wp. . ./EuBP_image_brochure-2011.pdf

Index